Electrochemistry of Zirconia Gas Sensors

Electrochemistry of Zirconia Gas Sensors

Serge Zhuiykov

CRC Press
Taylor & Francis Group
Boca Raton London New York

CRC Press is an imprint of the
Taylor & Francis Group, an **informa** business

CRC Press
Taylor & Francis Group
6000 Broken Sound Parkway NW, Suite 300
Boca Raton, FL 33487-2742

First issued in hardback 2019

© 2008 by Taylor & Francis Group, LLC
CRC Press is an imprint of Taylor & Francis Group, an Informa business

No claim to original U.S. Government works

ISBN-13: 978-1-4200-4761-5 (hbk)

Library of Congress Cataloging-in-Publication Data

Zhuiykov, Serge.
    Electrochemistry of zirconia gas sensors / Serge Zhuiykov.
        p. cm.
    Includes bibliographical references and index.
    ISBN-13: 978-1-4200-4761-5 (alk. paper)
    ISBN-10: 1-4200-4761-2 (alk. paper)
    1. Gas detectors. 2. Zirconium dioxide. 3. Electrochemical analysis. I. Title.

TP754.Z58 2007
681'.2--dc22   2007000171

Visit the Taylor & Francis Web site at
http://www.taylorandfrancis.com

and the CRC Press Web site at
http://www.crcpress.com

Dedicated to my beloved parents, Alla and Ivan Zhuiykov

Table of Contents

Preface

Increasing energy demands and atmospheric pollution require identification and quantification of chemical species in many industrial applications involving high temperatures and chemical contaminants. With increased communication and transportation, the world has become metaphorically smaller. As a result, more countries and their people are becoming developed and the level of development is increasing in all countries and at an increasing rate in most countries. Consequently, there is a continuing need for the development of low-cost sensors for applications in such harsh industrial environments as the metal-processing, heat-treating, glass, pulp, ceramic, automotive, aerospace, power, combustion, petrochemical, chemical, and security industries. In many industries, the monitoring and control of combustion-related emissions are top priorities. On the other hand, the need for reliable sensors along with predicted emission-modeling tools has been driven by environmental legislation in most of the developed countries. Therefore, sensors capable of providing emission profiles for such gases as carbon monoxide (CO), carbon dioxide (CO_2), nitrogen oxides (NO_x), oxygen (O_2), hydrocarbons (H_xC_y), and volatile organic compounds (VOCs) will allow for feedback control systems, resulting in lower pollutions and more efficient use of fuels.

The ability to develop single-gas or multigas sensors to the requirements of different industrial applications expands in size and in complexity the sensor's parts that can be produced, as well as increasing structural efficiency and design flexibility, optimizing materials compatibility, maximizing sensor properties, and minimizing cost. Unfortunately, it has to be admitted that *the advances in sensor development have not kept pace with the advances in materials.*

Today, principally new materials often require totally new methods and processes. Traditional sensor materials or improved versions of them usually simply require improved versions of traditional manufacturing processes to reach new levels of performance. Such improvements may be achieved in the sensor development process itself, or through development of special procedures for employing a traditional process. These facts are directly related to the development of solid electrolyte sensors in general and zirconia-based sensors measuring specific gaseous species in particular.

More types of oxides used for electrodes of the zirconia-based sensors both enable and encourage more possible combinations. Greater diversity in zirconia structures and types of oxide electrodes leads, in turn, to more incompatibilities in chemical, physical, electrochemical, and mechanical properties. The irony is that the more diversity achieved with advanced solid electrolyte and electrode materials, the bigger the challenges that arise for their joining. Beyond sheer diversity, modern designs tend to place higher demands on the sensor materials being selected and used. Modern design approaches also demand state-of-the-art techniques and tools allowing higher operating temperatures and providing greater durability in hostile environments. To address these demands, carefully created composites and structures of zirconia must be preserved in the sensor structures, and the exotic combinations of electrode materials have to be used in the hybrid structures of electrodes to optimize their performance, manufacturing efficiency, reliability, life, cost, and environmental compatibility. In short, modern electrode materials and structures being produced from them are being pushed to new limits.

Consequently, this book addresses a range of different stages of development of zirconia-based sensors for gaseous and molten metal environments, focusing in accessible form from analysis of interaction at the measuring environment–zirconia sensor interface to the reliability testing of the sensors. Furthermore, the coverage focuses on different fundamental aspects of the electrochemistry and physical chemistry of zirconia, mathematical modeling, optimization parameters, and structures of the electrode materials. The book will fill the gap among pure academic research of the zirconia-based (O_2, CO_2, SO_2, NO_x, and C_xH_y) gas sensors: their electrochemistry, the development of the sensing electrodes, mathematical modeling and optimal design, an explanation of the influence of the double electrical layer on the sensor output signal, and the applied, technological, down-to-earth approaches adopted by the vast majority of the industrial companies working in this field.

Electrochemistry of the triple-phase-boundary (TPB): gas-electrolyte-semiconductor electrode is as yet not entirely understood. Unfortunately, in the gas sensors industry as well as in the advanced materials development field, the inalterable methods used early in the 21st century are still nothing more than empirical.

As far as TPB electrochemistry is concerned, modern surface science still unambiguously throws crossword puzzles at us. In these puzzles one discovered letter offers only the clue to other words while each empty square awaits its letter. Notwithstanding the brilliant advances of the recent years, many empty squares have yet to be filled before it would be possible to decipher the expression, "*Electrochemistry of gas-zirconia-semiconductor TPB*." This book attempts to conquer this. If this crossword puzzle will be solved, we would acquire absolute power over the TPB reactions and the power to control them.

This book is addressed principally to scientists, applied researchers, and production engineers working in the fabrication, design, testing, characterization, and analysis of new materials for the zirconia-based sensors. It can be useful for students studying materials science and engineering, for those working in the analysis and characterization of new ceramic materials, and for those developing various technologies of sensor fabrication. The chapters include a large number of literature citations, which will be of interest to those seeking more information on the fundamental aspects of zirconia electrolytes, semiconductor oxides, applied technologies, and sensor performance in different measuring environments.

Acknowledgments

While it may be unusual to say that the long hours of research and writing for this book gave me as much pleasure as discovering something completely unexpected, I often felt that they did. I am grateful to many other people who directly and/or indirectly helped me during preparation of this book, including Professor Janusz Nowotny, director of the Centre for Materials Research in Energy Conversion, University of New South Wales, Australia, who spent a tremendous amount of time teaching me about work function; and Professor Norio Miura, Art, Science and Technology Center for Cooperative Research (KASTEC), Kyushu University, Japan, for his invaluable help in all aspects related to solid electrolyte gas sensors and in providing testing facilities, materials, and sensors for meaningful test results to be obtained. Great thanks to Dr. Vlad Plachnitsa, KASTEC, Japan, for his help with conducting some experiments; and to lifelong friends Dr. Radislav A. Potyrailo, GE Research, United States, and Dr. Vlad Maksutov, Australian Locum Medical Service Pty. Ltd., Australia.

I also want to acknowledge the support from the management of the Commonwealth Scientific Industrial Research Organization (CSIRO), CMMT (CSIRO Manufacturing and Materials Technology) Division, Australia; and from Dr. Ivan Cole, deputy chief of the CMMT Division.

My special thanks to Fire Science Team Leader Vince Dowling, CMMT, CSIRO, who encouraged the goals of *Electrochemistry of Zirconia Gas Sensors*. Thanks to my colleagues at CMMT, CSIRO, Dr. Donavan Marney, Dr. Greg Griffin, Glenn Bradbury, David O'Brian, Alex Web, William Mikus, and Rex Pollard, who helped me to carve out the time needed to write as efficiently as needed for this book. I also want to thank Eugene Kats, who took red pen to more than one chapter. Without their contribution, it wouldn't be possible to realize many research ideas, which have resulted and been discussed in the book.

My special thanks to David Whittaker, executive officer of the ActivFire® Listing Scheme, CSIRO, who has sharpened my style significantly. Special thanks also to Teresa Fitzgerald for her uncompromised way to be herself in any circumstances.

I wish also to express my gratitude to the Fire Protection Association of Australia and especially to Barry Lee. Special thanks to Ralph Garbutt of Resource Risk Management and to Dr. Alex Nosenkov of Technologies Australia Group Pty. Ltd. I am also grateful to Standards Australia International for the special honor to represent Australia and to be a head of the Australian delegation at the International Standards Organization (ISO) TC21/SC8 Technical Committee.

Also, great thanks to the group at CRC Press, Catherine Giacari and Allison Shatkin, who believed in solid-state sensors and who are just great to work with.

Very special thanks to my parents, Ivan and Alla Zhuiykov, who brought wisdom into my life; my wife, Tatiana, for her constant love, support, encouragement, and patience; and my children, Michael, Slava, and Maxim.

Dr. Serge Zhuiykov
CSIRO, CMMT

About the Author

Dr. Serge Zhuiykov is a senior research scientist at the Manufacturing and Materials Technology Division of the Commonwealth Scientific Industrial Research Organization (CSIRO), Australia. As a research scientist, he possesses combined academic and industrial experience from working at different universities in Australia, Japan, and Europe and in industrial environments for more than 16 years after completion of his Ph.D. He is also chairman of the *FP-011-02* Technical Committee of Standards Australia International and a head of the Australian delegation at the International Standards Organization (ISO) TC21/SC8 Technical Committee, since 2005. He is the author of more than 100 scientific publications, including 14 international patents, and is the recipient of numerous awards. His current research concentrates on the development, design, and evaluation of various new functional materials for solid-state gas sensors and for fire detection and suppression systems.

1 Introduction to Electrochemistry of Solid Electrolyte Gas Sensors

1.1 ELECTROCHEMISTRY OF ZIRCONIA SOLID ELECTROLYTES AS THE BASIS FOR UNDERSTANDING ELECTROCHEMICAL GAS SENSORS

1.1.1 SOLID OXYGEN-IONIC ELECTROLYTES

Solid electrolytes based on stabilized zirconia have been studied since the discovery of electrolytic oxygen evolution from ZrO_2-Y_2O_3 solid solutions (Nernst glower) by Nernst in 1899. Perhaps this was the first finding which clearly illustrated that ionic conductivity exists in the solid state. The ionic conductivity in those solid solutions occurs via oxide ionic vacancies (V_O) generated due to charge compensation. Stabilized zirconia ceramics have been the subject of extensive scientific research during the last 30–35 years owing to the diverse technological applications. These materials have been used in fuel cells, oxygen separators, oxygen pumps, and electrochemical gas sensors, throughout which there have been two primary areas of research: (1) optimization of the manufacturing technology and structure of the stabilized zirconia, and (2) development of knowledge relating to the solid electrolyte–electrode interface. While it is well recognized that both directions reached the level of their maturity, there still seems to be some dispute regarding the explanation of complexity of physical and electrochemical processes on the electrolyte-electrode interface.

The first comprehensive review of the properties of zirconia ceramics was published in 1970 [1]. Since then, the number of publications has been constantly growing, and now it can be concluded that the task of development of the zirconia-based solid electrolyte sensors for measurement of oxygen and oxygen-containing gases is almost completed. Although research of zirconia/electrode interfaces is still developing, the first breakthrough for the zirconia-based gas sensors occurred recently owing to the introduction and commercialization of *in-situ* λ-sensors based on yttria-stabilized zirconia (denoted as YSZ) for the detection of equilibrium oxygen partial pressure in automotive exhausts [2]. Despite this attention, a number of questions remain to be resolved, including the temperature and concentration dependence of the structural disorder and its interrelation with the high ionic conductivity which underlies many of the industrial uses of these electrolytes.

The vast majority of the zirconia-based solid electrolytes can be represented by two types of oxide solid solutions: $ZrO_2 + M_1O$ or $ZrO_2 + R_2O_3$ (R: rare-earth element). These solutions have a centered cubic lattice (fluorite type), and they can appear when the dimensions of the base cation M^{4+} and the substitutive cations (M_1^{2+}, R^{3+}) are close enough. In general terms, the zirconia-based solid solutions can be expressed as follows:

$$ZrO_2 + xM_1O, \qquad (1.1)$$

$$ZrO_2 + xR_1O_{1.5}, \qquad (1.2)$$

$$ZrO_2 + xR_2O_3, \qquad (1.3)$$

where x is the mole part of the substitutive ion or the mole part of R_2O_3 solid solution. The mole part of $R_1O_{1.5}$ can be shown as

$$[R_1O_{1.5}] = [2\ R_2O_3]/(1 + [R_2O_3]). \qquad (1.4)$$

Zirconia-based ceramics can be stabilized by adding such binary oxides as CaO, MgO, Sc_2O_3, and Y_2O_3 to zirconia. However, Y_2O_3 doping is particularly effective in producing high ionic conductivity and is the most widely used [3, 4]. The published literature in relation to the structural properties of stabilized zirconia is extensive, and therefore only a brief summary of properties regarding gas-sensing principles is presented in this chapter. A ZrO_2-Y_2O_3 partial phase diagram [5, 6], shown in Figure 1.1, must be considered for better understanding various properties of the YSZ-based solid electrolytes and their behavior during longtime exposure to high temperatures. This diagram illustrates the effect of Y_2O_3 on the stability ranges of the three zirconia polymorphs—monoclinic (M), tetragonal (T), and cubic (C)—as well as the regions where the stable solid solutions with the fluorite structure can be defined.

Pure zirconia, ZrO_2, changes its crystalline structure at least three times during heating from the normal temperature up to the melting point. Pure ZrO_2 adopts the monoclinic baddelyite structure (M-ZrO_2, space group $P2_1/c$) under ambient conditions, with the Zr^{4+} in a distorted sevenfold coordination. On increasing the temperature, it transforms to a tetragonal distorted fluorite structure (T-ZrO_2, $P4_2/nmc$) at T ~1097°C, with Zr^{4+} surrounded by eight anions, but with two slightly different Zr^{4+}-O^{2-} distances [7]. Perfect eightfold coordination is achieved at T ~2371°C, with a transformation to a cubic fluorite structured phase (F-ZrO_2, $Fm3m$), followed by melting at T ~2715°C. The ambient temperature structure of yttria, Y_2O_3, has space group $1a\overline{3}$ and can also be derived from that of fluorite by removal of $^1/_4$ of the anions [8]. There are two sixfold-coordinated and symmetry-independent Y^{3+}, which have the anion vacancies arranged along either a face diagonal or a body diagonal. The only known ordered phase in the ZrO_2-Y_2O_3 system is $Zr_3Y_4O_{12}$. This adopts a rhombohedral structure which is also strongly related to that of fluorite [7]. It is important to note that the tetragonal-to-monoclinic transformation is well-known to

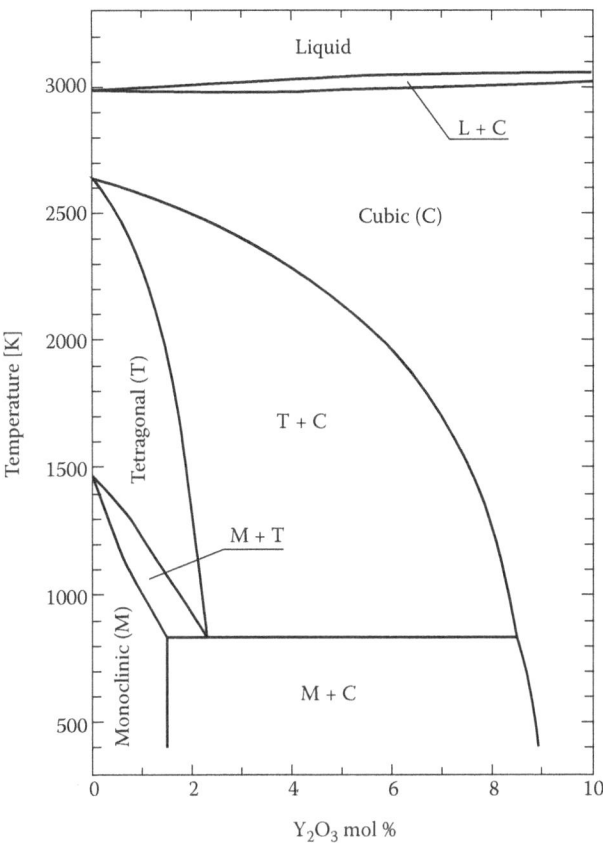

FIGURE 1.1 Partial phase diagram for ZrO_2–Y_2O_3. (From Nowotny, J. et al., Charge transfer at oxygen/zirconia interface at elevated temperatures. Part 1: Basic properties and terms, *Adv. Appl. Ceramics* **104** (2005) 147–153. With permission.)

result in microstructural degradation owing to a volume expansion of 3–5% [9]. At low Y_2O_3 concentrations ($x < 0.15$), there are regions of the crystal ~20 angstrom units (Å) in size which contain relatively few oxygen vacancies, causing the lattice to undergo a slight tetragonal distortion of the type observed in the tetragonal phase of $(ZrO_2)_{1-x}$ $(Y_2O_3)_x$ at $x < 0.09$. The oxygen vacancies are preferentially arranged in pairs on nearest-neighbor anion sites in the ‹111› fluorite directions, with a cation located between them and extensive relaxations of the surrounding nearest-neighbor cations and anions. As the yttria content increases, these ‹111› vacancy pairs pack together in ‹112› directions to form aggregates, whose short-range defect structure resembles the long-range crystal structure of the ordered compound $Zr_3Y_4O_{12}$ and other anion-deficient fluorite-related systems. The aggregates are typically ~15 Å in diameter, though both their size and density increase slightly with x. As temperature increases, these aggregates remain stable up to the melting point. There is also an increasing number of single vacancies and ‹111› vacancy pairs (with surrounding relaxation fields) as x increases, and these isolated clusters become mobile at

T > 727°C and result in the high ionic conductivity of the material. In light of these observations, it has been proposed [10] that the anomalous decrease in the ionic conductivity with increasing x is a consequence of the decreasing mobility of the isolated defects, possibly owing to blockage by an increasing number of the static aggregates.

1.1.2 Transport Properties

The electrochemical performance of zirconia is determined by the mechanism and kinetics of charge transfer between the zirconia lattice and the oxygen-containing environment. It has been generally assumed that the charge transfer between oxygen and zirconia occurs on the triple-phase boundary (TPB): zirconia, gas, and electrode [5, 11, 12]. The charge transfer in ionic crystals is usually carried out by ions. Their transference is possible due to the availability of some point defects in the lattice, representing the deviation of periodic atomic structure of this lattice from the ideal one.

Point (microscopic) defects in contrast from the macroscopic are compatible with the atomic distances between the neighboring atoms. The initial cause of appearance of the point defects in the first place is the local energy fluctuations, owing to the temperature fluctuations. Point defects can be divided into *Frenkel defects* and *Schottky defects*, and these often occur in ionic crystals. The former are due to misplacement of ions and vacancies. Charges are balanced in the whole crystal despite the presence of interstitial or extra ions and vacancies. If an atom leaves its site in the lattice (thereby creating a vacancy) and then moves to the surface of the crystal, it becomes a Schottky defect. On the other hand, an atom that vacates its position in the lattice and transfers to an interstitial position in the crystal is known as a Frenkel defect. The formation of a Frenkel defect therefore produces two defects within the lattice—a vacancy and the interstitial defect—while the formation of a Schottky defect leaves only one defect within the lattice, that is, a vacancy. Aside from the formation of Schottky and Frenkel defects, there is a third mechanism by which an intrinsic point defect may be formed, that is, the movement of a surface atom into an interstitial site. Considering the electroneutrality condition for the stoichiometric solid solution, the ratio of mole parts of the anion and cation vacancies is simply defined by the valence of atoms (ions). Therefore, for solid solution M_nX_m, the ratio of the anion vacancies is equal to m/n.

Usually the concentration of Frenkel and Schottky defects is relatively small. The maximum mole part of these defects does not exceed several tenths of a percent. Thus, the electroconductivity of such solid solutions is minimal even at the temperatures close to their melting point.

Apart from the point defects, there are impurity defects in ionic crystals due to some impurities in raw materials. The impact of impurity segregation on ionic conductivity of the solid electrolytes will be considered in detail in section 1.4 of this chapter. The vacancies, developed in the solid solutions during the substitution of the main ion (M in the solid solution $M(M_1)O_{2-x}$) by the ion substituent (M_1) of the different valence, have special meaning for solid electrolytes among impurity defects. In this case, the vacancies must appear from one of the solid-state sublattices at the preservation of its crystal structure. Most of the solid oxygen-ionic electrolytes represent themselves as solid solutions with this particular type of defect structure.

These solid solutions have several types of charge carriers. They can be represented by anions, cations, electrons, and holes. The Ohm law is fair for each of them. The full current represents itself as a summary of the partial currents by n particles:

$$I = \sum_n In = U \sum_n \sigma_n , \qquad (1.5)$$

where σ_n is the partial electroconductivity by the nth particle, which is equal to product of charge (q_n), concentration (c_n), and mobility (B_n) of this particle:

$$\sigma_n = q_n c_n B_n . \qquad (1.6)$$

The temperature dependence of σ_n can be determined by the following equation:

$$\sigma_n = \sigma_0 \exp (-E_n/RT), \qquad (1.7)$$

where σ_0 is the constant and E_n is the transference activation energy. R is the gas constant, and T is the absolute temperature. Portion of the partial conductivity

$$t_n = \sigma_n/\sigma \qquad (1.8)$$

is the transference number by n particles.

The flow of the charged particles in the chemical (μ) and electrical (φ) fields can be described by the Wagner equation:

$$j_n = -\frac{\sigma_n}{q_n^2}\left(\frac{d\mu}{dx} + q_n \frac{d\varphi}{dx}\right) . \qquad (1.9)$$

Considering the Nernst–Einstein correlation, establishing the dependence of the coefficient of diffusion (D) on transport characteristics,

$$D_n = RT \, \sigma_n/q_n^2 \, c_n, \qquad (1.10)$$

(1.9) can be transformed into the following equation [13]:

$$j_n = -\frac{D_n c_n}{RT}\left(\frac{d\mu}{dx} + q_n \frac{d\varphi}{dx}\right) . \qquad (1.11)$$

The last correlation is acceptable to any particles, including the neutral atoms.

The diffusion of charged particles plays a very important role in solid electrolytes. Sometimes, it can be referred to as the *Wagner diffusion mechanism* [14]. In accordance with the electrostatic laws, the following condition of electroneutrality can be fulfilled in any element of the volume of solid state:

$$\sum_n q_n c_n = 0. \tag{1.12}$$

This leads to the important feature of the charged particle's diffusion within the solid state, manifested in the fact that the transfer of charged particles does not accompany the proceeding of electric current. For the particles of two different types, the correlation (1.12) can be transformed:

$$q_1 dc_1/dx + q_2 dc_2/dx = 0. \tag{1.13}$$

Consequently, the correlations for the particle's flow in accordance with the first Fick law are given as follows:

$$j_1 = - D_{12} \, dc_1/dx \text{ and } j_2 = - D_{12} \, dc_2/dx, \tag{1.14}$$

with the total effective coefficient of diffusion (D_{12}) represented as

$$D_{12} = t_1 D_2 + t_2 D_1. \tag{1.15}$$

The correlation (1.15) plays an important role in the analysis of processes stipulated by the small electron conductivity of solid electrolytes, such as the electrolytic permeability of the solid state or the establishment of the chemical potential profile of the elements of solid electrolytes.

1.2 ELECTROPHYSICAL PROPERTIES OF SOLID ELECTROLYTES

It is evident that in crystals of the perfect structure (ideal situation), when they have no vacancies ($c_{V^{..}} = 0$), it is impossible to have direct transfer of the electrical charges. Only the appearance of vacancies allows the possibility for transfer of oxygen ions from the engaged assembly into the vacant one. This reaction is usually accompanied by the transfer of both charge and mass. In this case, the probability of "jumping over" the oxygen ions is proportional to the concentration of vacancies. Subsequently, the particles, responsible for the current transfer in the solid electrolyte systems, are the oxygen vacancies. Therefore, the transfer mechanism of the oxygen ions in the solid electrolyte systems should be named the *vacant mechanism*. It means that the transfer of oxygen vacancies in the electrical field occurs from anode to cathode. As a result, vacancies behave as particles with charge $q = - (2e)$. The ratio between ionic and vacant mechanisms of the current transfer in solid electrolytes is similar to the ratio between electron and hole conductivities. In both cases, the second ones are additional to the first ones.

It is following from the above that the concentration of the oxygen vacancies, that is, $c_n = c_{V'}$, can be used as the concentration of the ionic current carriers in equation (1.6). Based on the ionic conductivity theory, considering the appearance of couple complexes of the admixture cation-vacancy type, as well as interactions of nonassociated defects, it is expected that the electroconductivity can be a growth monotony with the increase in concentration of the oxygen vacancies up to 50%. However, in reality, the electroconductivity/composition dependence has its maximum at about ~8–10% of vacancies [15]. When the high-temperature cubic zirconia is stabilized by doping with alkali or rare-earth oxides, the dopant cations, substituting the Zr^{+4} sites in the crystal structure, stipulate the appearance of oxygen vacancies in order to maintain the charge neutrality of the crystal. For example, 8 mol % Y_2O_3 in stabilized zirconia contains about 4 mol % oxygen vacancies [16]. Such a high vacancy concentration facilitates the selective O_2 diffusion via a vacancy diffusion mechanism. Furthermore, the electrical conductivity remains predominantly ionic even at elevated temperatures with practically no concomitant electronic conductivity over a wide range of oxygen activities [17]. In other words, the average ionic transference number remains near unity (i.e., $\bar{t}_i > 0.99$), indicating that almost all current through YSZ is carried by the oxygen ions.

High conductivity in the solid state has been stipulated not only by high concentration of the current carriers, but also by their mobility (see Equation (1.6)). High conductivity in the solid solutions has been provided by their crystalline structure and the optimal correlation between the radiuses of the main cations and anions, as well as by the ratio between radiuses of the cation substituent and the main cation (see Figure 1.2). Thus, two factors—high concentration of oxygen vacancies and their substantial mobility—provide superior ionic conductivity of the solid solutions transferring them into the solid electrolytes.

The main principle, determining the application of solid electrolytes in gas sensors, is based on type of conductivity and on stability of the electrophysical properties at alteration parameters of the measuring environment. In one's turn, type (character) of conductivity depends on structural characteristics of the solid electrolyte, working temperature, and measuring partial pressure.

Based on the Wagner method [16], the required character of conductivity can be achieved by the introduction of admixture into the basis oxide, which has a common anion with the basis oxide, and the cation has less valence. Type and quantity of the defects are stipulated by the admixture and its concentration. As a rule, the majority of well-known solid electrolytes with pure oxygen-ionic conductivity have a fluorite CaF_2 crystalline structure [17].

Appearance of the electronic conductivity in oxide electrolytes is connected to the development of conductive electrons in the lattice of solid solution, forming as a result of the reduction of cations $Me^z \rightarrow Me^{z+1} + \Theta$, where Θ is an electron. Moreover, a part of the anion is removed from the solid oxide lattice (n type of conductivity).

Appearance of the current carriers as electron holes is connected to the oxidation of cations, that is, with their loss of electrons $Me^z \rightarrow Me^{z-1} + \oplus$, where $\oplus$ is the hole.

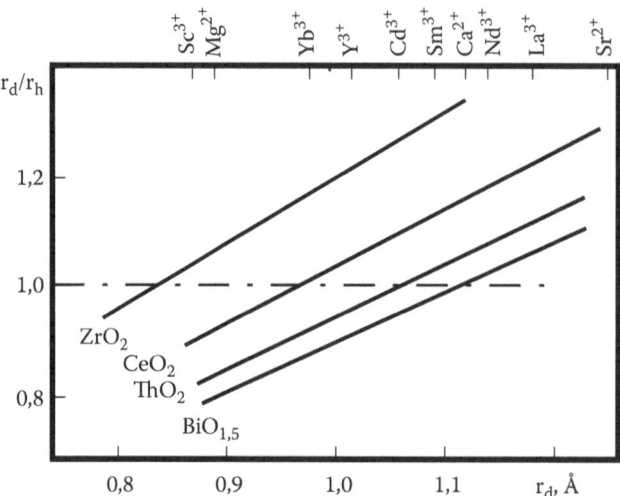

FIGURE 1.2 Ratio range between dimensions of the cation substituent (r_d) to the main cation (r_h), at which the fluorite type of structure exists.

In spite of all this, the admixture conductivity is stipulated by excess of the negative component (p type of conductivity).

Thus, the total conductivity of oxide electrolytes can be described as follows [17]:

$$\sigma = \sigma_{ion} + \sigma_\ominus + \sigma_\oplus.$$

and the relative conductivity can be expressed as an ionic transference number

$$t_{ion} = \frac{\sigma}{\sigma_{ion} + \sigma_\ominus + \sigma_\oplus}. \tag{1.16}$$

Interaction of the solid electrolyte lattice with the oxygen-containing environment can be expressed by the following equations [18–20]:

$$\tfrac{1}{2}\,O_2 + V_{\ddot{O}} \leftrightarrow O^{2-} + 2\oplus; \tag{1.17}$$

$$\tfrac{1}{2}\,O_2 + V_{\ddot{O}} + 2\ominus \leftrightarrow O^{2-}. \tag{1.18}$$

In the simplest case, behavior of defects in the solid electrolyte is described by the mass law in force. Considering independence of the concentration of vacancies and anions from the oxygen partial pressure in gas (P_{O2}), the following correlation of conductivity in the solid electrolyte can be obtained from the equilibrium reactions (1.17) and (1.18):

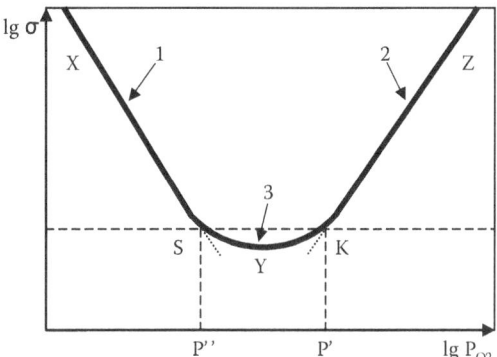

FIGURE 1.3 Dependence of electronic conductivity in solid oxygen-ionic electrolytes on the oxygen partial pressure. 1: region of an electronic conductivity of p type; 2: region of an electronic conductivity of n type; and 3: region of pure oxygen-ionic conductivity. (From Zhuiykov, S., Electron model of solid oxygen-ionic electrolytes used in gas sensors, *Int. J. Applied Ceramic Techn.* **3** (2006) 401–411. With permission.)

$$\sigma = \sigma_{ion} + k_1 P_{O2}^{1/n} + k_2 P_{O2}^{-1/m}. \qquad (1.19)$$

Values n and m are determined by the type of the dominated defects within the levels of the oxygen partial pressure. Conductivity dependence on the oxygen partial pressure (P_{O2}) is shown in Figure 1.3 for the constant temperature. The curve XYZ is typical for pure oxides (ZrO_2, ThO_2), which possess the mixed conductivity in the wide P_{O2} range excluding the region around the minimum point, where $t = 1$. The $XSKZ$ curve characterizes the dependence of total conductivity (1.19) on the oxygen partial pressure for the mixed zirconia-based oxide electrolytes. Section SK corresponds to the pure oxygen-ionic conductivity region ($\bar{t_i} > 0.99$) at P_{O2} changes within $P_\theta < P_{O2} < P_\oplus$. Pressures P_θ and $P_\oplus$ are characterized as the beginning of the electron and hole conductivities in the solid electrolyte, respectively.

The existence of the relatively large region of the pure oxygen-ionic conductivity in the zirconia-based electrolytes provides their wide use in the solid electrolyte oxygen sensors [21]. So far, most of the solid electrolyte sensors accepted by industry are based on dioxide zirconia (ZrO_2-Y_2O_3) possessing a pure oxygen-ionic conductivity within the wide temperature range of 380–1400°C and measuring oxygen partial pressures from 10^{-25} up to 10^5 Pa [3, 22, 23]. However, it has to be admitted that the conductivity character substantially depends on such technological parameters of material as an admixture concentration, the average crystalline size, and so on [24].

The *emf*, generated by the following solid electrolyte cell,

$$Pt\ (\mu_1)\ \|\ \text{solid electrolyte}\ \|\ (\mu_2)\ Pt, \qquad (1.20)$$

is inseparably connected to chemical potential μ and to the ionic transference number t by the following Wagner equation [16]:

$$E = \frac{1}{4F}\int_{\mu_2}^{\mu_1} t(\mu)\,d\mu \; . \tag{1.21}$$

In case $t = 1$, considering that $\mu = \mu_0 + RT \ln P$, the following equation takes place:

$$E = \frac{\mu_2 - \mu_1}{4F} = \frac{RT}{4F}\ln\frac{P_2}{P_1} \; . \tag{1.22}$$

The correlation between t and P_{O2} for $t < 1$ can be expressed as follows [25]:

$$t = \left[1 + \left(\frac{P}{P\oplus}\right)^{1/n} + \left(\frac{P}{P*}\right)^{-1/m}\right]^{-1} , \tag{1.23}$$

where $P*$ corresponds to $\sigma_{ion} = 0.50$.

For the zirconia-based electrolytes in the region where $P_{O2} < 10^5$ Pa, Equation (1.23) can be expressed as

$$t = \left[1 + \left(\frac{P}{P*}\right)^{-0.25}\right]^{-1} . \tag{1.24}$$

This is connected to the fact that for these electrolytes, the hole conductivity does not exist up to $P_{O2} = 10^5$ Pa. Then *emf* for the electrochemical cell (1.20) can be written as follows:

$$E = \frac{RT}{F}\ln\frac{(P*)^{0.25} + P_2^{0.25}}{(P*)^{0.25} + P_1^{0.25}} \; . \tag{1.25}$$

The electrolyte dissociation from the lower partial pressure side does not initially result in the decrease of *emf* for cell (1.20) because the average ionic transference number plays the crucial role and can be given as follows [26]:

$$\overline{t} = \frac{1}{\mu_2 - \mu_1}\int_{\mu_1}^{\mu_2} t(\mu)\,d\mu \; . \tag{1.26}$$

After substitution of (1.24) into (1.26) and considering (1.22) for the zirconia-based electrolytes, the following Wagner appears:

$$E = \bar{t}\, E_0. \tag{1.27}$$

It can be observed from Equation (1.27) that the average ionic transference number indicates on the *emf* deviation of the electrochemical cell (1.20) from the thermodynamic *emf* at the presence of electronic conductivity in solid electrolytes.

At the critical partial pressure,

$$lg\, P^{**} = 23.5 - (^{60500}/_T), \tag{1.28}$$

the electronic conductivity appears in the ZrO_2-CaO electrolytes and the *emf* deviation of the cell (1.20) from the thermodynamical value is no more than 1% ($\bar{t}$ = 0.99) [27, 28]. Therefore, the measuring *emf* in this case can be represented as follows:

$$E^{**} = \bar{t}\,\frac{RT}{4F}\, In\,\frac{P_2}{P^{**}}. \tag{1.29}$$

After substitution of Equation (1.28) into Equation (1.29), the critical *emf* for the zirconia-based electrolyte at $\bar{t}$ = 0.99 for P_2 = 0.21•10^5 Pa is as follows:

$$E^{**} = 2.96 - 1.19 \cdot 10^{-3}\, T. \tag{1.30}$$

Consequently, the critical temperature for the zirconia electrolyte at $\bar{t}$ < 1 can be determined from the *emf* dependence on temperature, presented by Equation (1.29).

1.3 AGING OF SOLID ELECTROLYTES

Zirconia-based electrolytes are generally operated at the high-temperature range of 650–1200°C. Such electrochemical devices as oxygen sensors and solid oxide fuel cells can provide the highest total efficiency at a temperature of about 1000°C, where the ionic conductivity reaches the required high level. However, such high temperatures often lead to reactions between the components, thermal degradation, or thermal expansion mismatch. Aging of solid electrolytes is the process of changing their electrical conductivity during prolonged exposure to high temperatures. From the practical point of view, it is very important to know how the conductivity may decrease in time, referred to as *aging* as a result of longtime annealing at temperatures of ~1000°C and higher. The aging of a solid electrolyte system has an influence on the major characteristics of the gas sensor. The accuracy and stability of the sensor may be affected during the high temperature measurements as a result of aging. The aging process for stabilized zirconia was first investigated for solid electrolyte ZrO_2-CaO [29], and it was concluded that a defect-ordering process was taking place. Later, the aging was found and studied for ZrO_2-Y_2O_3, ZrO_2-Sc_2O_3 [30–33], and $Zr(Y, Sc)O_{2-x/2}$ [34].

It was reported that the initial decrease of conductivity was attributed to precipitation of tetragonal zirconia from the cubic matrix and further decreases to the ordering in the cubic phase [35]. However, even after annealing for 42 days (1000 hours), the slight relative X-ray diffraction (XRD) intensity changes of the cubic phase are detected. It was examined and reported that the short-range ordering of oxide ion vacancies around Zr ions, resulting from relaxation of the lattice distortion, is responsible for the decrease in conductivity with aging [36]. In addition, a noticeable contribution, which can be both positive and negative, to the aging intensity by the change in electrical conductivity of grain boundaries of polycrystalline solid electrolytes has been established [37].

The following equations have been proposed for describing the kinetics of the aging process [13]:

$$R_\tau = R_\infty - B \ exp \ (-k\tau), \tag{1.31}$$

$$R_\tau = R_0 + B\tau^{0.5}, \tag{1.32}$$

where R_0, R_τ, and R_∞ are the resistances of the electrolyte before aging, at the moment of aging τ, and at $\tau \to \infty$, respectively; and k and B are constants. Equation (1.31) determines the change of the current carriers at the monomolecular quasi-chemical reaction, and Equation (1.32) describes the change of the current carriers as a result of a stationary diffusion process at the growth of a new phase. However, only (1.32) confirms experimentally measured dependencies within the narrow time interval.

Results of the modern experimental research have shown that all solid electrolytes can be divided into two major groups by the character of the aging process. The aging kinetic has a different nature for single- and two-phase polycrystalline electrolytes. Based on the solid solution phase diagrams, these two groups correspond to the single- and two-phase solid solutions. Moreover, it is also dependent upon which stabilizing oxide (Y_2O_3, MgO, Yb_2O_3, Sc_2O_3, etc.) was selected for the preparation of a ZrO_2-based electrolyte.

1.3.1 SINGLE-PHASE SOLID ELECTROLYTES

Let us consider regularity of the aging processes for the single-phase solid electrolytes. It was concluded that each chemical composition of electrolyte has its own unique start aging temperature (T_{st}) [13, 33]. For example, the start aging temperature for the solid electrolyte $Zr(Y)O_{2-x/2}$ is increased gradually from 1027 to 1200°C when the parameter x was changed from $x = 0.091$ to $x = 0.248$. Therefore, each chemical composition of electrolyte has its own limiting of aging (Figure 1.4), which depends only on temperature, as is clearly shown in Figure 1.5. From the analysis of kinetic curves, the aging process in the close vicinity of T_{st} proceeds relatively quickly within dozens of hours. The aging speed is decelerated substantially with the decrease of the working temperature and can be represented by thousands and dozens of thousands of hours. The limiting deep of aging (R_∞/R_0) is grown with the decrease of temperature.

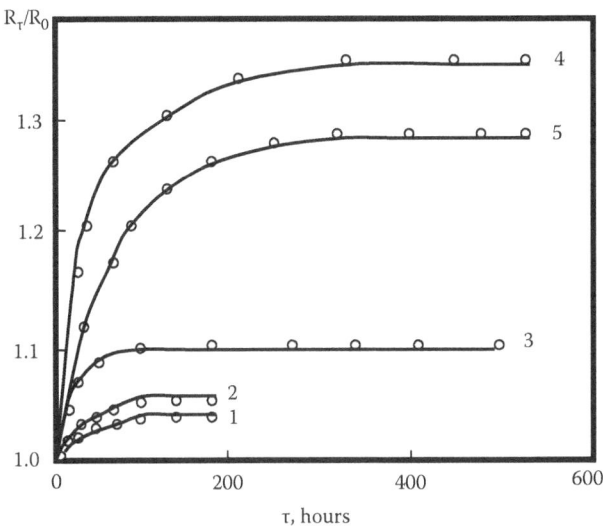

FIGURE 1.4 Kinetics of the aging process for electrolyte $Zr(Y)O_{2-x/2}$ in the single-phase region at $1100°C$: Y_2O_3, mol. %. 1: 10; 2: 12.1; 3: 15; 4: 20; 5: 33.

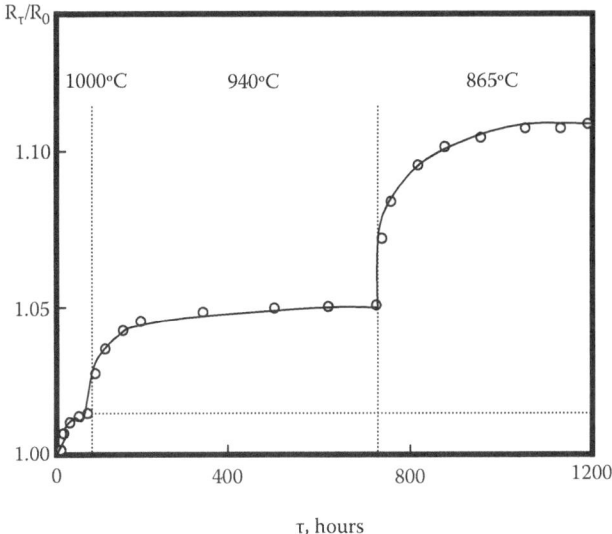

FIGURE 1.5 The aging of the solid electrolyte $Zr(Y)O_{1.909}$ at the consecutive temperature changes.

One of the reasons of aging in the single-phase electrolytes is the development of the ordering structure in the cation sublattice of the solid solution [30]. However, it is impossible to use ordinary XRD methods for investigating these ordering structures owing to the same value of the X-ray dissipation factor for Zr^{4+} and Y^{3+}.

Another possible reason of aging in such solid electrolytes can be the appearance and growth of the microdomains of a small size possessing equal to the lattice value. These microdomains of size 5–30 nm have been discovered by the electron-graphical method in the system ZrO_2-CaO [38, 39]. However, no direct experimental data have been reported so far about their existence in the solid solutions ZrO_2-R_2O_3.

1.3.2 Two-Phase Solid Electrolytes

Solid electrolytes $Zr(R)O_{2-x/2}$ possess the maximum ionic conductivity corresponding to the lowest boundary of the solid solution of the fluorite type at $x = (0.16 - 0.20)$. If $x < 0.16$, the solid solution dissociates on two phases with a following sharp decrease of its electroconductivity. It was shown [33] that the correlation between the change in the phase composition and the decrease of electroconductivity in time does exist for the zirconia-based electrolytes.

In contrast from the single-phase electrolytes, the two-phase electrolytes are characterized by the deeper aging, and the duration of aging is usually much longer. The absence of equilibrium was reported for the zirconia-based electrolytes held at the temperature of 1100°C for 2000–3000 hours [13], for the ZrO_2-Sc_2O_3 electrolyte part of the aging curve has the S-shape form [34].

The aging of oxygen sensors based on HfO_2-ZrO_2-Y_2O_3 and ZrO_2-Y_2O_3 solid electrolytes with platinum electrodes and with 15 and 10 mol % of Y_2O_3, respectively, was investigated [40]. The sensors were annealed for 42 days (1000 hours) at 1000°C. It has recently been found [41] that the aging process does not occur before 1000°C in electrolytes doped with 10 mol % of Y_2O_3. Figure 1.6 illustrates how the conductivity of YSZ with different yttria content has changed after annealing for 30 days at 1000°C [42]. This investigation of the aging processes has shown that a ZrO_2 electrolyte containing 10 mol % of Y_2O_3 is much less affected by the aging process than the same electrolyte containing 8 mol % of Y_2O_3. With a lower than 10 mol % of Y_2O_3, the cubic solid solution gradually breaks up into two phases, thus leading to a drop in its electrical conductivity [43]. Therefore, zirconia electrolytes containing a 10 mol % of Y_2O_3 are more preferable to use in practical gas sensors working at the high-temperature range of 900–1200°C.

Figure 1.7 shows the results of the resistivity measurement for both electrolyte specimens before and after annealing. The deterioration of conductivity of the ZrO_2-Y_2O_3 sample is significant; however, the level of reduction is still such that the total conductivity remains high. It is also well-known that the rate of conductivity deterioration of ZrO_2-Y_2O_3 electrolytes is very rapid during the initial stages of annealing. Nevertheless, it slows down considerably and it appears that the deterioration will settle down after prolonged heat treatment. A number of features of these results require further elaboration. This is a trend in the amount of conductivity reduction with time which is related in some way to the amount of HfO_2 present in the HfO_2-ZrO_2-Y_2O_3 system. The HfO_2-ZrO_2-Y_2O_3 electrolyte showed a similar degradation rate in conductivity to the ZrO_2-Y_2O_3 electrolyte as a function of time, but the degradation rate decreased with an increase in the HfO_2 content. For example, the resistivity of the HfO_2-ZrO_2-Y_2O_3 electrolyte (15 mol % of Y_2O_3 and 65 mol % of HfO_2) at a temperature of 1000°C was increased by ~50% after 42 days of annealing.

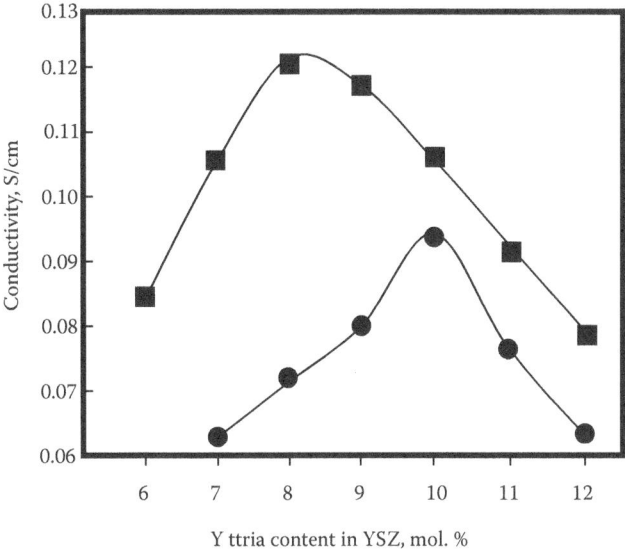

FIGURE 1.6 Conductivities of YSZ solid electrolytes at 1000°C versus yttria content: (■) before annealing; and (●) after 30 days of annealing. (Reprinted from Zhuiykov, S., Development of dual sulphur oxides and oxygen solid state sensor for "*in-situ*" measurements, *Fuel* **79** (2000) 1255–1265, with permission from Elsevier Science.)

In contrast, the resistivity of the ZrO_2-Y_2O_3 (10 mol % of Y_2O_3) electrolyte at the same temperature was increased by ~110%. This may suggest that the main cause of aging of solid electrolytes in the two-phase region appears to be the precipitation and growth of the second conductive phase. The kinetics of the aging process is determined by the kinetics of two processes running independently of each other: growth of the number of second-phase centers, and the growth of their bulk. In the case of polycrystalline solid electrolytes, the process of the second-phase growth is connected with the diffusion of solid solution components. Therefore, an increase in the HfO_2 content in the HfO_2-ZrO_2-Y_2O_3 system diminishes the degradation rate in conductivity of the composite electrolyte. These results also appear to indicate that the HfO_2-based electrolytes are more likely to be used in sensors measuring extremely low oxygen partial pressure at temperatures higher than 1000–1100°C [40].

1.4 AN ELECTRON MODEL OF SOLID OXYGEN-IONIC ELECTROLYTES USED IN GAS SENSORS

For the zirconia-based gas sensors, the low level of threshold temperature, when the zirconia electrolytes possess pure ionic conductivity, is approximately 500–550°C for polycrystalline structures [44–46] and around 380–420°C for single crystals [47, 48]. The conductivity of the YSZ-based electrolyte below these temperatures is compatible with the conductivity of isolators. Moreover, any reduction in operating

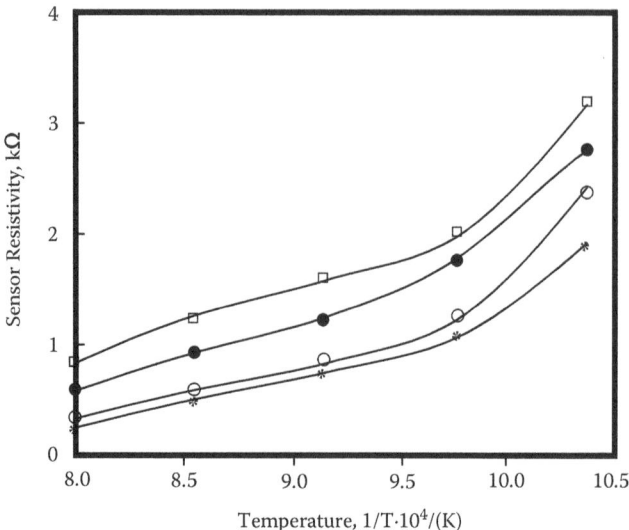

FIGURE 1.7 Measurements of the resistivity of solid electrolyte systems before and after annealing at 1000°C for 1000 hours: (*) ZrO_2–10 mol% Y_2O_3 before annealing; (○) HfO_2–ZrO_2–15 mol% Y_2O_3 before annealing; (□) ZrO_2–10 mol% Y_2O_3 after annealing; and (●) HfO_2–ZrO_2–15 mol% Y_2O_3 after annealing. (Reprinted from Zhuiykov, S., Investigation of conductivity, microstructure and stability of HfO_2–ZrO_2–Y_2O_3–Al_2O_3 electrolyte compositions for high-temperature oxygen measurement, *J. Europ. Ceram. Soc.* **20** (2000) 967–976, with permission from Elsevier Science.)

temperature of the oxygen sensor by the fabrication technology has a considerable effect on materials selection and on lifetime predictions for such devices. As a result of that, the electron processes, stipulated by the admixture and the structural defects, start to influence the properties of oxygen-ionic electrolytes at low temperatures.

In addition, modern solid electrolyte theory considers electrons only as a secondary charge carrier [23]. Furthermore, theory uses the definition *chemical oxygen potential* as a binding element of the electrode reactions for description of galvanic systems with oxygen-ionic electrolytes. However, the change of the chemical potential for oxygen vacancies within the zirconia-based electrolyte is restricted by the narrow region of homogeneity of the solid electrolyte, and the concept of *gaseous oxygen* for the single-crystal electrolytes has no physical sense. The chemical potential of electrons is much more preferable to use as a binding element for the electrode reactions on the boundaries of both polycrystalline and single-crystal electrolytes. In the electrochemical form, the chemical potential of electrons represents the Fermi level E_F, detaching, by the definition [49], the free conditions from the engaged ones at the absolute zero temperature. The electrical fields at the solid electrolyte-electrode interfaces depend on the chemical potential differences of both mobile species: ions and electrons. The mobility of ions results in variations of the chemical composition of compounds and even in the formation of new interfacial materials, which generate changes in the performance and cause short lifetimes [50]. Moreover, the electrical

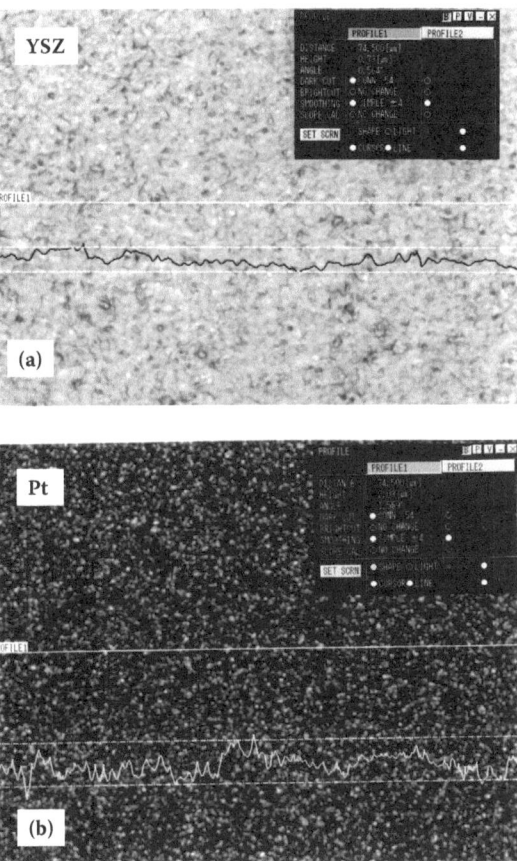

FIGURE 1.8 *a*: Surface SEM images of polished YSZ substrate; *b*: Pt electrode deposited on YSZ substrate; and *c*: three-dimensional image of the surface of the Pt electrode deposited on YSZ. (From Zhuiykov, S., Electron model of solid oxygen-ionic electrolytes used in gas sensors, *Int. J. Applied Ceramic Techn.* **3** (2006) 401–411. With permission.)

field at the contact between a metal electrode and a solid electrolyte is much stronger and extends over a much smaller regime than in the case of a semiconductor junction. This is owing to the much higher concentration of charge carriers in metals and solid ion-conductive electrolytes compared to semiconductors.

Figure 1.8 shows the roughness of the typical polished $(ZrO_2)_{0.9}(Y_2O_3)_{0.1}$ surface (*a*); the roughness of the Pt electrode, made of Pt paste, deposited on this surface (*b*); and the three-dimensional image of the surface of the Pt electrode deposited on YSZ (*c*). This figure clearly illustrates that the difference between the lowest and highest peaks for the polished YSZ substrate does not exceed 0.73 μm, and for the deposited Pt electrode on the YSZ substrate is about 2.03 μm. Magnified scanning electron microscopy (SEM) images of YSZ substrate and Pt electrode up to the same nanoscale, shown in Figure 1.9, indicate that the YSZ structure consists of small and large developed YSZ grains with an average size from 300 nm up to 1 μm. A

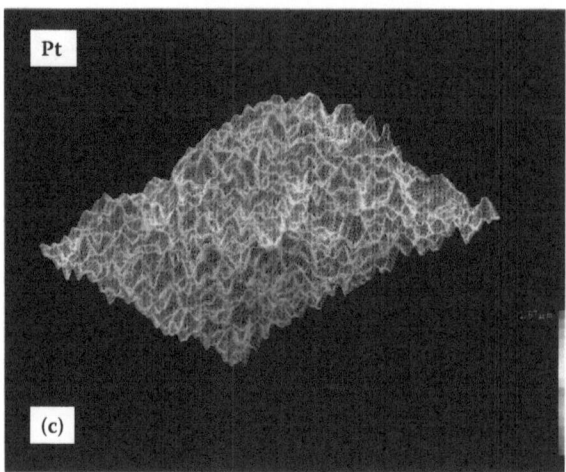

FIGURE 1.8 (Continued.)

similar situation has been observed for the Pt electrode (see Figure 1.9, *b*). The vast majority of Pt particles have an average size of about 1 μm. In addition, the TEM image, shown in Figure 1.10, indicates that the vicinity of interface between YSZ and NiO-sensing electrode in the YSZ-based NO_x planar sensor is only 2 nm [51, 52]. Both provided SEM and transmission electron microscopy (TEM) images, and referenced data support the aspect that the thickness of the space charge layer is usually in the range of a few nm instead of μm, which consequently results in much higher sensitivity in regard to the formation of interfacial layers.

From the solid-state zone theory point of view, solid electrolytes are dielectrics with the wide forbidden zone E = 4 – 6 eV [23]. In contrast from the ordinary dielectrics, which have a fixed Fermi level E_F in the middle of the forbidden zone, the position of the Fermi level for the solid electrolytes depends on the thermody-namical oxygen potential on crystalline boundaries. This is due to the fact that the solid electrolytes are not just dielectrics, but rather crystalline phases with ionic conductivity, stipulated by the high density of oxygen vacancies ($n_V = 10^{26} - 10^{28}$ m^{-3}), in the anion sublattice of the electrolyte.

Then, for example, if the oxygen partial pressure decreases in the measuring environment, the solid electrolyte loses part of the oxygen atoms in correspondence with the Equations (1.17) and (1.18) for the electrode reactions.

Partial dissociation of the electrolyte results in reduction of density of the electron conditions in the valent zone and, consequently, to the electron transfer into the conductivity zone. Owing to this, the Fermi level increases within the forbidden zone of the electrolyte. On the analogy of semiconductors, the solid electrolytes can be considered as admixture semiconductors in which the content of donor (or acceptor) admixture depends on the oxygen partial pressure in the analyzing envi-ronment [53].

The chemical potential of electrons μ_θ in dielectric material can be represented as $\mu_\theta = E_F - E_C$, where E_C is the bottom level of the conductivity zone. In this case,

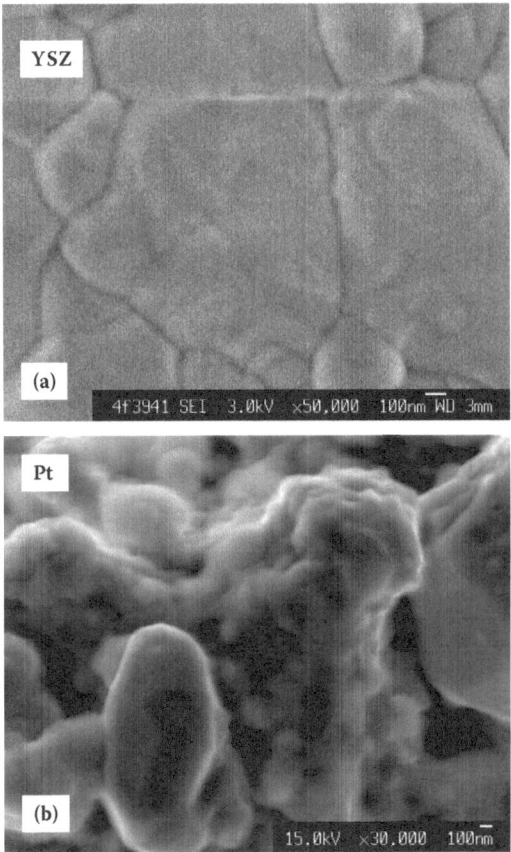

FIGURE 1.9 Nanoscaled SEM images of (*a*) YSZ electrolyte and (*b*) Pt electrode. (From Zhuiykov, S., Electron model of solid oxygen-ionic electrolytes used in gas sensors, *Int. J. Applied Ceramic Techn.* **3** (2006) 401–411. With permission.)

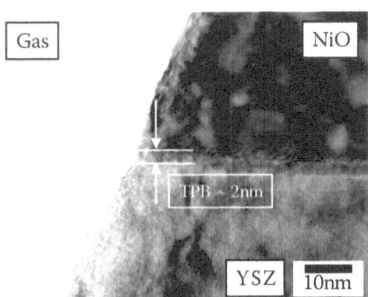

FIGURE 1.10 TEM image in the vicinity of interface between YSZ and the NiO-SE sintered at 1400°C. (From Zhuiykov, S., Electron model of solid oxygen-ionic electrolytes used in gas sensors, *Int. J. Applied Ceramic Techn.* **3** (2006) 401–411. With permission.)

from the thermodynamic equilibrium condition for reactions described by Equations (1.17) and (1.18), the following correlation can be easily obtained:

$$(E_F)_f = E_C + \mu^* - \tfrac{1}{2}\,\mu_V - \tfrac{1}{4}\,\mu_{O2}. \tag{1.33}$$

Here, index f designates the electrolyte surface, μ^* is the constant, and μ_V is the chemical potential of the oxygen vacancies. $\mu_{O2} = kT \ln P_{O2}$ represents the thermo-dynamical oxygen potential in the environment, where k, T, and P_{O2} denote the Boltzmann constant, temperature, and oxygen partial pressure, respectively.

Dependence of the Fermi level on the thermodynamical oxygen potential at the boundaries of solid electrolyte μ_{O2} is simple, owing to the small density of the electron conditions in the forbidden zone of dielectric and on account of high density of the vacant (V) groups in the anion sublattice of the electrolyte. Analytically, it can be shown as

$$\left| \frac{\partial \mu_v}{\partial (E_F - E_C)} \right| \ll 1 . \tag{1.34}$$

Both correlations (1.33) and (1.34) point out that the electrons play an important role in the electrophysics of solid electrolytes. They also point out the singleness of purpose of describing the properties of electrolytes within the solid-state zone theory.

Let us consider the contact phenomena on the boundary of the solid electro-lyte–metal interface, schematically shown in Figure 1.11. In this case, the work function of electrons from the metal χ_M is bigger than the work function of electrons from the oxide electrolyte χ_E. Electrons immediately transfer from the metal into the solid electrolyte after establishment of initial contact between electrolyte and electrode. This is caused by the electron diffusion driven by the electron concentra-tion gradient. Only under working conditions (i.e., low oxygen partial pressure) would the electrons produced by the oxygen oxidation at the surface be transported into the metal electrode. The Fermi level displaces to the middle of the forbidden zone; however, in contrast from semiconductors, the volume charge of the solid electrolyte is neutralized by the oxygen ions going away off the surface of the electrolyte. The oxygen vacancies, accumulated on the surface of the electrolyte, form the dense part of the electric double layer [54]. In the zirconia-based electro-lytes, as in the metals, the screening length $L_D = (\varepsilon kT/8\pi e^2 n_V)^{1/2}$, determining the thickness of the electric double layer, is equal by the order of magnitude to the interatom distance. This small screening length is stipulated by high density of the oxygen vacancies–charge carrier in the solid oxygen-ionic electrolytes.

The displacement of Fermi level E_F disturbs the thermodynamic equilibrium in the solid electrolyte-electrode system, expressed by Equation (1.33). Subsequently, the solid electrolyte dissociates in accordance with reaction (1.17) followed by the appearance of free electrons on its surface, increasing density of the oxygen vacan-cies in the double layer and rising of the Fermi level. The equilibrium restores (see Figure 1.11, b) when the contact potentials difference E_k becomes equal to

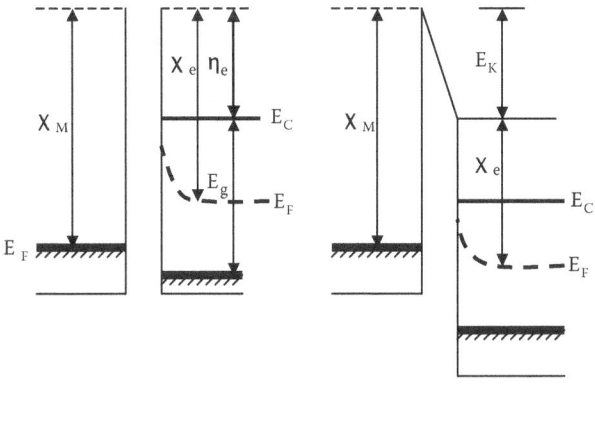

a   b

FIGURE 1.11 Schematic interpretation of the energetic structures for electrons in metal and solid electrolyte (*a*) before their contact and (*b*) after establishment of contact. (From Zhuiykov, S., Electron model of solid oxygen-ionic electrolytes used in gas sensors, *Int. J. Applied Ceramic Techn.* **3** (2006) 401–411. With permission.)

$$E_k = (\chi_M - \chi_E)/e, \tag{1.35}$$

where *e* is the charge of the electron.

Similar to semiconductors, the work function of electrons from the oxide electrolyte χ_E can be expressed by the Wagner equation [14]:

$$\chi_E \equiv \eta_e - E_F + E_C, \tag{1.36}$$

where η_e is an electron affinity of the solid electrolyte.

Then the linear dependence of the contact potentials difference E_k on the thermodynamic oxygen potential μ_{O2} in the measuring environment can be obtained from Equations (1.33) and (1.35):

$$E_k = (\chi_M - \eta_e + \mu^* - \tfrac{1}{2}\,\mu_V)/e - \mu_{O2}/4e. \tag{1.37}$$

Therefore, the output *emf*, which is equal to the difference of contact potentials difference, appears at the different oxygen partial pressures P_{O2} on the opposite boundaries of the solid electrolyte in open circuit (schematically shown in Figure 1.12):

$$E = E_k(0) - E_k(h) = [\chi_E(h) - \chi_E(0)]/e, \tag{1.38}$$

where *h* is the electrolyte thickness.

Correlation (1.38) with due regard for the Wagner Equation (1.36) can be transformed to

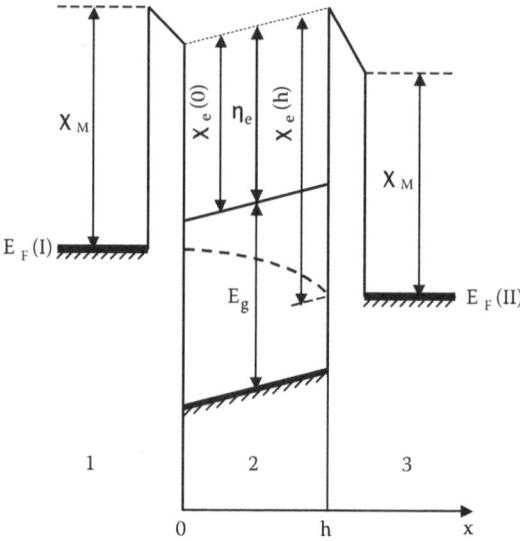

FIGURE 1.12 Equivalent scheme of the galvanic concentration element. *1*: first electrode (φ_1); *2*: solid oxygen-ionic electrolyte; and *3*: second electrode (φ_2). (From Zhuiykov, S., Electron model of solid oxygen-ionic electrolytes used in gas sensors, *Int. J. Applied Ceramic Techn.* **3** (2006) 401–411. With permission.)

$$E = [E_F (0) - E_F (h)]/e. \qquad (1.39)$$

The emerging gradient of the chemical potential of electrons brings to the diffusive charges transfer. As a result, the electrical field $d\varphi/dx$, which decreased the *emf* of circuit in accordance with (1.37), appears within the electrolyte. It is graphically shown in Figure 1.12 as a pitch of power levels of electrons. In one's turn the electrical field provokes the ionic current compensating diffusive transfer of the electrons and supporting the difference of the Fermi levels on the boundaries of the solid electrolyte. If the electrical field within the solid electrolyte cannot be ignored, then the difference of the electrostatic potentials,

$$\Delta\varphi = \int_0^h \frac{d\varphi}{dx}dx \; ,$$

should be added to Equation (1.38):

$$E = \left[\chi_E(h) - \chi_E(0)\right]/e + \int_0^h \frac{d\varphi}{dx}dx \; . \qquad (1.40)$$

If the Fermi level within the electrolyte can nominally be expressed by Equation (1.33), then Equation (1.40) can be transformed to the well-known Wagner's integral [16]:

$$E = \frac{1}{4e} \int_0^h t_i \frac{d\mu_{O2}}{dx} dx ,$$

where t_i is the ionic transference number. Such transformation indicates on correspondence of the presented model to the Wagner's solid electrolyte theory [14].

Based on the fact that $\Delta\varphi = -\Delta E_C$, the correlation (1.39), in difference from Equation (1.38), does not change in the presence of the electrical field within the solid electrolyte. Therefore, the output *emf* of the opened galvanic circuit can be determined by the difference in Fermi levels on the boundaries of the solid electrolyte regardless of the electronic-to-ionic conductivities ratio.

Maximum allowable *emf* of the opened circuit does not exceed the width of the forbidden zone divided on the electron charge: $E < \varepsilon_g/e$. This restriction is stipulated by the high density of the electron conditions in the conductivity zone and in the valent zone, in comparison with the forbidden zone of the electron energy in dielectric. In this case, insignificant shift of the Fermi level in the permitted zones leads to substantial growth of the electron-hole conductivity, acceleration of the diffusive transfer of electrons and holes through the electrolyte, extension of the internal electrical field $d\varphi/dx$, and, consequently, the reduction of the output *emf* of the galvanic circuit, determined by correlation (1.40).

Hence, from the electron model of the solid electrolyte following that the *emf* of the opened galvanic circuit, firstly, can be determined by the difference of the Fermi levels on the boundaries of solid electrolyte and, secondly, can be restricted by the ratio between the width of the forbidden zone and the charge of electron. Otherwise, the properties of the electrolyte can be changed by its dissociation. In view of the fact that the electrons have high density within the conductivity zone and within the valent zone by comparison to the forbidden zone, the correlation (1.34) for these parts of the energetic spectrum of electrons does not fulfill. Thus, in order to change the Fermi level within the conductivity zone, that is, break inequality,

$$E < \varepsilon_g/e, \tag{1.41}$$

it is necessary to change the electrolyte composition by its dissociation. In this scenario, the value of the oxygen partial pressure P_{O2}^0, corresponding to the minimum electron-hole conductivity of the oxide electrolyte, allows defining the lowest level of P_{O2}^m at the admitted oxygen pressures, at which the electrolyte can yet be used in oxygen sensors. Based on the presented electron model of oxygen-ionic electrolytes, the minimum electron-hole conductivity can be achieved when the Fermi level E_F disposes in the middle of the forbidden zone. If the Fermi level E_F corresponds to the oxygen partial pressure P_{O2}^0 (*T*), then from the correlation (1.41) with due regard for the polar correction V_p [55, 56] and the Nernst equation, the following inequality yields,

$$\left(\varepsilon_g - 2V_p\right)/2e \geq \frac{kT}{4e} \, ln\left(\frac{P_{02}^0}{P_{02}^m}\right), \tag{1.42}$$

where $P_{O2}'' = P_{O2}^0$ and $P_{O2}^{0'} = P_{O2}^m$. Then, the minimum measuring P_{O2}^m can be expressed as

$$- \, ln \, P_{O2}^m = - \, ln \, P_{O2}^0 \, (T) + (2 \, \varepsilon_g - 4V_p)/kT. \tag{1.43}$$

From the data [57] concerning determination of the *emf* of the dissociated electrolyte in relation to the air reference electrode, it can be found that for the electrons localized on the anion vacancies in electrolyte $(ZrO_2)_{0.9}(Y_2O_3)_{0.1}$, the polar correction is equal to ~0.4 eV.

Let us suppose that the polar correction V_p for the $(ZrO_2)_{0.85}(CaO)_{0.15}$ electrolyte has the same order of magnitude as for $(ZrO_2)_{0.9}(Y_2O_3)_{0.1}$. Then the temperature dependence of the minimum P_{O2}^m (T) for $(ZrO_2)_{0.9}(Y_2O_3)_{0.1}$ and $(ZrO_2)_{0.85}(CaO)_{0.15}$ electrolytes can be calculated by inserting the values of P_{O2}^0 (T), ε_g, and V_p, respectively, in Equation (1.43):

$$- (lg \, P_{02}^m)_Y = - \, 10 + 53 \, (10^3/T); \tag{1.44}$$

$$- (lg \, P_{02}^m)_{Ca} = - \, 5.4 + 70.5 \, (10^3/T). \tag{1.45}$$

Figure 1.13 shows the low boundary of the oxygen-ionic conductivity region based on Equation (1.44) for the $(ZrO_2)_{0.9}(Y_2O_3)_{0.1}$ electrolyte. The logarithm of the dissociation pressure for oxide Na_2O is also shown for comparison. It is evident from

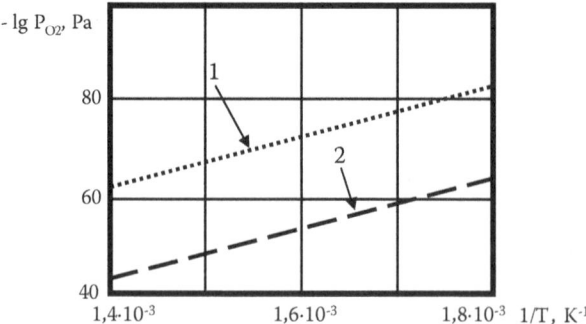

FIGURE 1.13 *1*: Low boundary of the oxygen-ionic conductivity for the solid $(ZrO_2)_{0.9}(Y_2O_3)_{0.1}$ electrolyte; and *2*: and the temperature dependence of logarithm pressure of dissociation for oxide Na_2O. (From Zhuiykov, S., Electron model of solid oxygen-ionic electrolytes used in gas sensors, *Int. J. Applied Ceramic Techn.* **3** (2006) 401–411. With permission.)

this figure that the YSZ allowing control of the oxidizing potential not only oxygen saturated by admixture, but also deoxidized sodium in the low temperature region. Taking into consideration the data of solubility oxygen in sodium, the YSZ-based solid electrolyte sensors ensure control of the activity of the oxygen admixture in liquid sodium in the range of $a = 1 - 10^{-10}$.

Basically, the nature of the electron conductivity in the solid electrolytes depends on two mechanisms of charge transfer: (1) transfer of electrons in the conductivity zone, separated from the valent zone in the solid electrolyte by the wide enough forbidden zone; and (2) the spasmodic transitions of electrons from one local level to another within the forbidden zone.

Apparently, the admixture atoms of the metals with alternative valence, local levels of which are adjacent to the valent zone of electrolytes, bring the greatest contribution into the electron conductivity of solid electrolytes. This fact, in one's turn, increases the density of electrons on these levels simply because the equilibrium between electrons within the conductivity zone and on the local levels can be expressed by the following law in force:

$$\frac{n^e_-}{N_e} = \frac{n^z_-}{n} exp \frac{\Delta G_e}{RT} , \tag{1.46}$$

where n^e_- is the density of electrons on the local levels, N_e is the density of the local levels, and n^z_- is the density of electrons within the conductivity zone. ΔG_e is the activation energy of the electron transfer from the local level into the conductivity zone, $n = (mkT/2\pi h)^{3/2}$, and m is the mass of the electron.

Consequently, it can be concluded that the "deeper" the level within the forbidden zone, the larger the value of ΔG_e and, therefore, the higher the density of electrons on local levels.

In the close vicinity between atoms forming the local levels in the energetic spectrum of electrons, the value of the potential barrier between them is less ΔG_e on some value ΔG_z (see Figure 1.14). The total conductivity of electrons on the local levels and in the conductivity zone, considering that both uniformity and isothermal conditions of the polycrystalline solid electrolyte are met, can be described by the following equation:

$$\sigma_- = F\left[n^z_- u^z_- + n^e_- u^z_- \cdot exp\left(\frac{\Delta G_z - \Delta G_e}{RT}\right)\right], \tag{1.47}$$

where u^z_- is the mobility of electrons. If n^e_- can be substituted in this equation from the correlation (1.46), the conductivity turned out,

$$\sigma_- = Fn^z_- u^z_- \left(1 + \frac{N^*_e}{n} exp \frac{\Delta G_z}{RT}\right), \tag{1.48}$$

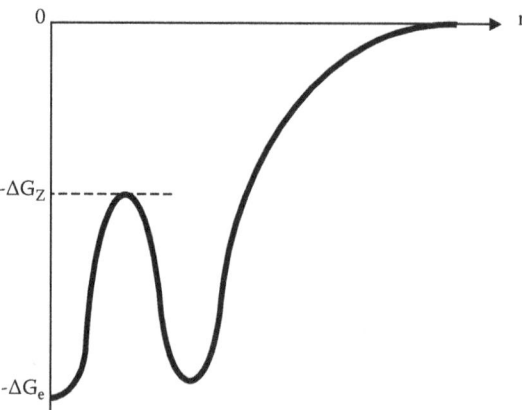

FIGURE 1.14 Potential energy of electrons on the local level. (From Zhuiykov, S., Electron model of solid oxygen-ionic electrolytes used in gas sensors, *Int. J. Applied Ceramic Techn.* **3** (2006) 401–411. With permission.)

where N^*_e is the density associations of two and more atoms of metal, forming the local levels of electrons.

As long as we have an equilibrium between electrons on the local levels and electrons in the conductivity zone, that is, $\mu^z_- = \mu^e_-$, the ionic transference number (t_i) can be described by the following equation:

$$t_i = \left\{ 1 + \left[\frac{n}{2N} \left(\frac{n_{\ddot{o}}}{N} \right)^{-3/2} \cdot \frac{1 + \dfrac{N^*_e}{n} exp\,\dfrac{\Delta G_z}{RT}}{1 + \dfrac{n_{O''}}{n_{\ddot{o}}} \cdot \dfrac{u_{O''}}{u_{\ddot{o}}}} \times \right.\right.$$

$$\left.\left. \times exp\left(-\frac{\Delta G^0}{4RT} \right) \frac{u^z_-}{u_{\ddot{o}}} \right] P_{02}^{-1/4} \right\}^{-1}$$ (1.49)

where N is the number of groups of oxygen sublattices in the unit of volume and $n_{O''}$, n are the density of the oxygen vacancies and intergroup oxygen ions, respectively, corresponding to the electroneutrality condition; $n_{\ddot{o}} - n_{O''} - (n\_/2) = n_{bO''}\, n_-$ is the density of electrons, and n_{bo} is the density of surplus oxygen vacancies. The second item in Equation (1.49) represents the conductivities ratio $\sigma_-/(\sigma_{\ddot{o}} + \sigma_{O''})$. Therefore, by designating the multiplier independent on the oxygen partial pressure, as P_-, the correlation (1.49) can be transformed into the well-known equation for the oxygen ions transference number:

$$t_i = \left[1 + \left(\frac{P_-}{P_{02}} \right) \right]^{-1}.$$

Nonregulating of such ionic crystals as solid electrolytes is insignificant at low temperatures, that is, $n_{\ddot{O}} \gg n_{O''}$. Moreover, it is known that the ratio of the mobility of vacancies and electrons in the conductivity zone can be linked to the temperature:

$$\frac{u_-}{u_{\ddot{O}}} \sim exp\left(\frac{const}{T}\right).$$

Then the functional dependence of the logarithm P_- on temperature can be expressed as

$$- lg \; P_- = A + B \; (10^3/T) - 4lg \; [1 + c^* \; exp \; (A'_m + B'_m \; (10^3/T))],$$

where A, B, A'_m, and B'_m are the constants and c^* is the concentration of associations for two admixture atoms, forming the local levels in the energetic spectrum of electrons.

Thus, as long as associations are in equilibrium with the atoms making them available, then at the condition of observance of law in force, the correlation $c^* \sim c$ takes place, where c is the concentration of appropriate admixture atoms. Consequently,

$$- lg \; P_- = A + B \; (10^3/T) - 4lg \; [1 + c^2 \; exp \; (A_m + B_m \; (10^3/T))]. \qquad (1.50)$$

Equation (1.50) evidently expresses an extra contribution of the admixture atoms into the electron mechanism of conductivity, which leaves a restriction on the area of the primary anion conductivity [58].

Therefore, even small traces of technological admixtures in the solid oxygen-ionic electrolytes (for example, iron, vanadium, and titanium), which usually accumulate on the grain boundaries, can substantially influence the limits of the practical applicability of electrolytes by temperature and by the level of partial pressure.

The quadratic dependence of the technological admixture in Equation (1.50), relating to the complex influence of the admixtures on the conductivity of the solid electrolyte, has expressed at least in their twin interactions. Since the requirements for the close vicinity of two or more atoms, forming the local energetic levels, are relatively easy to meet, there are many possibilities for uninterruptible spasmodic electron transfer from one level on another. Otherwise, the transference of electrons would be carried out only within the conductivity zone. In fact, the height of the potential barrier for the electron displacement from one potential level to another depends substantially on the distance between levels. Taking into account that the surrounding ions are exerting a screening influence on the electrons transfer in the crystalline structure, only the close vicinity with a view to the bonded couple of atoms decreases the height of potential barrier.

Using the assumptions given above, it was possible to determine the binary metal compositions with unlimited solubility, which is illustrated in Figure 1.15. These binary metal compositions are vital for the selection of materials for a sensing electrode of the mixed-potential zirconia-based gas sensors [23, 24]. This is because

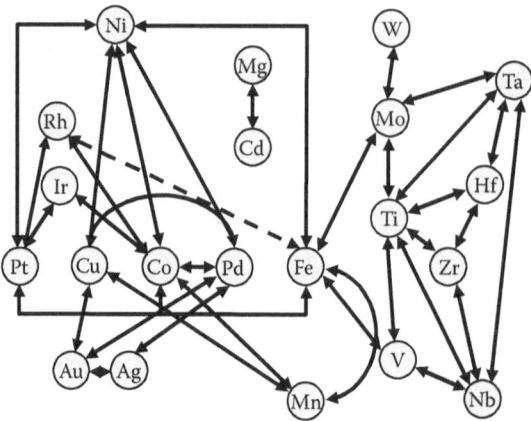

FIGURE 1.15 Binary metallic compositions with unlimited solubility; the solid line represents systems with transfer without breach of homogeneity, and the dotted line represents systems with solubility within the wide temperature range.

the ideal sensing electrode is likely to be a mixture of two or more oxides or one oxide doped by the admixture of another metal or metal oxide, which usually enhances sensitivity to the measuring gas or, alternatively, suppresses unwanted sensitivity to another gas [25]. The metallurgical stability of such a composite system is determined by the character of the phase equilibrium of metals, speed of their interdiffusion, and adhesion to each other. Basically, there are two approaches to the selection of the multiphase system nowadays: (1) the selection of materials developing a continuous row of the solid solutions with high speed of the diffusive "eroding" boundaries at the immediate contact (see Figure 1.15), and (2) the binary compositions with the minimum speed of the interdiffusion, which do not form intermediate phases in the working temperature range of the gas sensors (see Figure 1.16).

Verification of the proposed model can be illustrated by the investigation of the influence of the iron admixture on the conductivity of the YSZ electrolyte at the low threshold temperature. The following electrochemical cell has been investigated:

$$Pt \mid Fe, FeO \parallel YSZ \text{ solid electrolyte} \parallel O_2 \ (P_{O2} = 0.21 \times 10^5 Pa) \mid Pt.$$

Five samples of the YSZ with different concentrations of iron, impregnated into the electrolyte, have been studied. For investigation of the influence of the iron admixture on the electrophysical properties of the solid electrolyte, four samples from five were saturated by a FeCl solution of different concentrations (1, 3.8, 7.5, and 15%), followed by the annealing of all samples at 1600°C for 3 hours. The fifth sample had the minimum iron admixture (0.001%). After annealing, all samples had a density of 5.4 g/cm³ with zero open porosity. Figure 1.17 shows experimental *emf* measurements for all samples of the prepared electrolyte. The thermodynamical *emf* ($\bar{t} = 1$) corresponding to Equation (1.22) is shown as well. The results of the experimental measurements in this figure clearly show that the more iron admixture

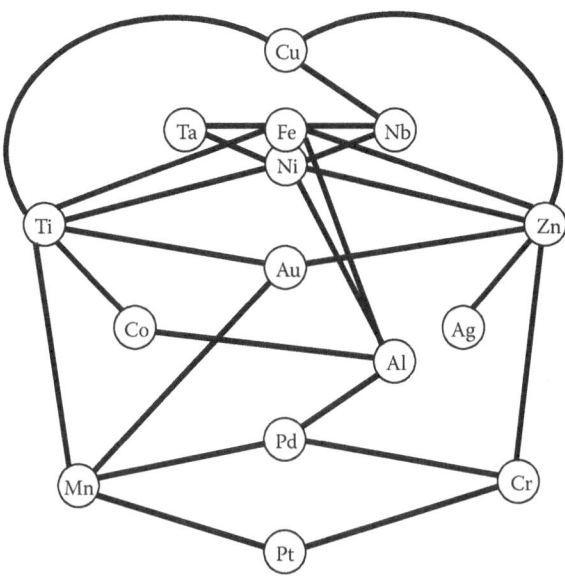

FIGURE 1.16 Binary metallic compositions with minimum speed of interdiffusion of metals.

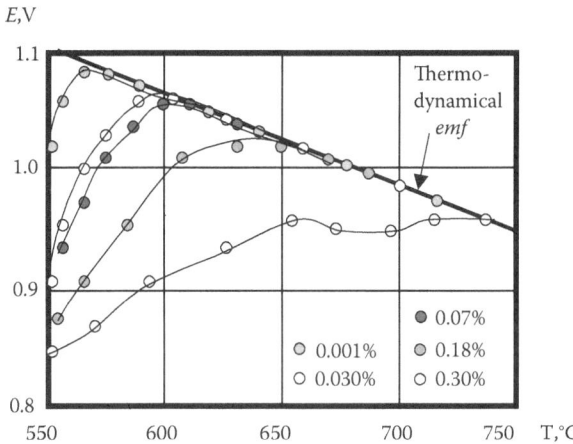

FIGURE 1.17 Calculated and measured dependences of the zirconia electrochemical cells with different percentages of the iron admixture on temperature. (From Zhuiykov, S. [52]. With permission.)

the solid electrolyte possesses, the higher the threshold temperature is required for zirconia to be a pure ionic conductor. The functional dependence of the threshold temperature (T_t) on the logarithm of the iron admixture concentration (C_{Fe}) in the solid electrolyte can be expressed as follows:

$$1000/T_t = 0.83 - 0.43 \; lg \; C_{Fe}.$$

This temperature characterizes the transfer from the mix of electrolyte electroconductivity into the oxygen-ionic conductivity by correlation $E/E_0 = 0.98$. As a result,
these experiments concluded that the decrease of percentage of the iron admixture
in zirconia from 0.3 to 0.001% allows decreasing the lower level of threshold
temperature T_t from 740 to 560°C.

From a structural perspective, the degree of interface modification is responsible
for regulating the kinetics of the *emf* measurement as a function of its influence on
the structural gas permeability of electrodes, which controls the initial rate of physic-
chemical reactions of adsorption–desorption and diffusion on electrodes. In addition,
the sensitivity of the zirconia gas sensors is highly sensitive to chemistry and can
be lost by minor changes either in the phase purity or at the presence of metallic
admixtures in the ceramic ionic conductors.

Consequently, the proposed model allows the necessary information regarding
the electrolyte–metal electrode interface and about the character of the electronic
conductivity in solid electrolytes to be obtained. To an extent, this is additionally
reflected by the broad range of theoretical studies currently published in the scientific
media and is inconsistent with some of the research outcomes relative to both
physical chemistry of phenomena on the electrolyte-electrode interfaces and their
structures. Partially, this is due to relative simplifications of the models, which do
not take into account multidimensional effects, convective transport within interfaces, and thermal diffusion owing to the temperature gradients. An opportunity may
exist in the further development of a number of the specific mathematical and
numerical models of solid electrolyte gas sensors matched to their specific applications; however, this must be balanced with the resistance of sensor manufacturers
to carry out numerous numbers of tests for verification and validation of these models
in addition to the technological improvements.

1.5 ELECTRODE PROCESSES IN SOLID ELECTROLYTE SENSORS

1.5.1 ELECTRODE REACTION WITHIN THE TRIPLE-PHASE BOUNDARY

If gaseous, electrochemically active components of the measuring environment are
not dissolved in the electrode, then the electrode process will consist of the following
stages (also shown in Figure 1.18). They are adsorption–desorption of electrochemically active gaseous components on gas-electrolyte (GE) and gas-metal (GM) interfaces, ionization reaction (with electron transfer) on the metal-electrolyte (ME) and
gas-electrolyte interfaces, and mass-transfer processes on all boundaries of three
phases (gas-metal, gas-electrolyte, and metal-electrolyte). Furthermore, mass transfer of electrons and holes on the surface electrolyte layer may also occur. It is evident
that the quantity of the current in the stationary state is equal to the quantity of the
nonmetal component adsorbing on the gas-metal and gas-electrolyte surfaces as a
result of ionization of this component on the ME and GE surfaces.

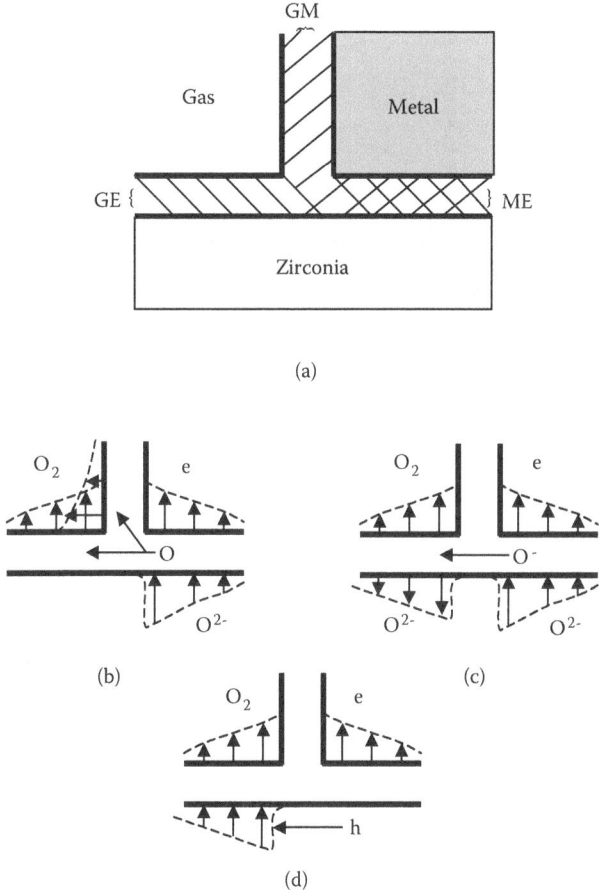

FIGURE 1.18 Model of the electrode structure (*a*), and schemes of the development of electrode process in the TPB at the surface diffusion of (*b*) oxygen atoms, (*c*) subions, and (*d*) electron holes. (From Zhuiykov, S. [52]. With permission.)

Mass transfer, basically taking place by the diffusion on the interphase surfaces, stipulates the transfer of electrochemically active components from the places of adsorption to the places of the electrochemical reaction. A scheme of the electrode process development allows various scenarios determined by the combination of the separate stages. For example, electrode system $M, O_2 | O^{2-}$ variants, considering the extension of the reaction zone in the contact of the TPB by the surface diffusion of oxygen atoms, subions (O^-), and electron (holes), come to the following (scheme for anodic reaction):

1. Discharge of the oxygen ions on the metal-electrolyte boundary to the atoms (reaction DR; see Figure 1.19); diffusion of the oxygen atoms on the metal-electrolyte, gas-metal, gas-electrolyte boundaries (Stages D_1,

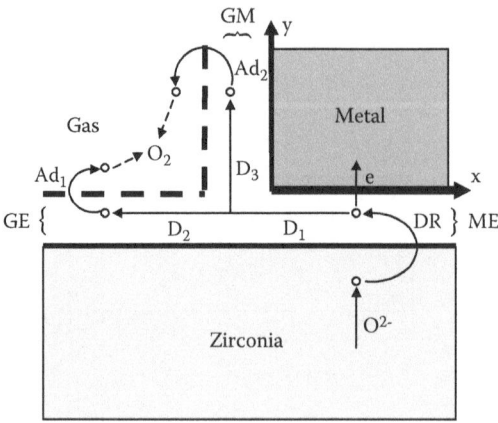

FIGURE 1.19 Scheme of the anodic process at the TPB for oxygen sensor without consideration of surface diffusion of electrons and holes. DR: discharge reaction with an electron transfer; D_1, D_2, D_3: diffusion of oxygen by TPB interfaces; and Ad_1, Ad_2: reactions of oxygen desorption.

D_2, and D_3, respectively); and desorption from the gas-metal and gas-electrolyte surfaces (reactions Ad_1 and Ad_2).

2. Discharge of the oxygen ions O^{2-} to subions O^-; diffusion of subions O^- by the metal-electrolyte and gas-electrolyte interfaces; and disproportion of subions O^- on O^{2-} and O on the gas-electrolyte interface with following desorption of oxygen molecules into the gaseous phase and transfer of ions O^{2-} into the zirconia electrolyte. This case can take place at the surface diffusion of subions (see Figure 1.18, c).

3. Discharge of oxygen ions on the gas-electrolyte surface with the following electron withdrawal (holes admission) on the surface layer of solid electrolyte and the removal of molecular oxygen from the surface of the electrolyte as a result of desorption (see Figure 1.18, d). It is also possible that the diffusion on the gas-metal interface can take place, with desorption following.

There is no doubt that the variants described above cannot comprehend all the possible ways of the reaction zone extension, even for the relatively simple electrode system. It is possible that some of the electrode processes can take place simultaneously on the gas-electrolyte, gas-metal, and metal-electrolyte interfaces. The removal of oxygen in the second variant, for instance, can be represented by the following reactions: diffusion of subions along the metal-electrolyte interface, and diffusion of oxygen atoms on the gas-metal interface. Prior to this, the oxidation reaction of subion O^- to atom O should take place with the transfer of electrons into the metal.

If the speed of the reactions on the exchange interfaces (reactions Ad and DR; see Figure 1.19) as well as the equations describing transfer of the electrochemically active component on the gas-metal, gas-electrolyte, and metal-electrolyte interfaces are known, then such electrochemical characteristics of the electrode reaction as polarization curves, equations of the transference processes at the changing of electrode potential, and so on can be calculated. As an example, the calculation method of stationary polarization characteristics for the system, representing itself as an oxygen metal electrode contacting to the zirconia electrolyte, can be shown in the next section.

1.5.2 DIFFUSION OF OXYGEN ATOMS

Let's consider the electrode system O_2, $Pt\,|\,Zr(Y)O_{2-x/2}$ at the diffusion of oxygen atoms on metal with the following two assumptions: the electrode represents itself the dense platinum stripe with width $2x_0$, attached to the zirconia electrolyte, and the coefficient of oxygen diffusion in the adsorption layer is independent from the oxygen concentration. We will deduce an equation of polarization curve for this system.

Assume that the electrode process develops by the scheme illustrated in Figure 1.19: adsorption-desorption reaction of oxygen takes place only on the metal surface (the electrolyte surface is not taken into consideration for the simplicity of calculations) by the dissociative mechanism and adsorption isotherm corresponding to the Henry equation. Then,

$$j(y) = j^0 \ (c_0^2(y) - P_{O2}), \tag{1.51}$$

where j^0 is the exchange current for reaction Ad at $P_{O2} = 1$, expressed in the electrical values; and $c_0(y)$ is the ratio between the local concentration of oxygen atoms at the point y on the gas-metal surface to the equilibrium concentration of oxygen atoms at $P_{O2} = 1$. Considering Fick's second law, the diffusion of oxygen atoms at the gas-metal layer can be written as

$$j(y) = D_{GM} \ (d^2c_0(y)/dy^2), \tag{1.52}$$

where D_{GM} is the coefficient of diffusion of atomic oxygen on the GM surface.

The process, shown in Figure 1.19, usually can be split into two stages: on the first stage the accumulation of oxygen on the metal occurs, and on the second stage the "throw off" of oxygen from the electrode takes place. This allows considering that the general electrode overpotential process also consists of two parts corresponding to the previous stages.

The solution of Equations (1.51) and (1.52) carries out at the following boundary conditions:

$$j_{GM} \ (\infty) = 0; \ dc_0(\infty)/dy = 0;$$

$$dc_0(0)/dy = I/D_{GM}, \tag{1.53}$$

where I is the electric current from the electrode on the unit of the TPB length. The following equation will be a solution of Equations (1.52) and (1.53):

$$I = I^0_{GM} \, [exp \, (6F\eta_1/RT) - 3exp \, (2F\eta_1/RT) + 2]^{1/2}, \tag{1.54}$$

where $I^0_{GM} = (3/2 \, j^0_{GM} \, D_{GM} \, P_{O2}^{3/2}]^{1/2} = j^0_{GM} \, P_{O2} \, \lambda_{GM}$ [13], η_1 is the overpotential of the stage 1 equal to

$$\eta_1 = \frac{RT}{2F} ln \frac{c_0(0)}{c_0(0,p)} = \frac{RT}{2F} ln\theta \, , \tag{1.55}$$

where $c_0(0)$ is the concentration of oxygen atoms on metal at $y = 0$, $c_0(0, p)$ is the concentration of oxygen atoms on metal at the equilibrium with P_{O2} in the gas phase, and λ_{GM} is the deepness of the surface diffusion saturation on the gas-metal boundary,

$$\lambda_{GM} = (3/2 \, D_{GM}/j^0 \, P_{O2}^{1/2}]^{1/2}. \tag{1.56}$$

For the second stage of the electrode process, using Fick's second law and the ionization reaction $O + 2e^- \leftrightarrow O^{2-}$, the local current density on the ME interface can be written considering that the activity of oxygen ions in zirconia is a constant value:

$$j_{ME}(x) = -D_{ME} \frac{d^2c_0(x)}{dx^2} = j^0_{ME} P_{O2}^{\alpha/2} \left[exp \frac{2\alpha F\eta}{RT} - \frac{c_0(x)}{c_0(0)} exp\left(-\frac{2\beta F\eta}{RT}\right) \right], \tag{1.57}$$

where j^0_{ME} is the density of the exchange current at $P_{O2} = 1$, η is the overpotential of the transition reaction, and α and β are transference coefficients. In case of unlimited width of contact between metal and electrode in calculation on the unity of the TPB, the solution of (1.57) is as follows:

$$I = I^0_{ME} \left[exp \frac{(1+\alpha) F\eta_2}{RT} - \theta \, exp\left(\frac{\beta F\eta_2}{RT}\right) \right], \tag{1.58}$$

where $I^0_{ME} = (j^0_{ME} \, D_{ME} \, P_{O2}^{(1 + \alpha/2)})^{1/2} = j^0_{ME} \, P_{O2}^{\alpha/2} \, \lambda_{ME}$, η_2 is the overpotential of the second stage of the reaction $[\eta_2 = (RT/2F)ln(c_0(\infty)/c_0(0))]$, and λ_{ME} is the deepness of the surface diffusion saturation on the metal-electrolyte boundary,

$$\lambda_{ME} = (D_{ME} \, P_{O2}^{(1 + \alpha)/2}/j^0_{ME})^{1/2}. \tag{1.59}$$

Dependence of the total overpotential of the electrode reaction on current can be found by the mutual solution of Equations (1.54) and (1.58). However, this solution cannot be found by analytical calculation. The curves' turn of this dependence is defined by the I^0_{GM}/I^0_{ME} ratio. The calculated polarization curves are presented in Figure 1.20. The cathode current was restricted by the value of I^0_{GM}, and the polarization curves become similar to the Taffel dependencies with the increase of η.

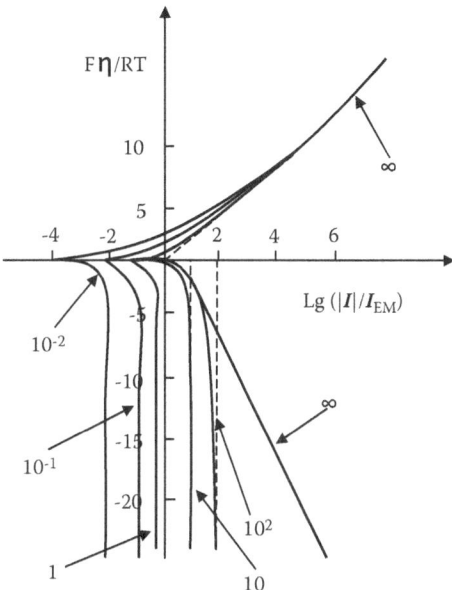

FIGURE 1.20 Polarization curves of the oxygen electrode at the surface diffusion of oxygen atoms; digits on curves represent I^0_{GM}/I^0_{ME} ratio.

Both correlations (1.54) and (1.58) are related to the electrodes with the uniform structure, when the width of both gas-metal and metal-zirconia surfaces is much bigger than the values for λ_{GM} and λ_{ME}. In the case of correspondence of their values, the speed of the electrode process will be predominantly determined by the exchange reactions, and Equation (1.58) will be as follows:

$$I = I^0_{ME} [\exp (2\alpha F\eta/RT) - \theta exp (-2\beta F\eta/RT) (\lambda_{ME}/x_0) th (x_0/\lambda_{ME})], \quad (1.60)$$

where x_0 is half of the width of the metal-electrolyte boundary. At the $\lambda_{ME} \gg \lambda_{ME}$, th $(x_0/\lambda_{ME}) = x_0/\lambda_{ME}$, and Equation (1.60) can be transformed into the equation of the delay discharge.

Two important conclusions follow from the correlations for the stationary polarization curve, which characterize in principle the processes described above and are independent of the reaction schemes at the TPB:

1. Each of the two stages of the above reaction represents itself as the interconnected processes consisting of the exchange and the interphase diffusion of the electrochemically active components. They are characterized by the exchange currents, which are proportional to the average geometrical value of the exchange current reactions and the coefficient of diffusion (Equations (1.54) and (1.58)). The fact is pointed out that it is impossible to determine characteristics of the separated stages of the

electrode process (adsorption–desorption, ionization, and diffusion on each of the interfaces) by only the results of polarization measurements.

2. The nature of the polarization process is the concentration–activation one, it being known that the correlation of these overstrain types change by the deepness of electrode. The consecutive result of this change is the decrease of oxygen activity on the surface of the electrode down to 10^{-20} – 10^{-30} at the cathode polarization. Such substantial alterations of P_{O2} on the surface of metal can cause the irreversible changes in the structure and properties of electrodes.

1.5.3 ROLE OF THE ELECTRIC DOUBLE LAYER IN ELECTRODE REACTIONS

The double-layer model on the TPB is used to visualize the ionic environment in the vicinity of a charged surface: gas-zirconia-SE (sensing electrode). It can be either a metal under potential or due to ionic groups on the surface of the TPB [54]. It is easier to understand this model as a sequence of steps that would take place near the surface if its neutralizing ions were suddenly stripped away. One of the first principles which we must recognize is that matter at the boundary of three phases possesses properties which differentiate it from matter freely extended in either of the continuous phases separated by the interface. When referring to a gas-zirconia-SE interface, it is perhaps relatively easy to visualize a difference between the interface and the gas. Where we have a charged surface, however, there must be a balancing countercharge, and this countercharge will occur on the TPB. The charges will not be uniformly distributed throughout the gas phase, but will be concentrated near the TPB. Thus, we have a small but finite volume of the gas phase which is different from the extended gaseous-measuring environment. This concept is central to electrochemistry, and reactions within this interfacial boundary that govern external observations of electrochemical reactions. It is also of great importance to solid-state chemistry, where ionic particles with different charges play a crucial role. There are several theoretical treatments of the solid-gas interface. We will look at a few common ones, not so much from the position of needing to use them, but more from the point of what they can inform us about the charged interface.

1.5.3.1 Helmholtz Double Layer

This theory is a simple approximation that the surface charge is neutralized by opposite-sign counter-ions placed at an increment of d away from the surface [59]. The surface charge potential is linearly dissipated from the surface to the counter-ions satisfying the charge. The distance, d, will be that to the center of the counter-ions, that is, their radius. The Helmholtz theoretical treatment does not adequately explain all the features, since it hypothesizes rigid layers of opposite charges. This does not occur in nature, as is schematically shown in Figure 1.21.

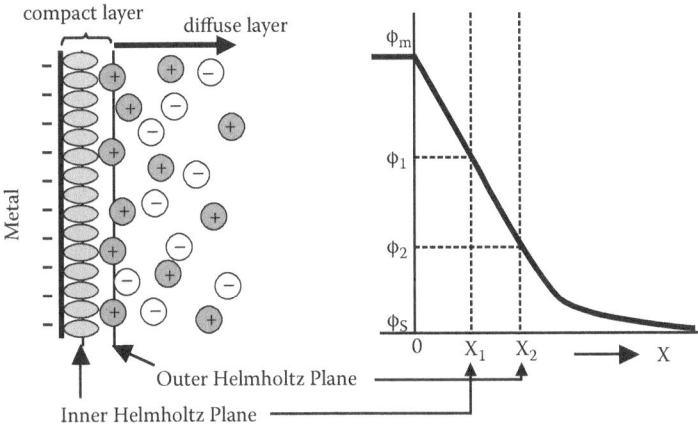

FIGURE 1.21 Model of the Helmholtz electric double layer.

1.5.3.2 Gouy–Chapman Double Layer

Gouy suggested that interfacial potential at the charged surface could be attributed to the presence of a number of ions of a given sign attached to its surface, and to an equal number of ions of opposite charge in the gas. In other words, counter-ions are not rigidly held, but tend to diffuse in the gas phase until the counterpotential set up by their departure restricts this tendency. The kinetic energy of the counterions will, in part, affect the thickness of the resulting diffuse double layer [60]. Gouy and, independently, Chapman developed theories of this so-called *diffuse double layer*, in which the change in concentration of the counter-ions near a charged surface follows the Boltzmann distribution:

$$n = n_o exp(-ze\Psi/K_B T), \quad (1.61)$$

where n_o is the bulk concentration, z is the charge on the ion, e is the electron charge, and K_B is the Boltzmann constant. However, in relation to the YSZ-metal-gas interface, there is a discrepancy since derivation of this form of the Boltzmann distribution assumes that activity is equal to molar concentration. This may be an acceptable approximation for the gas phase, but will not be true near a TPB. Now, since we have a diffuse double layer, rather than a rigid double layer, we must concern ourselves with the volume charge density rather than surface charge density when studying the coulombic interactions between charges. The volume charge density ρ of any volume i can be expressed as

$$\rho_i = \Sigma z_i en_i.$$

The coulombic interaction between charges can, then, be expressed by the Poisson equation. For plane surfaces, this can be expressed as

$$d^2\Psi/dx^2 = -4\pi\rho/d,$$

where Ψ varies from Ψ_0 at the surface to 0 in the gas phase. Thus, we can relate the charge density at any given point to the potential gradient away from the TPB. Combining the Boltzmann distribution with the Poisson equation and integrating under appropriate limits yields the electric potential as a function of distance from the TPB. The thickness of the diffuse double layer,

$$\lambda_{double} = [\varepsilon_r kT/(4\pi e^2 \Sigma n_{io} z_i^2)]^{1/2},$$

in other words, the double-layer thickness, decreases with increasing valence and concentration.

The Gouy–Chapman theory describes a rigid charged surface, with a cloud of oppositely charged ions in the gas phase, the concentration of the oppositely charged ions decreasing with distance from the TPB. This is the so-called diffuse double layer. This theory is still not entirely accurate. Experimentally, the double-layer thickness is generally found to be somewhat greater than calculated. This may relate to the error incorporated in assuming activity equals molar concentration when using the desired form of the Boltzmann distribution. Conceptually, it tends to be a function of the fact that both anions and cations exist in the gas phase, and with increasing distance away from the TPB the probability that ions of the same sign as the surface charge will be found within the double layer increases as well.

1.5.3.3 Stern Modification of the Diffuse Double Layer

The Gouy–Chapman theory provides a better approximation of reality than does the Helmholtz theory, but it still has limited quantitative application. It assumes that ions behave as point charges, which they cannot, and it assumes that there is no physical limit for the ions in their approach to the TPB, which is not true. Stern, therefore, modified the Gouy–Chapman diffuse double layer. His theory states that ions do have finite size, so they cannot approach the TPB closer than a few *nm* [54, 60]. The first ions of the Gouy–Chapman diffuse double layer are in the gas phase but not at the TPB. They are at some distance δ away from the zirconia-metal-gas interface. This distance will usually be taken as the radius of the ion. As a result, the potential and concentration of the diffuse part of the layer are low enough to justify treating the ions as point charges. Stern also assumed that it is possible that some of the ions are specifically adsorbed by the TPB in the plane δ, and this layer has become known as the Stern layer. Therefore, the potential will drop by $\Psi_0 - \Psi_\delta$ over the "molecular condenser" (i.e., the Helmholtz plane) and by Ψ_δ over the diffuse layer. Ψ_δ has become known as the zeta (ζ) potential.

The double layer is formed in order to neutralize the charged surface and, in turn, causes an electrokinetic potential between the TPB and any point in the mass of the oxygen ions. This *emf* difference is on the order of millivolts and is referred to as the *surface potential*. The magnitude of the surface potential is related to the TPB charge and the thickness of the double layer. As we leave the surface, the potential drops off roughly linearly in the Stern layer and then exponentially through the diffuse layer, approaching zero at the imaginary boundary of the double layer. The potential curve is useful because it indicates the strength of the electrical force

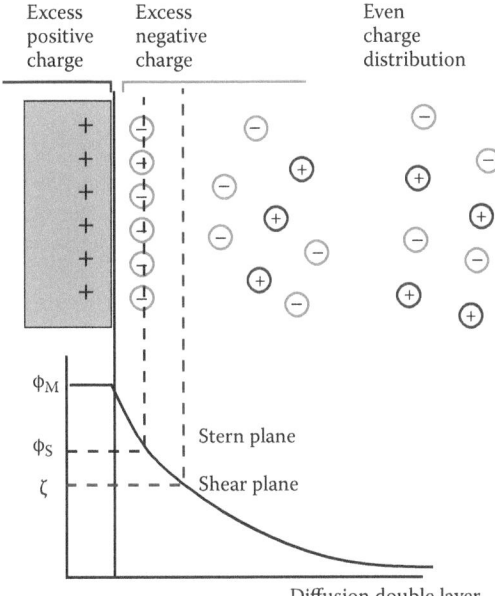

FIGURE 1.22 Interpretation of the change in charge density through the diffuse layer.

between particles and the distance at which this force comes into play. The particle's mobility is related to the dielectric constant and viscosity of the measuring gas and to the electrical potential at the boundary between the moving particle and the gas phase. This boundary is called the *slip plane* and is usually defined as the point where the Stern layer and the diffuse layer meet. The relationship between zeta potential and surface potential depends on the level of ions in the gas phase. Figure 1.22 schematically represents the change in charge density through the diffuse layer. Although zeta potential is an intermediate value, it is sometimes considered to be more significant than surface potential as far as electrochemistry of solid electrolytes is concerned.

REFERENCES

1. Etsell, T.H. and Flenders, S.N., The electrical properties of solid oxide electrolytes, *Chem Rev.* **70** (1970) 339–376.
2. Zhuiykov, S. and Nowotny, J., Zirconia-based sensors for environmental gases: A review, *Materials Forum* **24** (2000) 150–168.
3. Kudo, T. and Kawamura, J., Fast ionic conductors, in *Materials for Energy Conversion Devices*, Eds. S.S. Sorrell, J. Nowotny, and S. Sugihara, Cambridge, Woodhead Publishing, 2005, Chapter 7.
4. Badwal, S.P.S., Electrical conductivity of single crystal and polycrystalline yttria-stabilized zirconia, *J. Mater. Sci.* **19** (1984) 1767–1776.
5. Nowotny, J. et al., Charge transfer at oxygen/zirconia interface at elevated temperatures. Part 1: Basic properties and terms, *Adv. Appl. Ceramics* **104** (2005) 147–153.

6. Stubican, V.S., Hink, R.C., and Ray, S.P., Phase equilibria and ordering in the system ZrO_2-Y_2O_3, *J. Amer. Ceram. Soc.* **61** (1978) 17–21.

7. Goff, J.P. et al., Defect structure of yttria-stabilized zirconia and its influence on the ionic conductivity at elevated temperatures, *Physical Review B* **59** (1999) 14202–14220.

8. Kopp, A. et al., in *Science and technology of zirconia*, Eds. S.P.S. Badwal, M.J. Bannister, and R.H.J. Hannik, Lancaster, PA, Technomic Publishing, 1993, 567–575.

9. Standard, O.C. and Sorell, C.C., Densification of zirconia: Conventional methods, in *Zirconia engineering ceramics*, Ed. E. Kisi, Zurich, Trans Tech Publications, 1998, 251–300.

10. Goff, J.P. et al., Defect structure of yttria-stabilized zirconia and its influence on the ionic conductivity at elevated temperatures, *Physical Review B* **59** (1999) 59–67.

11. Badwal, S.P.S., Zirconia-based solid electrolytes: Microstructure, stability and ionic conductivity, *Solid State Ionics* **52** (1992) 23–32.

12. Badwal, S.P.S. and Ciacchi, F.T., Microstructure of Pt electrodes and its influence on the oxygen transfer kinetics, *Solid State Ionics* **18–19** (1986) 1054–1059.

13. Perfiliev, M.F. et al., *High-temperature electrolysis of gases*, Moscow, Science Publishing, 1988, 232.

14. Wagner, C., Beitrag zur Theoric des Aulaufvorgaugs. I, *Zeitschrift fur Phys. Chemie., Abt. B* B **21** (1933) 25–41.

15. Nakamura, A. and Wagner, J.B., Defect structure, ionic conductivity and diffusion in yttria stabilized zirconia and related oxide electrolytes with fluorite structure, *J. Electrochem Soc.* **133** (1986) 1542–1548.

16. Wagner, C., Beitrag zur Theoric des Aulaufvorgaugs. II, *Zeitschrift fur Phys. Chemie., Abt. B,* B **32** (1936) 447–462.

17. Patterson, I.W., Conduction domains for solid electrolytes, *J. Electrochem. Soc.* **118** (1971) 1033–1039.

18. Zhuiykov, S., Mathematical modeling of YSZ-based potentiometric gas sensors with oxide sensing electrodes: Part II: Complete and numerical models for analysis of sensor characteristics, *Sens. Actuators B, Chem.* **120** (2007) 645–656.

19. Tuller, H.L., Defect engineering: Design tools for solid state electrochemical devices, *Electrochemica Acta* **48** (2003) 2879–2887.

20. Marion, F., Brug, H., and Oehlig, I., Suv l'application des properties semi – conactrices des oxides a'la mesure et la regulation des tenenrs en oxides (de plusieurs bars a 10^{-16} bar), *Chimic Analytique* **51** (1969) 284–289.

21. Patterson, I.W., Mixed conduction in $Zr_{0.85}Ca_{0.15}O_{1.85}$ and $Th_{0.85}Y_{0.15}O_{1.925}$ solid electrolytes, *J. Electrochem. Soc.* **114** (1967) 752–758.

22. Nowotny, J. et al., Charge transfer at oxygen/zirconia interface at elevated temperatures. Part 2: Oxidation of zirconia, *Adv. Appl. Ceramics* **104** (2005) 154–164.

23. Chebotin, V.N. and Perfilev, M.F., *Electrochemistry of Solid Electrolytes*, Moscow, Chemistry Publishing, 1986, 312.

24. Nowotny, J. et al., Charge transfer at oxygen/zirconia interface at elevated temperatures. Part 3: Segregation induced interface properties, *Adv. Appl. Ceramics* **104** (2005) 165–173.

25. Schmalzried, H., Uber Zirkondioxid als Elektrolyt fur electrochemische Untersuchunger bei hoheren Temperaturen, *Zeitschrift fur Elektrochemic* **66** (1962) 572–576.

26. Schmalzried, H., Thermodynamic and kinetic aspects of interface reactions, *Reactivity of Solids* **8** (1990) 247–268.

27. Talanchuk, P. M. et al., *Semiconductor and Solid-Electrolyte Sensors*, Kiev, Technika Publishing, 1992, 220.

28. Zhou, M. and Ahmad, A., Synthesis, processing and characterization of calcia-stabilized zirconia solid electrolytes for oxygen sensing applications, *Materials Research Bulletin* **41** (2006) 690–696.

29. Tien, T.Y. and Subbarao, E.C., X-ray and electrical conductivity study of the fluorite phase in the system ZrO_2–CaO, *J. Chem. Phys.* **39** (1963) 1041–1047.

30. Suzuki, Y. and Takashi, T., Time dependence of conductivity of Y_2O_3 – stabilized zirconia, *Nippon Kagaku Kaishi* **8** (1975) 260–265.

31. Suzuki, Y. and Takashi, T., Stability of fluorite-type cubic solid solution in sintered samples of the ZrO_2–Y_2O_3 system, *Nippon Kagaku Kaishi* **11** (1977) 1610–1613.

32. Zhang, T.S. et al., Aging behaviour and ionic conductivity of ceria-based ceramics: A comparative study, *Solid State Ionics* **170** (2004) 209–217.

33. Vlasov, A.N. and Perfiliev, M.V., Ageing of ZrO_2-based solid electrolytes, *Solid State Ionics* **25** (1987) 245–253.

34. Marakami, Y., Nagano, I., and Yamamoto, H., Phase equilibria and phase change during ageing in the ZrO_2–Sc_2O_3 system, *J. Mater. Sci. Lett.* **16** (1997) 1686–1688.

35. Hattori, M. et al., Effect of aging on conductivity of yttria stabilized zirconia, *J. Power Sources* **126** (2004) 23–27.

36. Zhang, T.S. et al., Effects of dopant concentration and aging on the electrical properties of Y-doped ceria electrolytes, *Solid State Science* **5** (2003) 1505–1511.

37. Haering, C., Roosen A., and Schichl, H., Degradation of the electrical conductivity in stabilised zirconia systems: Part I: Yttria-stabilised zirconia, *Solid State Ionics* **176** (2005) 253–259.

38. Allpress, J.G. and Rossell, H.J., A micro-domain description of defective fluorite-type phases $Ca_xM_{1-x}O_{2-x}$ (M – Zr, Hf, x = 0.1 – 0.2), *J. Solid State Chem.* **15** (1975) 68–78.

39. Hudson, B. and Moseley, P.T., On the extent of ordering in stabilized zirconia, *J. Solid State Chem.* **19** (1976) 383–389.

40. Zhuiykov, S., Investigation of conductivity, microstructure and stability of HfO_2–ZrO_2–Y_2O_3–Al_2O_3 electrolyte compositions for high-temperature oxygen measurement, *J. Europ. Ceram. Soc.* **20** (2000) 967–976.

41. Raeder, H., Norby T., and Osborn, P.A., Ageing of yttria-stabilized zirconia electrolytes at 1000°C, in *Ceramic Processing Science and Technology*, Ed. H. Hausner, Columbus, OH, Amer. Ceram. Soc., 1995, 719–723.

42. Zhuiykov, S., Development of dual sulphur oxides and oxygen solid state sensor for "*in-situ*" measurements, *Fuel* **79** (2000) 1255–1265.

43. Blumenthal, R.N., A technical presentation of the factors affecting the accuracy of carbon/oxygen probes, in *Proc. of the Second Int. Conf. on Carburising and Nitriding with Atmospheres*, Cleveland, OH, 6–8 December 1995, 17–23.

44. Hibino, T. et al., Zirconia-based potentiometric sensors using metal oxide electrodes for detection of hydrocarbons, *J. Electrochem. Soc.* **131** (2001) H1–H5.

45. Dubbe, A., Fundamentals of solid state ionic micro gas sensors, *Sens. Actuators B: Chem.* **88** (2003) 138–148.

46. Maskell, W.C., Progress in the development of zirconia gas sensors, *Solid State Ionics* **134** (2000) 43–50.

47. Zhuiykov, S., Microstructure characterisation and oxygen sensing properties of Al_2O_3–ZrO_2–Y_2O_3 shaped eutectic composites, *Sensors and Materials* **12** (2000) 117–132.

48. Zhuiykov, S., Zirconia single crystal analyser for low-temperature measurements, *Process Control and Quality* **11** (1998) 23–37.

49. Landau, L.D. and Lifshits, E.M., *Statistical Physics*, Moscow, Science Publishing, 1976, 593.
50. Wepner, W., Interfacial processes of ion conducting ceramic materials for advanced chemical sensors, in *Proc. of 29th Int. Conf. of the American Ceramic Society*, Westerville, OH, 25–29 January 2005, **26** (2005) 15–24.
51. Elumalai, P. et al., Sensing characteristics of YSZ-based mixed-potential-type planar NO_x sensor using NiO sensing electrodes sintering at different temperatures, *J. Electrochem. Soc.* **152** (2005) H95–H101.
52. Zhuiykov, S., Electron model of solid oxygen-ionic electrolytes used in gas sensors, *Int. J. Applied Ceramic Techn.* **3** (2006) 401–411.
53. Mead, C.A. and Spitzer, W.G., Fermi level position at metal-semiconductor interfaces, *Physical Review* **134** (1964) A713–A716.
54. Kuzin, B.L. and Bronin, D.I., Electrical double-layer capacitance of the M, O_2, O^{2-} interfaces (M = Pt, Au, Pd, In_2O_3; O^{2-} = zirconia-based electrolyte), *Solid State Ionics* **136–137** (2000) 45–50.
55. Wright, D.A. et al., Optical absorption in current-blackened yttria-stabilized zirconia, *J. Mater. Sci.* **8** (1973) 876–883.
56. Kamiya, T. and Hosoto, H., Electronic structures and device applications of transparent oxide semiconductors: What is the real merit of oxide semiconductors? *Int. J. Appl. Ceram. Technol.* **2** (2005) 285–294.
57. Fabry, P., Kleitz, M., and Deportes, C., Sur l'utilization d'une electrode ponctuelle daus les allules a'oxyde electrolyte solide, *J. Sol. State Chem.* **6** (1973) 230–235.
58. Bonch-Bruevich, B.A. and Kalashnikov, S.G., *Physics of Semiconductors*, Moscow, Science Publishing, 1977, 627.
59. Lyklema, J., *Fundamentals of Interface and Colloid Science Vol. II: Solid-Liquid Interface*, London, Academic Press, 1995, 352.
60. Martin-Molina, A., Quesada-Perez, M., and Hidalgo-Alvarez, R., Electric double layers with electrolyte mixtures: Integral equations and simulations, *J. Phys. Chem. B* **110** (2006) 1326–1331.

2 Mathematical Modeling of Zirconia Gas Sensors with Distributed Parameters

2.1 COMPLETE MATHEMATICAL MODEL OF ELECTROCHEMICAL GAS SENSORS

Electrochemical gas sensors based on ceramic materials with high ionic conductivity have been technologically improved over the last twenty years and have been successfully used in many practical applications [1, 2]. A detailed mathematical model of the transport of both physical and electrochemical processes on the surface and within the sensing electrode (SE) as well as on the triple-phase boundary (TPB) YSZ-SE-gas can be a powerful tool for further development of these sensors. The activity in the field of computer-aided optimum design in engineering of the electrochemical gas sensors has also been increasing steadily over the last decade. A vast range of models exists today, varying in complexity and in the number of assumptions employed. There is no doubt that the considerable progress in computational modeling of the YSZ-based gas sensors over the last two decades has led to an improved understanding of the relevant physical, electrical, and chemical phenomena. However, the emphasis in a majority of the models has been either on the transport processes or on the electrochemical processes. Sometimes, the lack of complete understanding of the complexity of various reactions occurring on and within the SE and on the TPB has been substituted by the oversimplified sensing mechanism. A clear example of this approach is the zirconia-based mixed-potential-type gas sensors. These sensors with the oxide SE and Pt reference electrode (RE) have been developed for nitrogen oxide (NO_x), carbon monoxide (CO), and hydrocarbon (C_xH_y) detection at high temperatures of 500–900°C [3–28]. The oversimplified mixed-potential sensing mechanism has often been used in publications in order to explain the sensor behavior [11, 14, 18]. The domination of the mixed-potential theory has led to a situation where the potentially interesting results or phenomena obtained during experiments have not been reported simply because they could not be explained by the widely accepted mixed-potential theory. Furthermore, some of the results, which have recently been published, completely contradict the mixed-potential theory [29], showing that the sensor's output to both hydrocarbons CH_4 and C_3H_8 changes polarity from negative to positive for the NiO-SEs sintered at high temperatures of 1300°C and 1400°C. Fortunately, another gas-sensing mechanism, "differential electrode equilibria" [27, 28], has also been proposed to explain the NO_x sensitivity that is caused not only by the electrochemical reactions, but also by the different electrocatalytic activity and/or sorption-desorption behavior of two electrodes. Although the proposed mechanism has enhanced our knowledge in relation to the NO_x sensing, it does not explain the complexity of the electrochemical and physic-chemical processes on and within the SE. Both molecular and dissociative adsorption of gaseous NO_2, O_2, and H_2O are observed on many oxide and transitional metal SEs. As a consequence of dissociative adsorption, a variety of surface species such as $(NO)_2$, N_2O, and N and O adatoms have been found on the surfaces under different reaction conditions [30]. Apart from the competition between molecular and dissociative adsorption, the situation becomes even more complex when the surface topology changes during adsorption at high temperatures in excess of O_2. Consequently, the most important developments in the improvement of the mathematical models of the zirconia-based gas sensors must be based on understanding the mechanism of detailed electrochemical reactions and by accounting for the complex heat-mass-transfer processes occurring at the microscale level.

Recent modification of the design of the YSZ-based gas sensors has already been shifted toward the planar structure [2]. With the advent of micro- and nano-technology, that is, with reduction of the bulk volume in many sensors' design, it is evident that the role of interfaces and phase boundaries becomes increasingly important, especially in nonequilibrium thermodynamics and solid-state kinetics. From the designer's point of view, an adequate mathematical model of the specific gas sensor has to be numerically implemented in order to optimize characteristics of the sensor, as designers are constantly faced with the need to compromise between often antagonistic measures of design success: sensor performance, robustness to thermal shock, cost of development and manufacturing including applied technology for SEs and REs, cost of the sensor failure, and impact of the measuring environment. Expert knowledge, combined with the model-based decision-making computer tools, can provide a framework for optimization of a greater diversity of candidate solutions by assessing the variety of compromises at hand. The discretization of the mathematical model must be conducted extremely carefully to avoid influence of slight changes in accuracy. This is because these tiny changes can create an overwhelmingly different output. The formal name of this phenomenon is "sensitive dependence on initial conditions" [31]. Its informal and more popular name is *butterfly effect*. Simply stated, it means that the oversimplification in mathematical modeling and/or in making assumptions of both elementary and boundary conditions can have the power to transform the output signal of the sensor.

The proposed methodology for computer-aided optimal design in the development of YSZ-based gas sensors comprises three phases. Firstly, the complete mathematical model with distributed temporal and spatial parameters for electrochemical gas sensors is presented as a system of the differential equations in private derivatives of parabolic and hyperbolic types. The complexity of physical and chemical interactions, represented in this model, allows performing a mathematical description of the electrochemical gas sensors toward standardization of the calculating procedures. The complete mathematical model and the algorithm of transfer from the complete

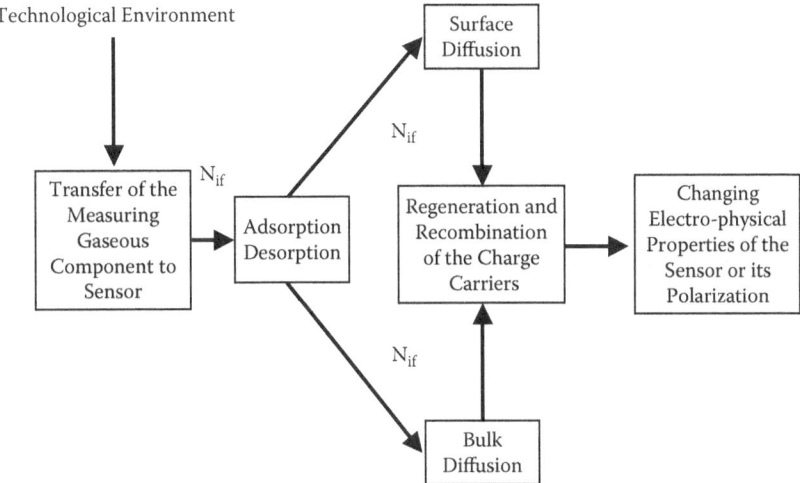

FIGURE 2.1 Summarized structural scheme of measuring transformations in the solid electrolyte gas sensors. (Reprinted from Zhuiykov, S., Mathematical modelling of YSZ-based potentiometric gas sensors with oxide sensing electrodes part II: Complete and numerical models for analysis of sensor characteristics, *Sensors and Actuators B, Chem.* 120 (2007) 645–656, with permission from Elsevier Science.)

model to models of the real gas sensors provide a decision-making tool for better optimal design of these sensors. Secondly, the general computational gas dynamics methodology is adopted to solve numerically the foregoing model equations. Thirdly, numerical simulations are carefully carried out to investigate the effects of changes in working temperature and in the gas concentration gradients on the sensor performance. This step can be repeated for each individual working temperature and for each measuring gas concentration.

Figure 2.1 illustrates the summarized structural scheme of the measuring transformations in the solid electrolyte gas sensors. As is clearly shown in this figure, the main measuring transformations are: transfer of the measuring gaseous component, its adsorption-desorption on the SE (RE), passage of the adsorbed gas through the phase boundary interface, surface and bulk diffusion in the SE toward the TPB, regeneration and recombination of the charge carriers, and, consequently, changing the electrophysical properties of the sensor or its polarization. For the solid electrolyte sensors, it corresponds to the change of electromotive force (*emf*). Other parameters can also be used as the output informative values for the zirconia-based gas sensors [2, 3].

The descriptive details of the physical and chemical processes and their a priori mathematical formulation for the YSZ-based electrochemical gas sensors allow one to combine them into the complete mathematical model, represented as a system of the differential equations in private derivatives of parabolic and hyperbolic types [32, 33]:

$$\frac{\partial \bar{U}(t,\bar{x})}{\partial t} = \sum_{i=1}^{n} \frac{\partial}{\partial x_i} \left\{ \sum_{j=1}^{n} \left[a_{i,j}\left(\bar{U},t,\bar{x},\bar{P_1}\right) \cdot \frac{\partial \bar{U}(t,\bar{x})}{\partial x_j} \right] \right\} +$$

$$+ \sum_{i=1}^{n} \left[b_i\left(\bar{U},t,\bar{x},\bar{F},\bar{P_1}\right) \right] \cdot \frac{\partial \bar{U}(t,x)}{\partial x_i} + \bar{F}(t,\bar{x}),$$

(2.1)

considering the elementary conditions

$$\bar{U}(t,\bar{x})\big|_{\Omega} = \bar{U}_0(\bar{x}),$$

(2.2)

and the boundary conditions of two types: Robin-type boundary conditions

$$\frac{\partial \bar{U}}{\partial n}\Big|_{\Sigma} = \bar{f_1}\left[\bar{U},\bar{P_2},\Psi\left(t,\bar{P_2}\right)\right]$$

(2.3)

or Dirichlet-type boundary conditions

$$\bar{U}\big|_{\Sigma} = \bar{f_2}\left[\bar{U},\bar{P_2},\Psi\left(t,\bar{P_2}\right)\right].$$

(2.4)

Here, $\bar{U}(t,\bar{x}) = [U_1, \dots , U_k]^T$ is the conditions' vector, components of which depend on the temporal coordinate $t \in (0, t_K)$ and the vector of spatial coordinate $x = (x_1, \dots , x_n) \in \Omega$, $\Omega \in R^n$, $n \le 3$. Spatial region Ω has a boundary Γ, and $\Sigma = (0, t_K) \times \Gamma$ represents itself as the sideway surface of the cylinder with base Ω and height t_K. $a_{i,j}$ ($\bar{U}$, t, $\bar{x}$, $\bar{P_1}$) and b_i ($\bar{U}$, t, $\bar{x}$, $\bar{F}$, $\bar{P_1}$) are the continuous nonlinear and positive functions of ($\bar{U}$, t, $\bar{x}$, $\bar{P_1}$) and ($\bar{U}$, t, $\bar{x}$, $\bar{F}$, $\bar{P_1}$), respectively, and $\bar{F}$ (t, $\bar{x}$) is the nonlinear function determining the influence of the external conditions on the boundary Γ. In this case, $\bar{F}$ and Ψ can be accidental or determining functions, and $\bar{f_1}$ and $\bar{f_2}$ are the nonlinear functions, calculating the character of the processes on the boundary of the region Ω. n is the external normal to the boundary Γ of the region Ω. $\bar{P_1} = (P_1' \dots , P_1^q)$ and $\bar{P_2} = (P_2', \dots , P_2^g)$ are the vectors of physical, chemical, thermodynamical, geometrical, and other parameters, respectively, determining the most important characteristics of the YSZ-based electrochemical gas sensors in the region Ω and on the boundary Γ.

The parts of every equation of the complete system (2.1) are as follows:

$$\sum_{i=1}^{n} \frac{\partial}{\partial x_i} \left\{ \sum_{j=1}^{n} \left[a_{i,j}\left(\bar{U},t,\bar{x},\bar{P_1}\right) \cdot \frac{\bar{U}(t,\bar{x})}{\partial x_j} \right] \right\}$$

is the diffusive part,

$$\frac{\partial \bar{U}\left(t, \bar{x}\right)}{\partial t}$$

is the nonstandard part,

$$\sum_{i=1}^{n} \left[b_i \left(\bar{U}, t, \bar{x}, \bar{F}, \bar{P_1} \right) \right] \cdot \frac{\partial \bar{U}\left(t, \bar{x}\right)}{\partial x_i}$$

is the convective part, and $\bar{F}$ $(t, \bar{x})$ is an additional term. The convective part determines the process of thermo-mass transfer. The variable term $\bar{U}$ $(t, \bar{x})$ can represent different physical parameters such as partial pressure or concentration of the measuring gas, enthalpy or temperature, and so on. Consequently, the functions $a_{i,j}, b_i, \bar{F}, \bar{f_1}$, and $\bar{f_2}$ can then be transferred as appropriate to each of these variable's physical value.

By using the deduction principle, the complete mathematical model of the electrochemical gas sensors with distributed parameters can be transformed to the mathematical models of the specific gas sensors, which is important for organization of their optimal design.

Based on analysis of equations (2.1)–(2.4) of the complete mathematical model, it can be confirmed that some of the parameters of the YSZ-based gas sensors (initial concentration of the charge carriers, concentration of the measuring gas, working temperature, etc.) are included directly into the system of equation (2.1). Other parameters, for example, thickness of the SE, are included into the boundary conditions (2.3) or (2.4). In this case, some of the functions ($\bar{f_1}$, $\bar{f_2}$, $\bar{\Psi}, \bar{U_0}$ ($\bar{x}$)), characterizing the interactions of the measuring environment with the SE (RE) of the sensor, must be taken into consideration at determination of the elementary and boundary conditions for the complete model (2.1)–(2.4).

Therefore, the complete mathematical model (2.1)–(2.4) can be used for the different tasks of the sensors' optimal design, including analysis the objects of measurement, development of engineering recommendations, organization of manufacturing technology, and selection of the rational conditions of use for the various measuring systems. Specifically, the algorithm, shown in Figure 2.2, provides the guidance for the transfer from a complete model with distributed parameters to the models of the real YSZ-based gas sensors. The main stages of this algorithm are as follows:

1. The minimum quantity of the interconnected variable parameters $\bar{U}$ ($t, \bar{x}$) is determined based on the preliminary analysis of the sensor. This quantity must be reflected in the model. The measured and referenced data, as well as data from the handbooks, can be used.
2. The main interchanges for the selected parameter, involving the measurement process within the sensor structure (Block 2), are considered. It should be determined whether the diffusion and/or thermoconductivity

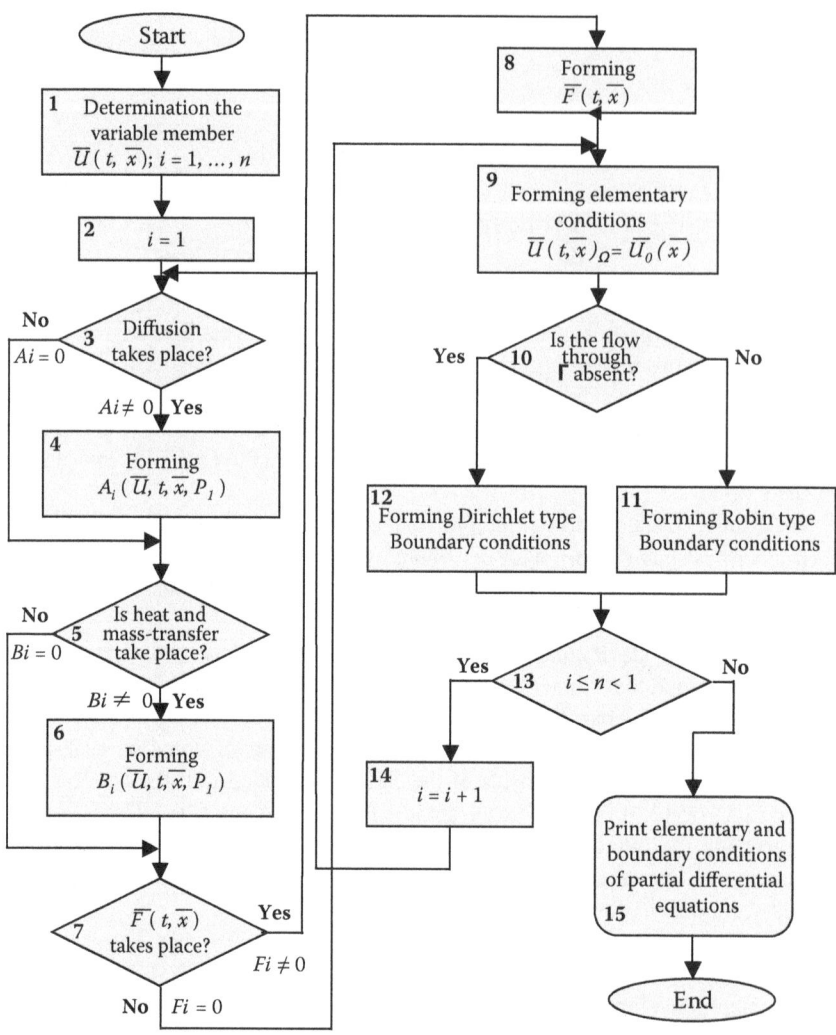

FIGURE 2.2 Algorithm of the transfer from the complete mathematical model with distributed parameters to the models of the real gas sensors. (Reprinted from Zhuiykov, S., Mathematical modelling of YSZ-based potentiometric gas sensors with oxide sensing electrodes part II: Complete and numerical models for analysis of sensor characteristics, *Sensors and Actuators B, Chem.* 120 (2007) 645–656, with permission from Elsevier Science.)

has an influence on the sensor output signal (Block 3). If these processes cannot be ignored, then the operator

$$A_i = \sum_{i=1}^{n} \frac{\partial}{\partial x_i} \left[a_i \left(\overline{U}, t, \overline{x}, \overline{P_1} \right) \cdot \frac{\partial}{\partial x_i} \right] \neq 0$$

and the transfer to Block 4, where the structure and coefficients of this operator will be determined, should be done. If the diffusion and/or thermoconductivity processes are negligible, then the operator $A_i = 0$ and the transfer to Block 5 takes place.

3. Block 4 is used for determination of the structure and parameters of the diffusive operator A_i on the basis of the calculative-experimental method.

4. The presence or absence of the thermo-mass transfer in the sensor structure is analyzed. If a thermo-mass-transfer process is absent, then the differential operator

$$B_i = \sum_{i=1}^{n} \left[b_i \left(\bar{U}, t, \bar{x}, \bar{F}, \bar{P_1} \right) \right] \cdot \frac{\partial}{\partial x_i} = 0$$

and the transfer to Block 7 occurs. If, in contrast, the thermo-mass transfer has taken place in the sensor structure, then $B_i \neq 0$ is established and the transfer to Block 6 should be conducted for further determination of the structure of this operator.

5. Structure of the differential operator $B_i \neq 0$ is determined (Block 6) on the basis of the physical laws, describing the thermo-mass-transfer processes (analytical method), or, alternatively, as a result of the structure-parametric identification tasks (experimental-calculative method).

6. Blocks 7 and 8 are used for forming function $\bar{F}$ $(t, \bar{x})$, determining the influence of the external conditions (deterministic or accidental). Deterministic part, as a rule, forms on the basis of the analytical approach; the accidental part forms on the experimental basis.

7. The elementary value of the variable member $\bar{U}$ $(t, \bar{x})_\Omega = \bar{U}_0$ $(\bar{x})$ on the region can be set by using Block 9.

8. Based on the analysis of the physical processes on the region boundary Γ, the presence or absence of the flow of variable member $\bar{U}_i$ (t, x), directing by the normal toward Γ, is determined (Block 10). If the flow of the variable member $\bar{U}_i$ (t, x) is absent, the transfer to Block 12 should be performed. Block 12 will form the Dirichlet-type boundary conditions. Otherwise, transfer to Block 11 takes place for development of the Robin-type boundary conditions.

9. By using Blocks 13 and 14, transfer to the next variable member $\bar{U}_{i+1}$ $(t, \bar{x})$ has occurred. The determination of the differential equations and of elementary and boundary conditions for the next variable member has to be performed again using Blocks 3–12.

10. After completion, the selection of all variable members (Block 13), the system of the partial differential equations, together with elementary and boundary conditions, is formed, calculated, and printed (Block 15).

The nature of the developing YSZ-based gas sensors is such that in the general outline, the main characteristics of these sensors (selectivity, sensor's response, sensitivity, etc.) are stipulated by the processes of transfer of the measuring component in the spatial region with simultaneous thermo-mass transfer with a gaseous environment. Therefore, the combination of the complete mathematical model and proposed algorithm provides a decision-making tool for the transference of the complete model into a mathematical model of the real YSZ-based gas sensors.

2.2 MODELING INTERACTIONS OF OXYGEN WITH THE ZIRCONIA SENSOR

The approach, described in the previous section, can be illustrated on the example of the zirconia-based potentiometric oxygen sensor with the Pt SE and metal-metaloxide (Me-MeO RE). Based on analysis of the electrochemical processes on the gas-electrode-YSZ interfaces, one of the key components of the mathematical model of the real YSZ-based sensors is modeling of the ion and electron transport. It is sometimes assumed that the YSZ is a sole contributor to ohm losses, and, therefore, the charge transport in the electrodes is often neglected. Nevertheless, electronic and ionic transports are fundamental processes of partial gas pressure measurement, since both electrons and ions must be present at the reaction site. The YSZ-based sensor is assumed to operate under steady-state conditions. Figure 2.3 illustrates the interaction scheme of the measuring gas with the YSZ-based oxygen sensor with the Me-MeO RE [32]. Stage 1 is diffusion of molecular oxygen in a gaseous measuring environment through the porous Pt SE, allowing for the effect of nonuniform gas pressure and sensor temperature, and Knudsen diffusion. Stage 2 is the adsorption-desorption and partial dissociation of oxygen molecules ($O_2 \rightarrow 2O^{2-} + 4e^-$) on the gas-SE-YSZ interface.

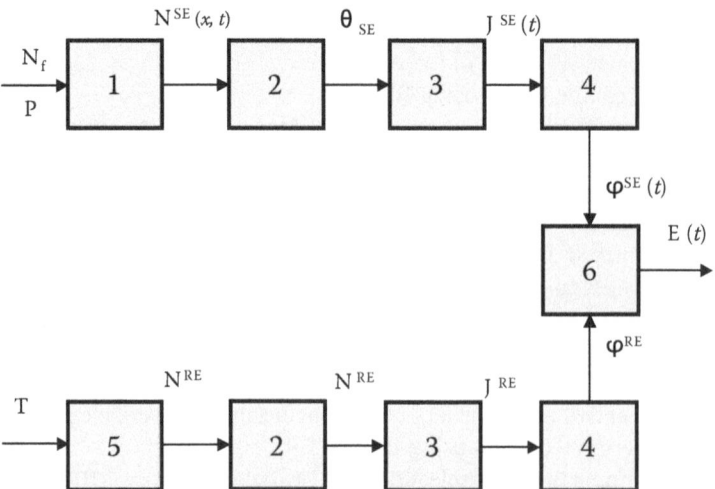

FIGURE 2.3 Schematic presentation of the interaction of the solid electrolyte oxygen sensor with gas environment: 1–6 stages of interaction. (From Zhuiykov, S., Mathematical model of electrochemical gas sensors with distributed temporal and spatial parameters and its transformation to models of the real YSZ-based sensors, *Ionics* **12** (2006) 135–148. With kind permission of Springer Science and Business Media.)

Moreover, the oxygen ions O^{2-} can subdissociate to subions O^-, which can diffuse on the metal-electrolyte and gas-electrolyte boundaries toward the gas-SE-YSZ interface. Stage 3 is the electrochemical reaction of oxygen ($O_2 + 4e^- \rightarrow 2O^{2-}$), which initiates changing the capacity of the double electric layer on the gas-SE-YSZ interface and, consequently, altering the potential of the SE (Stage 4). It should be noted that oxygen might be oxidized in different forms of O_2^-, O_2^{2-}, O^{2-}, O^-, and O depending on the working temperature, the oxygen partial pressure, and whether the SE is a reversible or irreversible electrode. The analogous processes occur on the RE (Stages 2–4) with the only difference being that the oxygen partial pressure is a characteristic of the solid-state RE, which represents a heterogeneous mixture of metal and its oxide decomposing under the temperature changes (Stage 5). The output informative parameter of the sensor on Stage 6 is the difference of potentials between the SE and RE: $E(t) = \varphi^{RE} - \varphi^{SE}(t)$.

In the case of the YSZ-based oxygen sensor with the Me-MeO RE, the electrochemical reactions take place at different kinetic rates on the dissimilar electrodes. On Stage 1, the dependence of the concentrations of physically adsorbed $N_{O2\ f}$ molecules on the partial pressure P_{O2} of the measuring gas can be determined. For Stage 1, the following assumptions should be taken into account:

- The surface of the SE is energetically uniform in the sense that the probability of adsorption and desorption is equal for all parts of the SE.
- The elementary adsorption process takes place at the interaction of gas molecules with the surface of the SE. Therefore, the concentration of adsorbed molecules is proportional to the number of gas interactions with the surface.
- The action radius of the adsorbing forces is relatively small, and therefore only molecules which interact with the clear surface can be adsorbed. The adsorbed molecules do not interact to each other.
- Adsorption takes place only within a monolayer. The number of adsorption centers is constant, independent of temperature, typical only for the current surface, and stipulated by its roughness.

In our case, the dissociative form of adsorption dominates for the oxygen sensor working at temperatures of 700–900°C, and consequently the adsorption isotherm differs from the linear Langmuir adsorption isotherm [34].

Based on the assumptions given above, the density of the adsorption stream is proportional to the probability of the existence of two free neighboring adsorption centers, and the density of the desorption stream is proportional to the probability of the existence of two engaged neighboring adsorption centers:

$$\frac{dN_{O2f}(t)}{dt} = \frac{1}{a}\left(1 - bN_{O2f}(t)\right)^2 P - c\left(bN_{O2f}(t)\right)^2, \tag{2.5}$$

where

$$\frac{1}{a} = \frac{\omega SN^*}{\sqrt{2\pi mkT}}, b = 1/N^*, c = \upsilon N^* e^{-Q/kt} .$$

N_{O2f} is the concentration of the physically adsorbed gaseous molecules of O_2 on the surface of the interactive system of the gas-oxide SE. m is the mass of the gas molecule. ω is the probability of adsorption ($\omega = 1$ at calculations [32]). S is the effective square of the adsorbed O_2 molecules. N^* is the number of adsorption centers. P is the partial pressure on the surface of the SE. $k = k_B$ is the Boltzmann constant. υ is the probability of desorption of adsorbed O_2 molecules. $\upsilon = 1/\tau_0$ and τ_0 are the minimum time for the gaseous component to be at the adsorption state. Q is the activation energy of adsorbed atoms, which is equal to the activation energy of adsorption of gaseous molecules plus the dissociation energy of the molecule: Q = $Q_{ads} + Q_{dis}$. $P/\sqrt{2\pi mkT}$ is the number of hits of O_2 molecules with the surface based on the kinetic theory of gases.

Initial conditions for Equation (2.5) are as follows:

$$N_{O2f}(0) = N'_{f0} = const. \tag{2.6}$$

Value N_{O2f} can be determined by the primary conditions of the sensor in the working chamber.

Diffusion of gaseous components may take place with three mechanisms: bulk, Knudsen, and surface diffusion [34]. Considering that the O_2 molecules of gas flow contact with the SE and reach the TPB among gas, YSZ, and the SE by diffusion in accordance with Fick's second law [35] the gas diffusion is expressed as a parabolic-type equation:

$$\frac{\partial N(x,t)}{\partial t} = D(N,x,T) \cdot \left(\frac{\partial^2 N(x,t)}{\partial x^2} + d \frac{N(x,t)}{k_B T} \cdot \frac{\partial \Psi(x,T)}{\partial x} \right), 0 < x < \delta; 0 < t < \infty ,$$

$$\tag{2.7}$$

where N (x, t) is the concentration of the gaseous components (oxygen molecules, atoms, and ions) in pores created by the grain boundaries in the thin-film SE. x is the spatial coordinate toward the surface of YSZ. D (N, x, T) is the coefficient of diffusion, which generally depends on the concentration of diffusing gas, spatial coordinate and temperature.

At a set temperature and concentration, the coefficient of diffusion D is constant and its temperature dependence, calculated by the Arrhenius law [36], can be ignored. However, in the presence of external forces F and F_1, stipulating the appearance of an additional flow of the gaseous components by electrodiffusion and by thermodiffusion (Soret effect [37]), respectively, the coefficient d in Equation (2.7) describes the force of the convective diffusion as follows:

$$d = \frac{D(N,x,T)}{k_B T} F + \frac{D(N,x,T)}{k_B T} F_1 . \tag{2.8}$$

The first and the second terms on the right-hand side of Equation (2.8) represent the effect of electrotransfer of the diffusing ions and the Soret effect (thermal diffusion due to temperature gradients), respectively. The electrical force is given by

$$F = (\zeta_{ef}/s) \, z^* \, q \, E_F,$$

where ζ_{ef} is the effective width of the grain boundary, and s is the average grain size. The effective charge of the ionized gas can be written as $z^*q = z - z'\sigma^*/\sigma^-$ in electron units, where z' is the number of conductivity electrons on one diffusive atom, σ^* is the integral cross-section of the dissipation of conductivity electron on the diffusive ion (it is equal at rough estimate to the square of the atom cross-section $\sigma^* = \pi r^2$; r is the radium of atom), $\sigma^- = z'q^2\rho/(2mE_F)^{1/2}$ is an average cross-section of dissipation, ρ is the specific electroresistance, and E_F is the tension of the electrical field.

The external force, stipulated by the Soret effect [37], can be calculated by the following equation:

$$F_1 = -\frac{Q_i^*}{T} \cdot \frac{\partial T}{\partial x} \,, \tag{2.9}$$

where Q_i^* is the heat of gas transfer, which is figured out by experiment (for example, for oxygen diffusion $Q^* = 12$ kJ/mol [32]) and $\partial T/\partial x$ is the temperature gradient.

Initial conditions for Equation (2.7) are as follows:

$$N_{O2f}(x, 0) = N'_{O2\,0} = const, \tag{2.10}$$

and, similarly to Equation (2.6), they can be determined by the primary conditions of the sensor in the working chamber.

The concentrations of O_2 on the gas-SE boundary at $x = 0$ are constant and equal to the gas concentrations in analyzing environments N, gas streams through the TPB ($x = \delta$), which are equal to zero since it has been assumed that there are no gas streams to the YSZ substrate or along boundaries between the film SE and the YSZ electrolyte:

$$N(0,t) = N = const; \quad \left.\frac{\partial N}{\partial x}\right|_{x=\delta} = 0 \,. \tag{2.11}$$

This is a very idealistic boundary condition because it has been presumed that any flows toward the SE and YSZ substrate or along the SE-YSZ interfaces are absent. However, considering the assumptions to the mathematical model given above, these flows can be ignored [32].

The next stage of oxygen interaction with the Pt SE of the YSZ-based sensor is the electrochemical reaction of oxygen ionization at the TPB. We assume that the oxygen atom accepts two electrons from the Pt SE at the TPB and as ion O^{2-} transfers through the double electrical layer on the boundary YSZ-SE. The reaction of the oxygen charge can be presented as follows: $O + 2e^- \rightarrow O^{2-}$. Corresponding to this reaction, the current charge density j can be determined by the kinetic equation $j = \vec{j} + \bar{j}$, where $\vec{j}$ and $\bar{j}$ are the current density of the direct and reverse reactions, respectively.

Opening the values of $\vec{j}$ and $\bar{j}$, the following correlation takes place:

$$j(t) = j_0^0 N^{1-\alpha}(t) N_{O2-}(t) \left\{ exp\left(\frac{-2\beta F \eta_e(t)}{RT}\right) - exp\left(\frac{2(1-\beta)F\eta_e(t)}{RT}\right) \right\},$$

(2.12)

where R is the universal gas constant; j_0^0 is the density of the standard exchange current, corresponding to the concentrations N and N_{O2-}; β is the transfer coefficient, characterizing the energy part of the double electrical layer; and η_e is the overstrain of the SE stipulated by deceleration of the electrochemical stage. η_e represents itself the deviation of potential from the equilibrium value at this moment of time.

The correlation for density of the standard exchange current can be expressed as follows [38]:

$$j_0^0 = 2Fk_1 \exp\left(-\frac{U_0}{RT}\right)\exp\left(-\frac{2\beta F\varphi_0}{RT}\right) = 2Fk_2 \exp\left(-\frac{U_0}{RT}\right)\exp\left[\frac{2(1-\beta)F\varphi_0}{RT}\right],$$

(2.13)

where k_1 and k_2 are constants of the speed of the direct and reverse reactions, respectively. U_0 is the jump of potential between the SE and YSZ, and φ_0 is the standard electrode potential.

During the transfer of oxygen ions through the SE-YSZ boundary, the double electric layer changes by changing the quantity of charges of opposite sign on both sides of the SE-YSZ boundary. The charge-discharge reaction of oxygen atoms takes place directly in this layer. Therefore, the distribution of the electrode potential and the exact location of the reacting particles influence both the value of the electrode overstrain and the speed of electrochemical reaction. The following assumptions were made at the analysis of the forming of the double electrical layer:

1. The existing double electrical layer can be described by the Stern model [38]. Based on this model, the dense Helmholtz layer does exist in the close vicinity of the TPB, and the oxygen ions O^{2-} are bound near the surface due to specially adsorbing and coulomb interactions. The rest of the ions, taking part in the double electrical layer, are distributed diffusively with a gradual decreasing of the charge density (see Figure 2.4).

2. Only particles in the close vicinity of the solid electrolyte surface take place in the electrochemical reaction. According to this fact, the surface layer possessing the defect structure in the double electrical layer can be

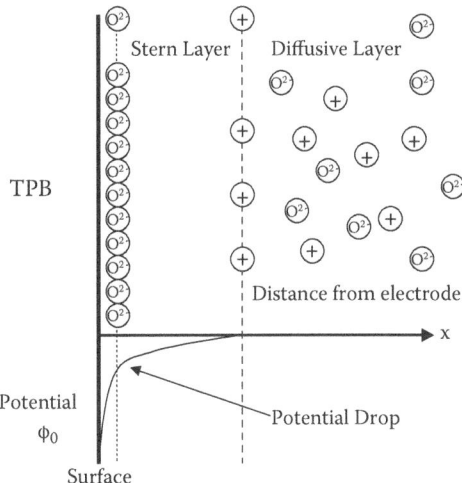

FIGURE 2.4 Schematic diagram of the electrical double layer. (Reprinted from Zhuiykov, S., Mathematical modelling of YSZ-based potentiometric gas sensors with oxide sensing electrodes part II: Complete and numerical models for analysis of sensor characteristics, *Sensors and Actuators B, Chem.* 120 (2007) 645–656, with permission from Elsevier Science.)

distinguished from the bulk layer. Thus, the capacity of the double electrical layer C can be calculated as follows:

$$\frac{1}{C} = \frac{1}{C'} + \frac{1}{C^D + C^S},$$

where C', C^D, and C^S are the capacity of the dense Helmholtz, diffusive, and surface layers, respectively.

The equivalent scheme of the boundary impedance of the SE-YSZ (Figure 2.5, *a*) can be represented as two parallel branches: one of each is characterized by the capacitance of the double electrical layer C. The second one, stipulated by the Faraday process, is characterized only by the transference resistance R_t for the SE, reversible to oxygen ions. The active resistance, equal to the ohm resistance of electrolyte R_{SE}, is connected in consecutive order. The scheme can be transformed to the more simple scheme, illustrated in Figure 2.5, *b*, where $R = (R_t \cdot R_{SE}/(R_t + R_{SE}))$. If the current is absent in the circuit, then $R = R_t$. Then the development of the double electrical layer on the SE-YSZ boundary can be considered as the process of charging the capacitor C by the current I through the resistance R_t. The value of current I can be calculated by Equations (2.12) and (2.13). In this case, the SE potential can be expressed as follows:

$$\varphi^{SE}(t) = \varphi_0^{SE} + \Delta\varphi(t), \tag{2.14}$$

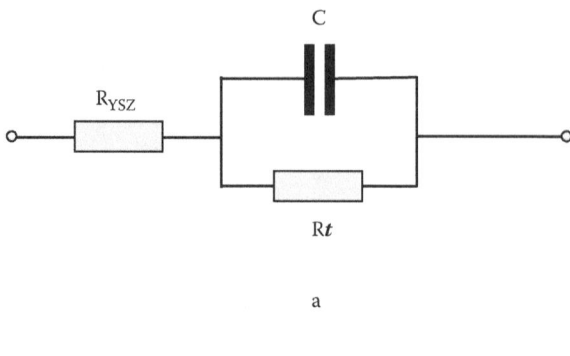

a

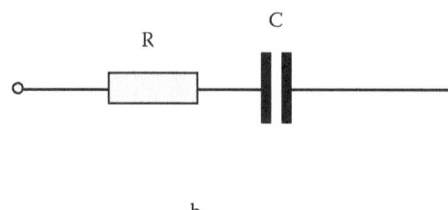

b

FIGURE 2.5 Equivalent scheme of the boundary impedance of the SE-YSZ (*a*) at the presence and (*b*) at the absence of the current in the circuit. (Reprinted from Zhuiykov, S., Mathematical modelling of YSZ-based potentiometric gas sensors with oxide sensing electrodes part II: Complete and numerical models for analysis of sensor characteristics, *Sensors and Actuators B, Chem.* 120 (2007) 645–656, with permission from Elsevier Science.)

where φ_0^{SE} is the value of potential at the initial moment of time, and $\Delta\varphi(t)$ is the changing of the SE potential during the charge reaction, which can be calculated as

$$\Delta\varphi(t) = \frac{1}{C}\int_0^t I(t)\,dt .\qquad(2.15)$$

$I(t)$ changes proportionally to the changing of overstrain on the SE $\eta_e(t)$:

$$I(t) = (S/R_{pol}) \bullet \eta_e(t),$$

where R_{pol} is the polarization resistance at the electrochemical overstrain. R_{pol} represents the kinetic analog of resistance in the Ohm law. $S = S_{SE} \bullet (1-\upsilon')$, where υ' is the porosity of the SE and S_{SE} is the area of the SE.

By the analog to the charge-discharge process of the capacitor, the value $\eta_e(t)$ changes in time as follows:

$$\eta_e\left(t\right) = \Delta\varphi\, exp\left[-\frac{S_{SE}\bullet\left(1-p\right)}{R_{pol}C}\bullet t\right],$$

where $\Delta\varphi$ is the deviation of the SE potential at the new equilibrium state φ_0^{SE} from the previous one φ_0. Based on the Nernst law, the equilibrium value of potential with respect to the air electrode is calculated as follows:

$$\varphi_0^{SE} = \frac{RT}{2F}\ln N_{eql},$$

where N_{eql} is the concentration of oxygen atoms on the zirconia-SE boundary at the equilibrium stage.

Considering Equations (2.14) and (2.15), the temporal change of the SE potential with the speed equal to the speed of electrochemical reaction is expressed by the following equation:

$$\varphi^{SE}\left(t\right) = \frac{RT}{2F}\ln N_{eql} + \left(\varphi_0 - \frac{RT}{2F}\ln N_{eql}\right)exp\left[-\frac{S_{SE}\bullet\left(1-p\right)}{R_{pol}C}\bullet t\right]. \qquad (2.16)$$

Consequently, the change of the SE potential $\varphi^{SE}\left(t\right)$ at the changing of the oxygen partial pressure P can be described by Equations (2.5)–(2.11) and (2.16).

The oxygen partial pressure P^{RE} on the RE is determined by the thermodynamic dissociation of the metal oxide $Me_lO_{2n} \leftrightarrow lMe + nO_2$ and is calculated by the following equation [39]:

$$P^{RE} = 10^5\, exp\,(\Delta G/nRT), \qquad (2.17)$$

where l and n are the coefficients, and ΔG is the standard Gibbs energy of formation of MeO. ΔG has the linear temperature dependence and has been tabulated for many metal oxides.

Considering the processes on the RE, it is assumed that the external environment does not affect the properties of the RE; therefore, P^{RE} = const at the fixed temperature. Furthermore, it is accepted that the processes at the RE are equilibrated. It means that the value of φ^{RE} is independent of time. Then the potential of the RE with respect to the air electrode is calculated as follows:

$$\varphi^{RE} = \frac{RT}{2F}\ln N_{eql}^{RE}, \qquad (2.18)$$

where N^{RE}_{eql} is the concentration of oxygen atoms on the YSZ-RE boundary at the equilibrium stage.

Therefore, the following system of equations is represented as the mathematical model of the YSZ-based O_2 sensor with the Pt SE and Me-MeO RE:

$$\frac{dN_{O2f}(t)}{dt} = \frac{1}{a}\left(1 - bN_{O2f}(t)\right)^2 P - c\left(bN_{O2f}(t)\right)^2 ,$$

where

$$\frac{1}{a} = \frac{\omega SN^*}{\sqrt{2\pi mkT}}, b = 1/N^*, c = \upsilon N^* e^{-Q/kt} .$$

$$\frac{\partial N(x,t)}{\partial t} = D(N,x,T) \cdot \left(\frac{\partial^2 N(x,t)}{\partial x^2} + d\frac{N(x,t)}{k_BT} \cdot \frac{\partial \psi(x,T)}{x}\right),$$

$$0 < x < \delta, \, 0 < t < \infty.$$

$$N_{O2f}(0) = N'_{f0} = const.$$

$$d = \frac{D(N,x,T)}{k_BT}F + \frac{D(N,x,T)}{k_BT}F_1 .$$

$$F = (\zeta_{ef}/s)\, z^*\, q\, E_F ,$$

$$F_1 = -\frac{Q_i^*}{T} \cdot \frac{\partial T}{\partial x} .$$

$$N_{O2f}(x, 0) = N'_{O2\,0} = const.$$

$$N(0,t) = N = const. \quad \left.\frac{\partial N_i}{\partial x}\right|_{x=\delta} = 0 .$$

$$j(t) = j_0^0 N^{1-\alpha}(t) N_{O2-}(t)\left\{exp\left(\frac{-2\beta F\eta_e(t)}{RT}\right) - exp\left(\frac{2(1-\beta)F\eta_e(t)}{RT}\right)\right\} .$$

$$j_0^0 = 2Fk_1\,exp\left(-\frac{U_0}{RT}\right)exp\left(-\frac{2\beta F\varphi_0}{RT}\right) = 2Fk_2\,exp\left(-\frac{U_0}{RT}\right)exp\left[\frac{2(1-\beta)F\varphi_0}{RT}\right] .$$

$$\varphi^{SE}(t) = \varphi_0^{SE} + \Delta\varphi\,(t),$$

$$\Delta\varphi(t) = \frac{1}{C}\int_0^t I(t)\,dt\ .$$

$$\varphi^{SE}(t) = \frac{RT}{2F}\ln N_{eql} + \left(\varphi_0 - \frac{RT}{2F}\ln N_{eql}\right)exp\left[-\frac{S_{SE}\bullet(1-p)}{R_{pol}C}\bullet t\right]\ .$$

$$P^{RE} = 10^5\ exp\ (\Delta G/nRT),$$

$$\varphi^{RE} = \frac{RT}{2F}\ln N_{eql}^{RE}\ .$$

Consequently, the sensor output parameter represents the electromotive force *emf* E (t), which is equal to the changing of the sensor potentials of electrodes, RE φ^{RE} and SE $\varphi^{SE}(t)$:

$$E\ (t) = \varphi^{RE} - \varphi^{SE}(t). \tag{2.19}$$

This model can be employed for analyzing the sensor performance at the different working temperatures and at the different measuring O_2 concentrations.

The final conclusions about such characteristics of the YSZ-based gas sensor as sensitivity, response and recovery time, and so on, which are stipulated by the mechanism of permeability of the measuring gas though the SE of the sensor, can only be done on the basis of comprehensive analysis of all thermodynamic and kinetic processes observed with the use of the mathematical model provided. The proposed algorithm of transfer from the complete mathematical model to the models of the real YSZ-based gas sensors has also been successfully employed for modeling the mixed-potential-type YSZ-based potentiometric NO_2 sensor [40, 41]. The designer must always seek a compromise between high sensitivity, low water vapor influence, long-term stability, and high selectivity of the YSZ-based sensors with the different materials of the SE (RE) and applied technologies for deposition of the SE on the YSZ substrate. This can lead to the conclusion that the development of the YSZ-based sensor with the metal or oxide SE can progress for more specific applications rather than target multipurpose devices designed for all markets. However, it should be noted that a success in this direction is very hard to predict because, in general, the optimal design is usually dependent on the designer's ability to select correctly both the optimizing parameters and the combination of the different metals or oxides together with necessary additives to the SE in order to achieve the optimal result attainment. However, acquaintance with the present decision-making tool will provide an opportunity to develop new YSZ-based gas sensors with greater success. New design and concepts for YSZ sensors, derived from improved theoretical understanding of the surface and bulk processes in the SE (RE), ensure that the development of the YSZ-based gas sensors will be more targeted on the specific applications and also will be simpler and relatively inexpensive.

2.3 MODELING INTERACTIONS OF VARIOUS GASES WITH NON-NERNSTIAN ZIRCONIA SENSORS

2.3.1 DESCRIPTION OF NON-NERNSTIAN BEHAVIOR

The situation when the typical zirconia-based oxygen electrochemical cell with Pt electrodes generates *emf* which differs from the Nernstian equation at a relatively low temperature, when at least two simultaneous oxidation/reduction reactions occur competitively on the SE at the presence of oxidizing/reducing gases, has been reported by many researchers [42–45]. This anomalous *emf* represents a mixed potential at the SE, which appears as a consequence of the coupling between electrochemical oxidation and reduction reactions [7, 46]. For example, in relation to NO_x sensing, the electrode reactions of NO_2 (NO) and O_2 proceed simultaneously at the TPB among gas, zirconia, and the oxide-SE, as was reported for the first time [47, 48], and are combined as follows:

$$\text{for } NO_2: NO_2 + 2e^- \rightarrow NO + O^{2-} \text{ and } O^{2-} \rightarrow {}^1\!/_2\, O_2 + 2\,e^- \qquad (2.20)$$

$$\text{for } NO: NO + O^{2-} \rightarrow NO_2 + 2e^- \text{ and } {}^1\!/_2\, O_2 + 2\,e^- \rightarrow O^{2-} \qquad (2.21)$$

Both NO_2 and NO tend to be partially reduced or oxidized, respectively, to form an equilibrium mixture of NO_2 and NO. Consistent with this tendency, each undergoes a cathodic or anodic reaction to be coupled with an anodic or cathodic reaction of O^{2-} or ${}^1\!/_2\, O_2$, respectively. Quantitative verification of the mixed-potential mechanism usually can be done by measuring polarization curves for the sensor in air, NO + air, and NO_2 + air at working temperatures. As the NO or NO_2 concentration in air increases, the polarization curves consequently shift upward or downward from that in air. The shifts of polarization curves are owing to the electrochemical reaction of NO or NO_2 in addition to that of oxygen. Thus, the mixed potential can be estimated from the intersections of these polarization curves. The estimated values then can be compared with those *emf* values which were experimentally observed for the same concentration of NO or NO_2 at the same temperature. Depending on the catalytic activity of the used SE materials, the electrodes behave differently in nonequilibrated gaseous environments. On the surface of the SE with high catalytic activity, the cathodic reaction of NO_2 may lead to the following decomposition reaction of NO_2 in the vicinity of the electrode:

$$NO_2 \rightarrow NO + {}^1\!/_2\, O_2. \qquad (2.22)$$

Consequently, high NO_2 sensitivity of the sensor will be lowered. To avoid NO_2 conversion on the SE at high working temperatures, the sintering temperature of the oxide SE can be increased to provide a relatively porous structure with larger grains and pores compared to the SE structure sintered at low temperatures [49]. Based on our previous results [15, 29], low catalytic activity of the gas-phase reaction (Equation 2.22) would lead to high NO_2 sensitivity. Schematically, the effect of the sintering temperature of the SE matrix on the catalytic activity to the gas-phase

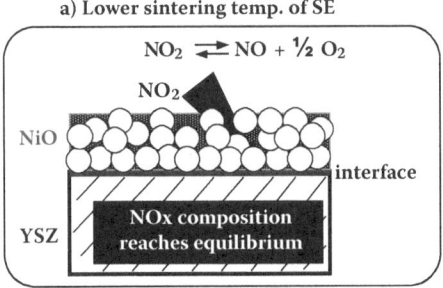

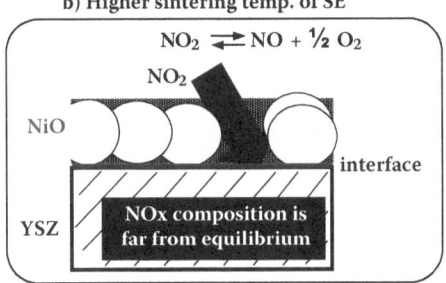

FIGURE 2.6 Effect of the sintering temperature of the SE matrix on the catalytic activity to the gas-phase decomposition reaction of NO_2. (Reprinted from Zhuiykov, S. and Miura, N., Development of zirconia-based potentiometric NOx sensors for automotive and energy industries in the early 21st century: What are the prospects for sensors? *Sensors and Actuators: B Chem.* 121 (2007) 639–651, with permission from Elsevier Science.)

decomposition reaction of NO_2 is shown in Figure 2.6. Larger pore sizes for the NiO-SE sintered at 1400°C compared to the NiO-SE sintered at 1100°C were confirmed by SEM investigation [49]. It is evident that NO_2 gas diffusing through the larger pores toward the TPB makes fewer contacts with the surface of SE grains. Thus, NO_2 can reach the TPB interface as such without serious decomposition to NO. This leads to the conclusion that in the case of porous structure of the SE, NO_x can be far from equilibrium when the measuring gas reaches the TPB after diffusing though the SE film. The confirmation of this explanation is shown in Figure 2.7, where the catalytic activity of the gas-phase decomposition of NO_2 (Equation 2.22) was measured for each of the NiO powder sintered at different temperatures. It can be clearly seen from this figure that the percentage of NO_2 conversion tends to be decreased as the sintering temperature is increased. These results indicated that the matrix of the SE and applied technology play significant roles in deciding the NO_2 sensitivity of the zirconia-based mixed-potential sensor.

Taking into account the results obtained by the different research groups, it can be concluded that the sensing mechanism of the solid electrolyte NO_x sensor with the oxide-SE is based on the mixed-potential model under the coexistence of NO_x and O_2. This mechanism is rather complex, and the NO_2 sensitivity can be indirectly determined by the following factors [50]:

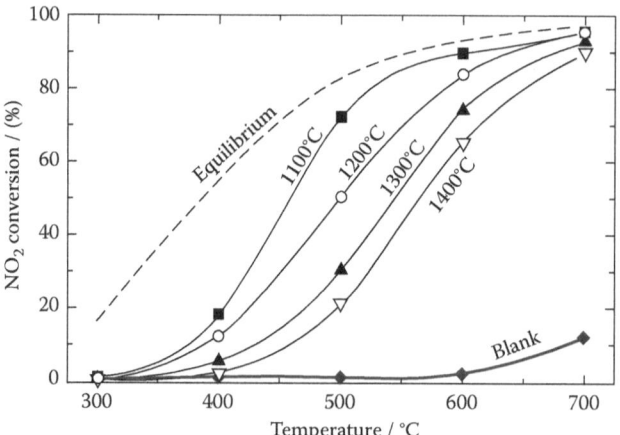

FIGURE 2.7 Temperature dependence of NO_2 conversion to NO for the gas-phase reaction (Equation 2.22) on each of the NiO powders sintered at different temperatures. (Reprinted from Zhuiykov, S. and Miura, N., Development of zirconia-based potentiometric NOx sensors for automotive and energy industries in the early 21st century: What are the prospects for sensors? *Sensors and Actuators: B Chem.* 121 (2007) 639–651, with permission from Elsevier Science.)

- Adsorption-desorption behavior of NO_2 on the SE. If the NO_2 adsorption is strong on the SE, this may lead to the high catalytic activity for cathodic reaction $NO_2 + 2e^- \rightarrow NO + O^{2-}$.
- Adsorption-desorption behavior of oxygen on the SE and oxygen-sensing performance of the SE. Strong oxygen adsorption not always directly correlated to the low catalytic activity for anodic reaction $O^{2-} \rightarrow \frac{1}{2} O_2 + 2 e^-$. However, the catalytic inactiveness of the SE toward O_2 is essential for establishment of a mixed potential responding to NO and NO_2 [51].
- Simultaneous changes in the capacitance of the double electric layer C_{dl} at interfaces YSZ-Pt-RE and YSZ-oxide-SE. The capacitance C_{dl} of Pt (Au)-RE-YSZ is usually independent of the temperature and P_{O2}, and has a value of about 30–150 $\mu F/cm^2$ [52]. In contrast, C_{dl} for YSZ-oxide-SE is dependent on the temperature and P_{O2}, and is approximately one order of magnitude lower than C_{dl} of SEs in the "metallic" state.
- The catalytic activity of the SE for the nonelectrochemical gas-phase decomposition reaction of NO_2: $NO_2 \rightarrow NO + \frac{1}{2} O_2$. The lower the catalytic activity, the higher the NO_2 sensitivity at high temperatures.

All of the above-mentioned factors entangle each other in a complicated manner and usually depend on the chemical composition and morphology of the SE. The last factor also depends on technology applied for making SEs.

2.3.2 NON-NERNSTIAN ZIRCONIA-BASED NO$_x$ SENSORS

The planar YSZ-based potentiometric NO_x sensor configuration is presented in Figure 2.8. This figure shows the views of the (*a*) front and (*b*) back sides of the

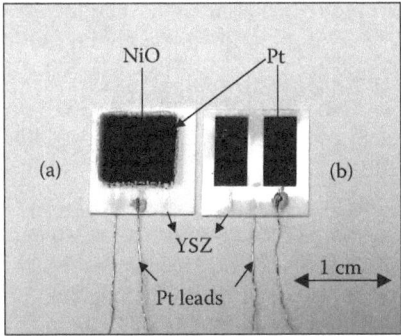

FIGURE 2.8 Front (*a*) and back (*b*) views of the planar non-Nernstian YSZ-based NO$_x$ sensor with the NiO-SE and Pt-RE. (Reprinted from Zhuiykov, S. and Miura, N., Development of zirconia-based potentiometric NO$_x$ sensors for automotive and energy industries in the early 21st century: What are the prospects for sensors? *Sensors and Actuators: B Chem.* 121 (2007) 639–651, with permission from Elsevier Science.)

sensor. This planar sensor was fabricated at KASTEC, Kyushu University, Japan, by using an YSZ plate (8 wt % Y$_2$O$_3$-doped, 10×10 mm; 0.2 mm thickness). Pt paste (Tanaka Kikinzoku, Japan) was printed on both sides of the YSZ plate and was fired at 1000°C for 2 hours in air. On one side of the YSZ plate, two rectangular Pt stripes were formed as the RE of the sensor; and on the other side, six narrow Pt stripes were formed as a base (current collector) for the thin-film NiO-SE. Both the NiO-SE and Pt-RE were exposed simultaneously to the sample gas or to the base gas. The difference in potential E (*emf*) between the NiO-SE (φ_1) and Pt-RE (φ_2) was measured as the sensor output signal. This planar sensor can be best described in the form of electrochemical cells as follows:

In a base gas O_2, NiO |YSZ| Pt, O_2 (2.23)

In a sample gas $NO_2 + O_2 + H_2O$, NiO |YSZ| Pt, $NO_2 + O_2 + H_2O$ (2.24)

One of the main assumptions for the thin-film YSZ-based sensors is the negligible interinfluence of the thin-film layers. This means that the mechanical, physical, chemical, and electrostatic components of their interactions are close to zero [53]. In order to achieve such conditions, the following requirements must be implemented: the purity of raw materials for thin films should be ultra-high (99.999%), and raw materials should have compatible coefficients of thermal expansion and parameters of the crystalline structure. They should also be characterized by the minimum value of the contact difference of potentials.

Modeling the interactions of various gases with the YSZ-based sensors, it is important to acknowledge that the state of a SE (RE) surface can determine various important characteristics of these sensors. Usually, statistical parameters of the surface height distribution function, such as the root-mean-square, slope, curvature, average height, average surface area, average roughness, and surface fractal number (dimension), have been used to characterize the surfaces of thin-film electrodes. However, experiments have shown that the surface topography cannot be adequately described by means of the obtained statistical parameters, and corrugation of the surface height of the SE may have a broad bandwidth [54]. This is because nonregularities are inherent in thin-film structures. These include the nonuniformity in pore distribution on the surface and within the thin-film electrodes, which may determine instability in various electrode processes such as adsorption-desorption, chemisorption, surface diffusion, diffusion by grain boundaries, molecular or Knudsen diffusion, and so on, or fluctuations in the SE electrical surface resistance [55]. Inadvertently, the tortuous pore structure slows down the dynamic performance of the sensors, and, consequently, the gas diffusion changes drastically within the structure of the SE over a relatively small distance. It has already been reported that the changes in sintering temperature and in the pore size and distribution have a substantial effect on the catalytic activity of the electrochemical reactions of oxygen and the gas-phase decomposition reaction of NO_2 on the YSZ-based planar NO_x sensors [9].

Physical and chemical properties of the surface of SEs in potentiometric gas sensors are sensitive to, and in some cases determined by, random roughness or surface disorder. Irregularities at an atomic level define electronic energy distribution at surface sites, and irregularities at a micrometer level determine the accessibility of particles to surface sites [56]. Recently, a new phenomenological method for the analysis of nonlinear system dynamics was proposed [57]. The methodology is based on a postulate that key information relating to the sensor system dynamics is contained in nonregularities of measured dynamic variables (temporal or spatial), as well as on the acceptance of a new scaling equation.

2.3.3 MATHEMATICAL FORMULATION OF ZIRCONIA-BASED NO_x SENSORS

One of the key components of the mathematical model of YSZ-based sensors is the modeling of the ion and electron transport. The scheme describing the interactions of the measuring gas with the planar potentiometric non-Nernstian YSZ-based NO_x sensor is shown in Figure 2.9. This scheme has been developed on the analysis of the electrode processes in the electrochemical systems [40] and is similar to the scheme for the Nernstian O_2 sensor (Figure 2.3). The main difference is the number of gaseous components reacting on and within the SE and the kinetics of the electrochemical reactions in the SE and on the TPB. Stage 1 is the multicomponent diffusion of gaseous species through the porous SE and RE, allowing for the effect of nonuniform gas pressure, sensor temperature, and Knudsen diffusion. This stage is the same as has previously been described for the Nernstian YSZ-based O_2 sensor. Stage 2 is the adsorption-desorption and partial dissociation of nitrogen dioxide molecules within the oxide SE ($NO_2 \leftrightarrow NO + \frac{1}{2} O_2$), accompanied by dissociation of water vapors ($H_2O + 2e^- \rightarrow H_2 + O^{2-}$) and oxygen molecules ($O_2 \rightarrow 2O^{2-} + 4e^-$) on the gas-SE-YSZ interface [41]. Furthermore, oxygen ions O^{2-} can subdissociate to subions O^-, which can diffuse on the metal-oxide-electrolyte and gas-electrolyte boundaries toward the gas-SE-YSZ interface. Stage 3 is the cathodic reaction of nitrogen dioxide ($2NO_2 + 4e^- \rightarrow 2NO + 2O^{2-}$) within the SE and on the gas-SE-YSZ interface, which changes the potential of the electrode. Stage 4 is the electrochemical

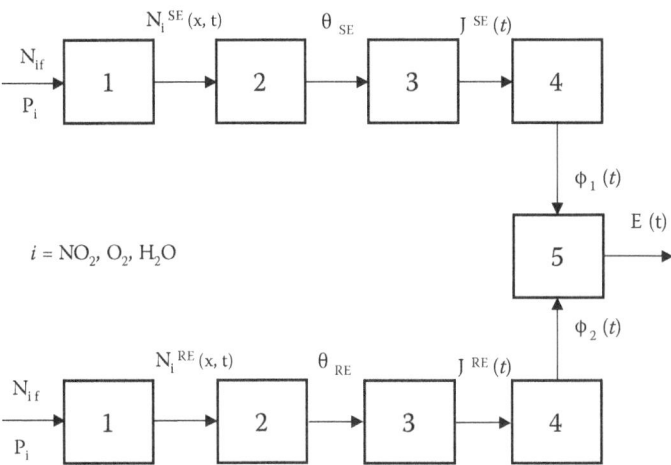

FIGURE 2.9 Schematic presentation of the interaction of the non-Nernstian solid electrolyte sensor with a gas environment: 1–5 stages of interaction. (Reprinted from Zhuiykov, S., Mathematical modelling of YSZ-based potentiometric gas sensors with oxide sensing electrodes part I: Model of interactions of measuring gas with sensor, *Sensors and Actuators B, Chem.* **119** (2006) 456–465, with permission from Elsevier Science.)

reaction of oxygen ($O_2 + 4e^- \rightarrow 2O^{2-}$), which initiates the change of capacitance of the double electric layer C_{dl} on the gas-SE-YSZ interface and, consequently, changes the potential of the SE (thin-film Pt-RE). The output parameter of the sensor in Stage 5 is the difference of potentials between the SE and RE: $E(t) = \varphi_1(t) - \varphi_2(t)$.

In the case of the planar non-Nernstian NO_x sensor, shown in Figure 2.8, the electrochemical reactions take place at different kinetic rates on the dissimilar electrodes. In Stage 1, the dependence of the concentrations of physically adsorbed $N_{NOx\,f}$ and $N_{O2\,f}$ molecules on the partial pressure P_{NOx} and P_{O2} of the measuring gases can be determined.

The dissociative form of O_2 and NO_2 adsorption is dominated for the NO_x sensor working at temperatures of 500–900°C. Desorption is accompanied by recombination of adsorbed atoms into molecules. The coordinate system and the sign convention for various fluxes in the electrochemical cell are illustrated in Figure 2.10 [40]. The species indices (1, 2, and 3) correspond respectively to NO_2, O_2, and H_2O, respectively, in the SE (RE).

The density of the adsorption stream is as follows:

$$\frac{dN_{if}(t)}{dt} = \sum_{i=1}^{3} \frac{1}{a}\left(1 - bN_{if}(t)\right)^2 P_i - c\left(bN_{if}(t)\right)^2,\qquad(2.25)$$

where

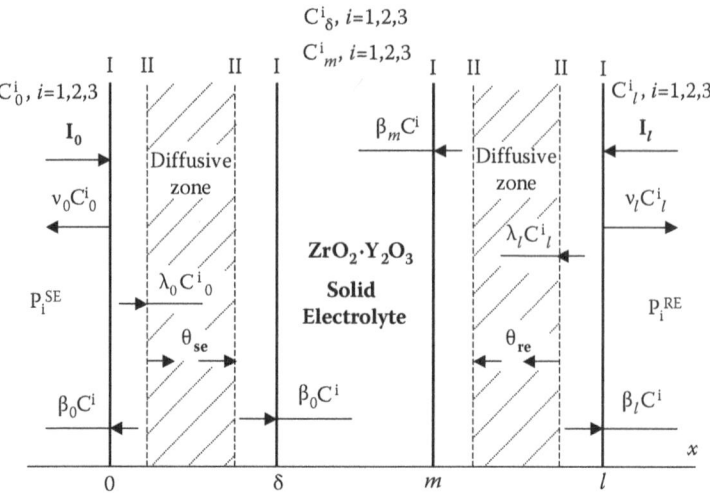

FIGURE 2.10 Coordinate system and the sign convention for various fluxes in the YSZ-based electrochemical cell. (Reprinted from Zhuiykov, S., Mathematical modelling of YSZ-based potentiometric gas sensors with oxide sensing electrodes part I: Model of interactions of measuring gas with sensor, *Sensors and Actuators B, Chem.* **119** (2006) 456–465, with permission from Elsevier Science.)

$$\frac{1}{a} = \frac{\omega S_i N_i *}{\sqrt{2\pi mkT}}, b = 1/N_i *, c = \upsilon_i N_i * e^{-Q/kt}, \text{ and } i = 2, 3 \, .$$

N_i* is the number of adsorption centers. P_i is the partial pressure ($i = 1, 2, 3$) of NO_2, O_2, and H_2O, respectively, on the surface of the SE. $k = k_B$ is the Boltzmann constant. υ_i is the probability of desorption of adsorbed NO_2, O_2, and H_2O molecules. $\upsilon_i = 1/\tau_{i\,0}$ and $\tau_{i\,0}$ are the minimum time for the gaseous component to be at the adsorption state. Q is the activation energy of adsorbed atoms, which is equal to the activation energy of adsorption of gaseous molecules plus the dissociation energy of the molecule: $Q = Q_{ads} + Q_{dis}$. $N_{i\,f}$ is the concentration of the physically adsorbed gaseous molecules (NO_2, O_2, and H_2O) on the surface of the interactive system of the gas-oxide SE. m is the mass of the gas molecule. ω is the probability of adsorption. S_i is the effective square of the adsorbed gas molecules (NO_2, O_2, and H_2O). $P_i /\sqrt{2\pi mkT}$ is the number of hits of NO_2, O_2, and H_2O molecules with the surface based on the kinetic theory of gases.

Initial conditions for Equation (2.25) are as follows:

$$N_{NO2\,f}(0) = N_{f0} = const. \tag{2.26}$$

$$N_{O2\,f}(0) = N'_{f0} = const. \tag{2.27}$$

$$N_{H2Of}(0) = N''_{f0} = const. \tag{2.28}$$

Values $N_{NO2\,f}$, $N_{O2\,f}$, and $N_{H2O\,f}$ are based on the primary conditions of the sensor in the working chamber, where NO_2 and H_2O are practically absent, and therefore for calculation it can be assumed that $N_{NO2\,f}(0) = 0$ and $N_{H2O\,f}(0) = 0$.

The gas mixture flow (molecules and atoms) within the porous oxide-SE (Pt RE), θ_{se} (θ_{re}), can be represented by the following equation:

$$\theta_{se} = -\frac{\alpha(x)}{k_B T}\frac{\partial\mu}{\partial x}, \tag{2.29}$$

where $\mu = k_B T \ln N(x, t) + \psi(x, T)$ is the chemical potential of dissolved gas. $N(x, t)$ is the concentration of atoms (molecules) of gas at point x at the moment t, α is the kinetic coefficient, and $\psi(x, T)$ is a function of spatial coordinate x and temperature T.

As mentioned above, diffusion of gaseous components may take place with three mechanisms: bulk, Knudsen, and surface diffusion. Each of the electrodes in the YSZ-based gas sensor is porous and may be described as a combination of two distinct layers (see Figure 2.10). A boundary functional layer **I** is for the critical electrochemical reactions to occur, and an "adjacent" porous diffusion layer **II** must conduct current (ions and electrons) through the metal-ceramic (SE) or metal (RE) matrix and allow for the diffusion of the chemical species. Layer **I** characterizes the surface concentrations C^i_0 (C^i_m) or C^i_δ (C^i_l), $i = 1, 2, 3$; and layer **II** characterizes the volume concentrations C^i, from which the process of random roaming begins. The performance of individual electrodes is influenced by the properties and composition of the constituent material as well as the microstructural parameters such as the particle size, the porosity and pore size, and the thickness. In the porous SE (RE), when the permeability becomes sufficiently small, the mean free path of the molecules becomes comparable to the pore size, and the collisions of molecules with the walls start to significantly affect the transport process via Knudsen diffusion. To assess the potential importance of this effect in modeling gas transport processes through the porous electrodes, consider that the molecules of the gas flow contact with the SE (RE) and reach the triple-phase boundary (TPB) among gas, YSZ, and electrode by diffusion that is expressed as the following equation:

$$\frac{\partial N_i(x,t)}{\partial t} = \sum_{i=1}^{3} D_i(N_i, x, T) \cdot \left(\frac{\partial^2 N_i(x,t)}{\partial x^2} + d\frac{N_i(x,t)}{k_B T} \cdot \frac{\partial\psi(x,T)}{x}\right), \tag{2.30}$$

$$i = 2, 3; \quad 0 < x < \delta; \quad m < x < l; \quad 0 < t < \infty,$$

where $N_i(x, t)$ is the concentration of the gaseous components (molecules, atoms, and ions) in pores created by the grain boundaries in the thin-film SE (RE). x is the spatial coordinate toward the surface of ZrO_2 (see Figure 2.10). $D_i(N_i, x, T)$ is the coefficient of diffusion, which generally depends on the concentration of diffusing gas, spatial coordinate, and temperature.

The presence of external forces F and F_1, stipulated by electrodiffusion and by thermodiffusion, respectively, can be expressed as follows:

$$d = \frac{D_i\left(N_i, x, T\right)}{k_B T} F + \frac{D_i\left(N_i, x, T\right)}{k_B T} F_1 \qquad (2.31)$$

The terms in Equation (2.31) are essentially similar to those presented in Equations (2.8) and (2.9). Initial conditions for Equation (2.30) are as follows:

$$N_{NO2\,f}\left(x,\,0\right) = N_{NO2\,0} = const. \qquad (2.32)$$

$$N_{O2\,f}\left(x,\,0\right) = N'_{O2\,0} = const. \qquad (2.33)$$

$$N_{H2O\,f}\left(x,\,0\right) = N''_{H2O\,0} = const. \qquad (2.34)$$

Similar to Equation (2.6), they can be determined by the primary conditions of the sensor in the working chamber. Therefore, for calculation purposes, it can be assumed that $N_{NO2\,0}\left(x,\,0\right) = 0$ and $N_{H2O\,0}\left(x,\,0\right) = 0$.

The concentrations of NO_2, O_2, and H_2O on the gas-SE boundary at $x = 0$ are constant and equal to the gas concentrations in measuring environments N_i, gas streams through the TPB ($x = \delta$; $x = l$), which are equal to zero since it has been assumed that there are no gas streams to the ZrO_2 substrate or along boundaries between the film SE (RE) and the substrate:

$$N_i\left(0,t\right) = N_i = const; \quad \left.\frac{\partial N_i}{\partial x}\right|_{x=\delta} = 0. \quad \left.\frac{\partial N_i}{\partial x}\right|_{x=l} = 0. \quad i = 2, 3. \qquad (2.35)$$

This is a very idealistic boundary condition because it has been presumed that any flows toward electrodes and the YSZ substrate or along the SE-YSZ (RE-YSZ) interfaces are absent. However, considering the assumptions to the mathematical model given above, these flows can be ignored [34]. For obtaining the boundary conditions to Equation (2.31) in the forms of $By = z$, $z = Cy$, $z \in \check{S}$, and $\check{S} = G + \Gamma$, the method, developed in [58], has been employed. The conditions on the surface phase distribution make sense, as the materials balance equations at the transference of molecules and atoms between these conditions and the diffusive zone (see, for example, the SE in Figure 2.9):

$$\frac{\partial C_0^i}{\partial t}\bigg|_{x=0} = I_0 - v_0 C_0^i - \xi_0 C_0^i + \rho_0 \left(C_0^i\right)^2; \quad 0 \le x \le \delta; 0 \le t \le \infty \quad i = 1,2,3.$$

$$\frac{\partial N_i}{\partial x}\bigg|_{x=\delta} = 0; \quad \lambda_0 C_0^i = \beta_0 C^i + \theta_{se};$$

$$\frac{\partial C_\delta^i}{\partial t}\bigg|_{x=\delta} = \xi_\delta C_\delta^i.$$

$$(2.36)$$

Here, I_0 is the flow of the gaseous molecules from the gas phase on the surface ($I_0 = 0$). $\lambda_{0(\delta)}$, $\beta_{0(\delta)}$, and v_0 are the constants of the speeds on the appropriate transfers on these surfaces. $\xi_{0(\delta)}$ is the constant of the speed for dissociation reaction of a gas molecule to atoms.

It is worthwhile to note that in the case of substantial deviation of the described gas-YSZ-SE system from the equilibrium state, the above-mentioned constants are also dependent on the gaseous flows θ_{se} and I_0, that is, $\lambda_0 = \lambda_0\ (\theta_{se}, I_0)$, $\beta_0 = \beta_0\ (\theta_{se}, I_0)$, $v_0 = v_0\ (\theta_{se}, I_0)$, and $\xi_{0(\delta)} = \xi_{0(\delta)}\ (\theta_{se}, I_0)$. At the equilibrium state, when $\theta_{se} = 0$, $\lambda_0{}^* = \lambda_0\ (\theta_{se}, I_0)$, $\beta_0{}^* = \beta_0\ (\theta_{se}, I_0)$, $v_0{}^* = v_0\ (\theta_{se}, I_0)$, and $\xi_{0(\delta)}{}^* = \xi_{0(\delta)}\ (\theta_{se}, I_0)$, the quantity Φ (in mol of gas on m³), dissolved within the SE, has been determined:

$$\Phi = \tfrac{1}{2} N_A^{-1/2} \left(2\pi m k_B T\right)^{-1/4} \left(\alpha_0 {}^*\beta_0 {}^{*2}\right)^{-1/2} \left(P_i^{SE}\right)^{-1/2} \delta^{-1} \int_0^l e^{-\omega(x)} dx; \quad i = 1,2,3\ .$$

$$(2.37)$$

Here, N_A is the Avogadro number, m is the mass of the gas molecule, and P_i^{SE} is the partial pressure of different gases (NO_2, O_2, and H_2O) on the SE:

$$\omega(x) = \frac{\left[\psi(x) - \psi(0)\right]}{k_B T}; \quad \alpha_{0(\delta)}{}^* = v_0 {}^* \xi_{0(\delta)}^{-1} \lambda_0 {}^{*-2}\ . \qquad (2.38)$$

Gas flows through the SE will be different ("ventilation" effect) and will depend on how the oxide SE was applied on the surface of YSZ and how it was heat-treated and sintered afterwards. The vast majority of the modern technologies will allow creating a uniform thickness of SEs (REs). However, the structural orientation of an oxide SE and, especially, its surface and bulk porosity will change from one technology to another and from one sintering temperature to another [9]. This fact, in turn, will influence characteristics of the YSZ–based sensor, such as sensitivity, reproducibility of measurements, response and recovery time, and so on. Let's indicate both necessary and sufficient conditions of existence of such a "ventilation" effect within the SE.

If the gas flow through SE does not depend on the applied technology and orientation of the oxide grains, then the value of θ_{se} should correspond to the correlation

$$((I_0 - \gamma_{(\delta)}\theta_{se})/\alpha_{(\delta)})^{1/2} - (\beta_{(\delta)}/\lambda_{\delta}) \cdot e^{-\omega\delta} \cdot (\gamma_{(0)}\theta_{se}/\alpha_{(0)})^{1/2} =$$

$$= ((\beta_{\delta}/\beta_0) \cdot e^{\omega\delta} + \beta_0 e^{-\omega\delta} F_{\delta} + 1)\,\theta_{se} \qquad (2.39)$$

at any values of the flow I_0. In Equation (2.39), $\omega_{\delta} = \omega(\delta)$, $\gamma_{(\delta)} = (\nu_0 + \xi_{0(\delta)})/2\,\xi_{0(\delta)}$, and

$$F_{\delta} = \int_0^{\delta} \frac{e^{-\omega(x)}}{D_i(N_i, x, T)}\,dx\,(i = 1, 2, 3)\ .$$

It is only possible at the simultaneous existence of the following conditions:

$$[\alpha_0\beta_0^2/(\alpha_{\delta}\beta_{\delta}^2)]^{1/2} = e^{-\omega\delta}; \qquad \gamma_0 = \gamma_{\delta}. \qquad (2.40)$$

If correlations (2.40) are performed (for example, the material of the SE is symmetrical), then the "ventilation" effect is absent. However, if one of the correlations (2.40) is violated, then the gas permeability of the SE will be strongly dependent on its structure. It is important to know that the "ventilation" effect can be seen in the close proximity of the equilibrium state when $\alpha_{0(\delta)} = \alpha_{0(\delta)}$ *, $\beta_{0(\delta)} = \beta_{0(\delta)}$*, and $\gamma_{0(\delta)} = \gamma_{0(\delta)}$ *. For this condition it is only necessary that γ_0 * $\neq \gamma_{\delta}$ *. Considering that $I_{\delta} = 0$ (connection of the SE to YSZ) and at the big diffusive resistance of the SE ($F_{\delta} \rightarrow \infty$), the following equation comes from Equation (2.39):

$$W = \theta_{se}/2I_0 = \tfrac{1}{2}N_A^{1/2}\left(2\pi m k_B T\right)^{1/4}\left(\alpha_0\beta_0^2\right)^{-1/2} D_i\left(N_i, x, T\right) H_{\delta}^{-1}\left(P_i^{SE}\right)^{-1/2},$$

$$H_{\delta} = \int_0^{\delta} e^{-\omega(x)}dx,\quad i = 1, 2, 3,$$

$$(2.41)$$

where W is the gas permeability of the SE [40].

When the gas flow θ_{se} reaches the TPB ($x = \delta$), the ionization of the atom or molecular gas takes place depending on the correlation of the speed of exchange reactions as well as the concentration and diffusion coefficients of various forms of the adsorbed gas particles. Different scenarios of extension of the electrode process zone are also possible.

Based on the fact that the potentiometric gas sensors establish thermodynamic equilibria at the interfaces with gases, let's consider this process for the YSZ-based sensor measuring the NO_2 concentration in a humidified atmosphere. The discharge reactions at the TPB (at $x \geq \delta$) can be shown as follows:

$$NO_2 + 2e^- \rightarrow NO + O^{2-}; \; O_2 + 4e^- \rightarrow 2O^{2-}; \; H_2O + 2e^- \rightarrow H_2 + O^{2-}. \; (2.42)$$

Corresponding to these reactions the current discharge density can be determined by the kinetic equation [40]:

$$j(t) = j_x^0(t) N^{1-\alpha}(t) N_x(t) \left\{ exp\left(\frac{-2\beta F\eta_e(t)}{RT} \right) - exp\left(\frac{2(1-\beta)F\eta_e(t)}{RT} \right) \right\},$$

$$(2.43)$$

where R is the universal gas constant; j_x^0 is the density of the standard exchange current, corresponding to the concentrations N and N_x; β is the transfer coefficient, and η_e is the overstrain of the SE stipulated by deceleration of the electrochemical stage.

Consequently, the sensor output, *emf* $E(t) = \varphi_1(t) - \varphi_2(t)$, where the change of the sensor potentials of the SE $\varphi_1(t)$ and RE $\varphi_2(t)$, can be determined by Equations (2.25)–(2.36) and (2.43).

2.4 NUMERICAL MATHEMATICAL MODELS OF ZIRCONIA GAS SENSORS

To numerically solve equations of the above mathematical models, the general computational gas dynamics is adopted in the present work. The general differential equations (2.7) and (2.31) are then discretized by the control volume-based finite difference method, and the resulting set of algebraic equations is iteratively solved. The numerical solver for the general differential equations can be repeatedly applied for each scale variable over a controlled volume mesh. This process must be conducted extremely carefully to avoid the influence of slight changes in the accuracy of discretization.

Let us introduce the difference net into the region $Q = [0, t_K] \times [0, z_K]$ considering that the coefficients of Equations (2.7)–(2.11) are inseparable within the reserved region $\Omega = [0, z_K]$ [59]:

$$\overline{\omega}_{\Delta t \Delta z} = \{ t_m = m\Delta t, \; m = \overline{0, M}, \; \Delta t = t_K/M, \; z_n = n\Delta z, \; n = \overline{0, N}, \; \Delta z = z_K/N \}.$$

Then,

$$N_i(m\Delta t, n\Delta z) = N^i_{m,n}, \; S_i \{ m\Delta t, n\Delta z, N_1, \ldots, N_K \} = S^i_{m,n}.$$

Differential equation of diffusion (2.7) then is discretized by using a double-layered by time and triple-layered by spatial coordinate centered finite difference scheme:

$$N^i_{m+1,n} = b^i_1(m) N^i_{m,n-1} + b^i_2(m) N^i_{m,n} + b^i_3(m) N^i_{m,n+1} + c^*_i(m) S^i_{m,n}, \; i = \overline{1, K},$$

$$(2.44)$$

where

$$b_1^i(m) = \left(\frac{b^i\{\cdot\}\Delta t}{\Delta z^2} - \frac{\Delta t}{2\Delta z}\right), \quad b_2^i(m) = \left(1 - \frac{2b^i\{\cdot\}\Delta t}{\Delta z^2} + c^i\{\cdot\}\Delta t\right),$$

$$b_3^i(m) = \left(\frac{b^i\{\cdot\}\Delta t}{\Delta z^2} + \frac{\Delta t}{2\Delta z}\right), \quad c^i_*(m) = \Delta t.$$

The following difference schemes were used for approximation of the boundary conditions for the equation of diffusion:

$$\frac{N_{m,n}^i - N_{m,n-1}^i}{\Delta z} - Y^i N_{m,n}^i = \gamma^i(m), \quad z = 0, \ t \geq 0, \ i = \overline{1,K},$$

$$\frac{N_{m,n+1}^i - N_{m,n}^i}{\Delta z} + U^i N_{m,n}^i = \psi^i(m), \quad z = z_k, \ t \geq 0, \ i = \overline{1,K}, \tag{2.45}$$

from where

$$N_{m,n-1}^i = (1 + Y^i \Delta z) N_{m,n}^i - \Delta z \, \gamma^i(m), z = 0, t \in (0, t_K),$$

$$N_{m,n+1}^i = (1 - U^i \Delta z) N_{m,n}^i + \Delta z \, \psi^i(m), z = z_K; t \in (0, t_K). \tag{2.46}$$

Scheme (2.44) can be rewritten in the vector-matrix form for $n = \overline{0,N}$, utilizing conditions (2.46) at the boundary points $n = 0$ and $n = N$:

$$\overline{N^i}(m+1) = \overline{B^i}(m) \, \overline{N^i}(m) + C_i(m), \quad i = \overline{1,K}, \tag{2.47}$$

$$\overline{B_i}(m) = \begin{bmatrix} x_1^i(m) & x_2^i(m) & 0 & \cdots & 0 & 0 & 0 \\ b_1^i(m) & b_2^i(m) & b_3^i(m) & \cdots & 0 & 0 & 0 \\ 0 & b_1^i(m) & b_2^i(m) & \cdots & 0 & 0 & 0 \\ \vdots & \vdots & \vdots & & \vdots & \vdots & \vdots \\ 0 & 0 & 0 & \cdots & b_1^i(m) & b_2^i(m) & b_3^i(m) \\ 0 & 0 & 0 & \cdots & 0 & x_3^i(m) & x_4^i(m) \end{bmatrix},$$

$$x_1^i(m) = [b_2^i(m) + b_1^i(m)(1 + Y^i \Delta z)], \quad x_2^i(m) = b_3^i(m),$$

$$x_3^i(m) = b_1^i(m), \quad x_4^i(m) = [b_2^i(m) + b_3^i(m)(1 - U^i \Delta z)],$$

$$\overline{N^i}(m) = [N_{m,0}^i, N_{m,1}^i, \ldots, N_{m,N}^i]^T,$$

where T is the transpose symbol

$$\bar{C}_i(m) = \begin{bmatrix} c_i^*(m) & S_{m,0}^i - b_1^i(m)\,\Delta z\,\gamma^i(m) \\ c_i^*(m) & S_{m,1}^i \\ \vdots & \vdots \\ c_i^*(m) & S_{m,N-1}^i \\ c_i^*(m) & S_{m,N}^i + b_3^i(m)\,\Delta z\,\psi^i(m) \end{bmatrix}.$$

The elementary conditions are expressed as follows:

$$\overline{N^i}(0) = \overline{N_0^i}; \quad \overline{N_0^i} = \left[N_{0,1}^i, N_{0,2}^i, ..., N_{0,N}^i \right]^T. \tag{2.48}$$

Naman's spectral steadiness criterion was used for analysis of the accuracy and steadiness of the finite difference scheme (2.44) and for estimation of the approximation error. According to this criterion, the scheme must be steady for any elementary conditions, including for the first harmonic $Y_{0,n} = e^{i\mu n}$, where μ is the harmonic's frequency ($0 \le \mu \le \infty$), $i = \sqrt{-1}$. Then $N_{n,m} = \lambda^m e^{i\mu n}$, where λ is the coefficient of change, which for the steadiness of difference scheme must be $|\lambda| \le 1$.

Let us define $\lambda^{m+1} e^{i\mu n} - \lambda^m e^{i\mu n} = h\,(\lambda^m e^{i\mu(n+1)} - 2\lambda^m e^{i\mu n} + \lambda^m e^{i\mu(n-1)})$, where $\lambda - 1 = 2h\,(\cos\mu - 1)$; $h = b^i\{\bullet\}\,(\Delta t/\Delta z^2)$. For the steadiness of scheme (2.44), it is necessary that h has to be within the following range: $0 \le h \le 0.5$. Consequently, at $b^i\{\bullet\}\,(\Delta t/\Delta z^2) \le 0.5$, the scheme (2.44) is a conditionally steady evident scheme; however, it is not a very accurate one. To improve its accuracy, the following conditions should be applied: $b^i\{\bullet\}\,(\Delta t/\Delta z^2) = 1/6$.

The decomposition function $N_{m,n}^i$ in the Taylor's row was used in the region point (m, n) for determination of the approximation error of the scheme (2.44):

$$N_{m,n}^i = \sum_{i,j=0}^{\infty} \frac{1}{i!\,j!} \cdot \frac{\partial^{i+j} N_{m,n}}{\partial t^i \partial z^j} (\Delta t)^i (\Delta z)^j. \tag{2.49}$$

By substituting (2.49) into (2.44) and the exclusion of the boundary points $z = 0$ and $z = z_K$ in interval $(0, z_K)$, we will receive

$$p_N = \frac{N_t^I}{2}\Delta t + 2\frac{N_z^{IV}}{24}\Delta z^2 + \frac{2N_z^{III}\Delta z^2}{6} + 0\left((\Delta t)^2\right) + 0\left[\Delta z\right]^4,$$

$$p_N = 0(\Delta t) + 0\left((\Delta z)^2\right). \tag{2.50}$$

The equation of diffusion (2.7) with the boundary conditions (2.11) has no analytical decision. It is possible, however, to find out the concentration of the measuring gas in the discrete points through the electrodes for nonidentical boundaries along spatial coordinate x by using the numerical model proposed. The accuracy of this calculation is also dependent on the level of discretization. Then the average values between the calculated and experimental concentrations can be defined as a single task for the accumulating or declining summary at the present moment of time:

$$\overline{K_{cal}}(t) = \frac{1}{N-2} \sum_{i=2}^{N-1} \left|\left(U_{cal\,i} - U_{exp\,i}\right)\right|,$$ (2.51)

where $U_{cal\,i}$ is the calculated value in the group i at the time t, and $U_{exp\,i}$ is the experimental value measured simultaneously. N is the common number of discrete points along the spatial coordinate x. It is rare that the value's order of the declining summary at the finite difference approximation can be defined accurately. Usually, the following situation takes place: the small step of the net, selected at the first place for good approximation of the differential equation, increases potential likelihood for the maximum deviation at the different equation decisions. In order to avoid that, the variation of the net steps along the temporal and spatial coordinates must be done. In our case, the following steps were chosen along the spatial coordinate x ($\Delta x = 0.2, 0.15, 0.1, 0.005$) and along the temporal coordinate t ($\Delta t = 0.01,$ $0.005, 0.0025, 0.00125$), respectively. The chosen net dimensions Δx and Δt are influenced by the accuracy of the numerical decisions in a different manner. The approximation deviation is decreased with the decrease of Δt in respect to the experimental value. Figure 2.11, a, shows the deviation profiles for the different values of Δt at fixed $\Delta x = 0.1$. It is clear from this figure that the decrease value Δt from 0.0025 to 0.00125 does not practically change the error profile. The effect from the changes in the net steps along the spatial coordinate Δx is more complex. Figure 2.11, b, shows the deviation profiles of approximation at $\Delta t = 0.005$ for three values Δx. The deviation constantly decreases at $\Delta x = 0.2$ (very rough net). If the step Δx can be decreased further down to 0.1, Figure 2.11, b, shows considerably lower deviation. However, at the next decrease step down to $\Delta x = 0.05$ (very small net), some increases of the deviation approximation are possible. Consequently, every step in the decrease of the spatial coordinate Δx must be accompanied by the appropriate step in the decrease of the temporal coordinate Δt to obtain convergent decisions. However, the calculation time increases substantially as the step in spatial coordinate Δx becomes too small. Hence, the compromise combination for the single measured task was found at the $\Delta x = 0.1$ and $\Delta t = 0.0025$ (Figure 2.11, c). Ratio $\Delta t / \Delta x = 0.025$ provides the sufficient decision accuracy and steadiness for calculations. It has been characterized that the deviation has a maximum value at the beginning of the transitional process for all calculated concentrations and working temperatures. It was stipulated by big initial concentration gradients at the boundary conditions.

In 2005, it was experimentally found that the slow recovery rate to NO_2 of the YSZ-based mixed-potential sensor with the NiO-SE can be significantly improved when the sample gas was humidified with 5 vol. % water vapor. These results were published in 2006 [13]. Figure 2.12 [41] illustrates numerical and experimental values for the response/recovery transients to 400 ppm NO_2 in 5 vol. % O_2 with a N_2 balance in the absence (a) and presence (b) of 5 vol. % H_2O at 850°C. Sample

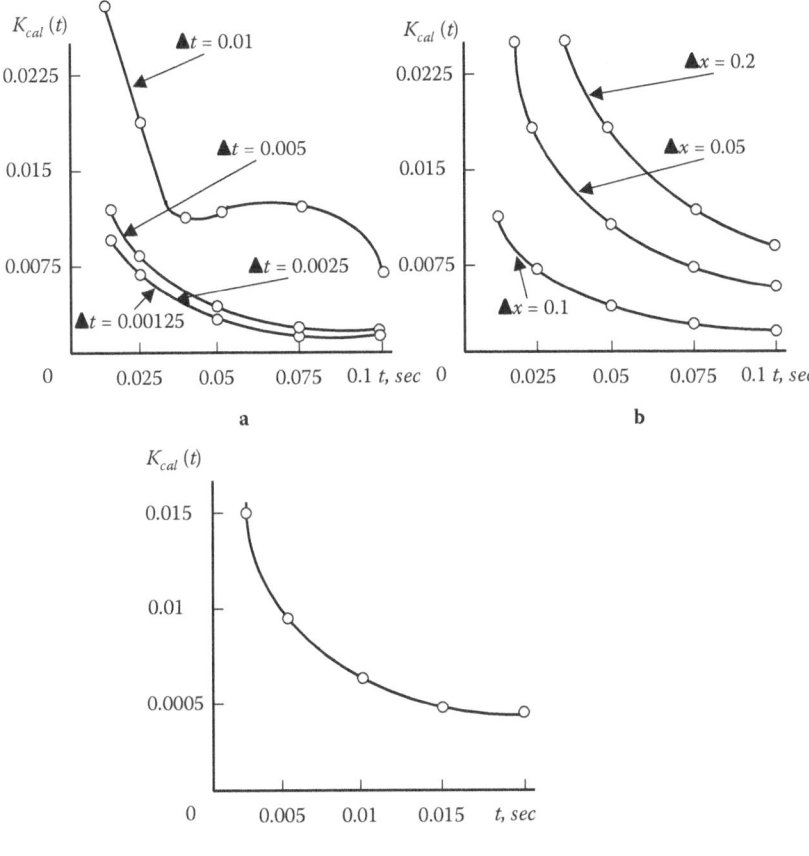

FIGURE 2.11 *a*: Deviation profiles for the different values of Δt at fixed $\Delta x = 0.1$; *b*: deviation profiles of approximation at $\Delta t = 0.005$ for three values Δx; and *c*: the compromise deviation profile for the single measured task at $\Delta x = 0.1$ and $\Delta t = 0.0025$. (Reprinted from Zhuiykov, S., Mathematical modelling of YSZ-based potentiometric gas sensors with oxide sensing electrodes part II: Complete and numerical models for analysis of sensor characteristics, *Sensors and Actuators B, Chem.* 120 (2007) 645–656, with permission from Elsevier Science.)

gas containing various NO_2 concentrations was prepared by diluting standard NO_2 gas with nitrogen. The NiO-SE was sintered on the YSZ surface for 2 hours at 1300°C. The thickness of the SE was 7 μm (as is clearly shown in Figure 2.13). The flow rate was constant during all measurement and was 100 cm³/min. YSZ sensors and all experimental measurements were obtained at KASTEC, Kyushu University, Japan. Figure 2.12 shows the changes in measuring *emf* of the planar sensor when 400 ppm NO_2 was introduced into the working chamber and changes in measuring *emf* when the sample gas was changed on the base gas. The deviation error between numerical and experimental values did not exceed 7% and decreased with the increasing of the working temperature. Noteworthy, the sensitivity of the sensor under wet conditions was higher than that under dry conditions, and the recovery

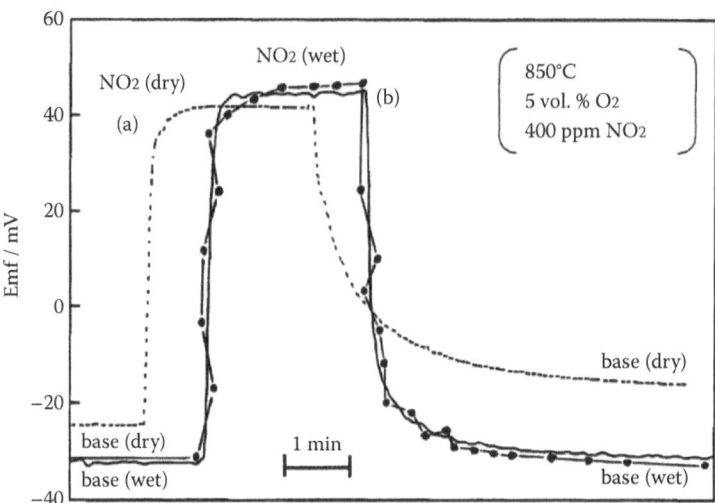

FIGURE 2.12 Numerical and measured response/recovery transients to 400 ppm NO₂ in 5 vol. % O₂ (N₂ balance) in the (*a*) absence and (*b*) presence of 5 vol. % water vapor at 850°C. (Reprinted from Zhuiykov, S., Mathematical modelling of YSZ-based potentiometric gas sensors with oxide sensing electrodes part II: Complete and numerical models for analysis of sensor characteristics, *Sensors and Actuators B, Chem.* 120, (2007) 645–656, with permission from Elsevier Science.)

rate was improved. Clearly, this feature cannot be explained in terms of the classical oversimplified mixed-potential theory. It was speculated that the exchange reaction between the adsorbed O_2 on the surface of NiO and the oxygen ions O^{2-} in the YSZ bulk proceeds more smoothly by the help of hydroxyl ions formed on the surface of NiO [13].

Following 2005–2006 publications, the comprehensive analysis of processes occurring at the TPB in the presence of water vapor for the YSZ-based NO₂ sensor with the NiO-SE has been done by using the complete mathematical model (2.1)–(2.4). This analysis has taken into account that the electrical fields at the solid electrolyte-electrode interfaces depend on the chemical potential differences of both mobile species: ions and electrons. The mobility of ions results in variations of the chemical composition of compounds and even in the formation of new interfacial materials, which generate changes in the performance and cause short lifetimes [60]. Moreover, the electrical field at the contact between a metal electrode and a solid electrolyte is much stronger and extends over a much smaller regime than in the case of a semiconductor junction. This is owing to the much higher concentration of charge carriers in metals and solid ion-conductive electrolytes compared to semiconductors. These assumptions suggest that the electric double layer at the TPB and the oxygen ions O^{2-} play a more significant role at the development of measuring potential on the NiO-SE interface than has been considered before. One of the first principles which must be recognized is that matter at the boundary of two phases possesses properties which differentiate it from matter freely extended in either of

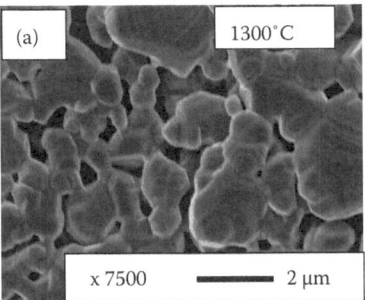

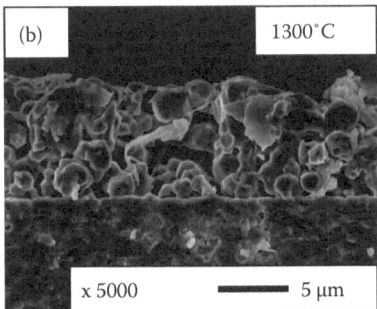

FIGURE 2.13 SEM images of the YSZ-based NO_2 sensor with the 7 μm NiO-SE sintered at 1300°C: (*a*) top view, and (*b*) cross-section view. (Reprinted from Elumalai, P. and Miura, N., Performances of NO_2 sensor using stabilized zirconia and NiO sensing electrode at high temperature, *Solid State Ionics* **176** (2005) 2517–2522, with permission from Elsevier Science.)

the continuous phases separated by the interface. Where we have a charged surface, however, there must be a balancing countercharge, and this countercharge will occur in the gas phase. The charge carried will not be uniformly distributed throughout the gas phase, but will be concentrated near the charged surface. Thus, we have a small but finite volume of the gas phase which is different from the extended gas. This concept is central to electrochemistry and reactions within this interfacial boundary that govern external observations of electrochemical reactions. The discharge reactions at the TPB in the presence of water vapor can be shown as follows:

$$NO_2 + 2e^- \rightarrow NO + O^{2-}; \quad O_2 + 4e^- \rightarrow 2O^{2-}; \quad H_2O + 2e^- \rightarrow H_2 + O^{2-}.$$

In all reactions given above, we have oxygen ions O^{2-} on the right part of the equations. Corresponding to these reactions, the current discharge density j can be determined by the kinetic equations (2.12)–(2.16). Furthermore, the capacitance of the electric double layer at the TPB among gas-oxide-SE-YSZ is different from that of the electric double layer at the TPB among gas-metal-SE-YSZ [38, 61]. In fact, for SEs, based on metal Pt or Au, the capacitance of the electric double layer is about 50–150 $\mu C/cm^2$ and basically independent of temperature and P_{O_2} fluctuations. However, for

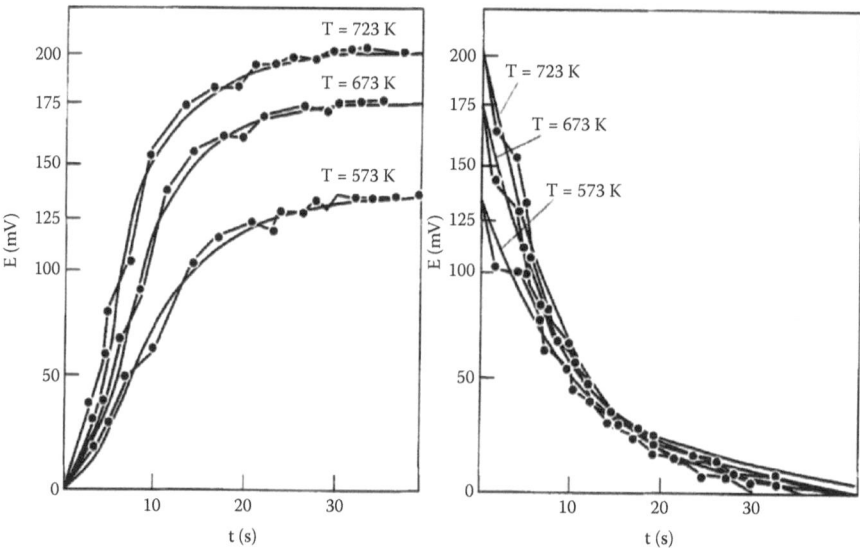

FIGURE 2.14 Response/recovery time of the hydrogen sensor based on the $(NH_4)_4Ta_{10}WO_{30}$ solid electrolyte ($C_1 = 10$ ppm; $C_2 = 100,000$ ppm). (From Zhuiykov, S., Hydrogen sensor based on a new type of proton conductive ceramic, *Int. J. Hydrogen Energy* **21** (1996) 749–759. With permission.)

semiconductor oxides, which are usually employed as SEs in the mixed-potential-type NO_x sensors, the capacitance of the electric double layer, firstly, is usually about an order of magnitude less than that for the noble metals and, secondly, depends on the temperature and P_{O2} level. Therefore, the presence of water vapor changes the concentration of oxygen ions O^{2-} and ultimately increases the capacitance of the electric double layer, which promotes faster kinetics for the recovery rate.

Another example of using the double-layered by time and triple-layered by spatial coordinate centered finite difference scheme (2.44) for calculation of the response/recovery transients of the hydrogen solid electrolyte sensor based on $(NH_4)_4Ta_{10}WO_{30}$ at the different temperatures is shown in Figure 2.14 [62, 63]. Analysis of both Figure 2.12 and Figure 2.14 has led to the conclusion that the majority of the physic-chemical and electrochemical processes, such as adsorption–desorption, diffusion, and changing the capacitance of the double electrical layer, exist in the most solid electrolyte gas sensors. The rate and kinetics of the above processes are based on the nature of the material of the SE (RE) and measuring gas, the TPB, the working temperature, and the coexistence of the oxygen partial pressure in the measuring gas.

The presented numerical approximation is preferable in comparison with other approximations because it provides high enough accuracy and can be easily executed on the ordinary PC. Moreover, from the experimental research point of view, one of the consequences of dependences of equilibrium values $\alpha_0^*\beta_0^{*2}$ on the flows is the possibility of deviation of the sorption isotherm from the parabolic pattern (see (2.37)). In this case, when the parameters shown above depend on θ_{se} ("nonequilibrium" system), new quality effects can appear at a mass transfer of the gas through the SE. Specifically, the gas permeability of the "thick" electrodes can be determined by the status of their surfaces, which would consequently influence the sensitivity of the sensor. The high surface-to-volume ratio of the SE provides better sensitivity of the sensor toward the measuring gas concentration. However, the reproducibility of the measurement would not be stable enough. On the other hand, in the case of the higher sintering temperature, the larger grains are present at the TPB. Hence, the larger grains may produce fewer reaction sites at the TPB during NO_2 measurement. Thus, in the case of the higher sintering temperature of SEs, the catalytic activity to the cathodic reaction of NO_2 [9] would be lower compared with the high surface-to-volume ratio, which is usually observed in the SEs sintered at lower temperatures.

Another modification of the sensor characteristics can be achieved by using technologies for making a nanoparticle SE. Figure 2.15 illustrates the cross-sectional SEM image of the YSZ-based sensor with the nanostructured NiO-SE. As is clearly shown in this figure, the thickness of the NiO-SE can decrease to 10–15 nm. The shift from microsize toward nanosize in the thickness of the SE is allowed to decrease the response/recovery time of the sensor and, sometimes, to enhance the NO_2 sensitivity by using nanoparticle seeds of BaO, La_2O_3, and CuO [10]. However, in spite of numerous publications about using nanotechnology for the YSZ-based gas sensors, to the best of our knowledge, none of these sensors so far has been involved in wide practical use. The reason is simple. Most YSZ sensors are working at high temperatures. The size of nanograins of the SE is growing in time at high temperatures and changing the TPB, and consequently all sensor characteristics will be entirely different in 6–12 months' time. Thus, no long-term stability (at least 12 months) has been reported so far. In the vast majority of publications, "long-term stability" has usually been reported for 3–4 months, and no further publications have followed. These facts have also shown that there is no universal technology or technological decisions for simultaneous improvement of all sensor characteristics.

Therefore, the surface and bulk modifications of microstructure of the SE in any particular case can change the metrological characteristics of the YSZ-based gas sensor. This fact has been proven by various publications, where the same material of the SE has been sintered at the different conditions, and subsequently, such characteristics of the gas sensor as sensitivity and response/recovery time at the same temperature differ significantly from one publication to another. The efficiency of such modifications can be determined by changing the value for $\alpha_0\beta_0^2$ in nonequilibrium conditions from its equilibrium condition $\alpha_0^*\beta_0^{*2}$. In case of $\alpha_0\beta_0^2 = \alpha_0^*\beta_0^{*2}$, the surface modifications cannot change the permeability of the "thick" electrodes because the value of $\alpha_0^*\beta_0^{*2} = \alpha_\delta^*\beta_\delta^{*2} \exp(-2\omega_\delta)$ can only be determined by the concentration of the measuring gas within the SE and does not connect to the state of the interphase boundaries.

When the superpermeability of the measuring gas through the material of the SE takes place, the big value of $W \leq 0.1 \div 1$ can be achieved by different ways, including realization of "diffuse" rate in the SE if the conditions $D(N, x, T) H^{-1}_\delta (\alpha_0\beta_0^2)^{-1/2} \sim N_A^{-1/2} (mk_BT)^{-1/2} (P_i^{SE})^{1/2}$ are fulfilled. In this situation, the establishment of the time for gas flow diffusing through the SE can differ significantly from the typical time for the diffusion processes value $\delta^2/(6D(N, x, T))$.

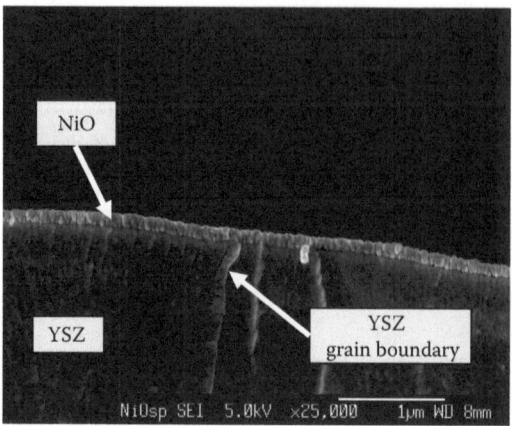

FIGURE 2.15 SEM image of cross-sectional view of the YSZ-based non-Nernstian potenti-ometric NO_2 sensor with the nanostructured NiO-SE. (Reprinted from Zhuiykov, S., Mathe-matical modelling of YSZ-based potentiometric gas sensors with oxide sensing electrodes part II: Complete and numerical models for analysis of sensor characteristics, *Sensors and Actuators B, Chem.* (in press), with permission from Elsevier Science.)

2.5 IDENTIFICATION PARAMETERS OF MATHEMATICAL MODELS

Although it is essential to validate mathematical models of gas sensors, this process is an extremely challenging one. High temperatures coupled with small size make it difficult to probe and measure parameters of interest even in a single test chamber. Many researchers validate models with single electrochemical performance data, which are usually the polarization curves (current versus voltage). This is a good practice. However, it must be done with care of other parameters which cannot be validated. The problem with this approach is that several combinations of parameters can yield similar results. Models often have some significant simplifications, and it is possible that different effects cancel each other out. More importantly, caution must be taken when a model is validated with a polarization curve because it is very easy to adjust parameters within the model so that the model matches the data.

Another approach to identification parameters of the presented model is based on validation of the coefficient of diffusion D (x). Both external and internal factors can influence this coefficient substantially. Therefore, its precise mathematical description cannot be obtained on the basis of a priori data, but only by using the special mathematic methods of data processing of the conditions of measurement u (t, z) at a set temperature. These data can be determined by the limited number of measurements in points x_j, $j = 1, \ldots , M$ of the spatial region $[0, x_u]$ and temporal region $[0, T]$; $(x_j \in [0, x_k])$, $(t \in [0, T])$, $j = 1, \ldots , M$, where M is the total number of measurement points. By taking these features into account, the equation for measurement can be presented in the following form:

$$z (t) = H (t) u_M (t), \qquad (2.52)$$

where $z(t) = [z_1(x^1, t), z_2(x^2, t), ..., z_m(x^m, t)]^T$ is the M-measured vector of measurement, $u_M(t) = [u(x^1, t), u(x^2, t), ..., u(x^m, t)]^T$ is the M-measured vector of conditions in the points of measurement, and $H(t) = \mathrm{diag}[H(x^1, t), H(x^2, t), ..., H(x^m, t)]$ is the diagonal matrix scaled $(M \times M)$, characterizing the mode of measurement. Suppose that the solution of Equations (2.30)–(2.37) exists and it is the sole one. Then the identification of the coefficient of diffusion $D(x)$ can be formulated in the following way.

It is necessary to calculate the real coefficient of diffusion $D(x)$ of Equations (2.30)–(2.37) on the basis of the discrete data in the space of measurement $z_j(x)$ ($j = 1, ..., M$), which causes the minimum to the quadratic function:

$$J = \int_0^{x_k} \int_0^T [u(t,x) - z_j(t,x)]^2 dx\, dt \to min \qquad (2.53)$$

considering the restrictions $D_l \le D \le D_h$, where D_l and D_h are the lowest and the highest limits of measurement of the coefficient of diffusion, respectively.

The formulated task is related to the class of problems on conditional extremes with due regard for Equations (2.30)–(2.37) as restrictions. It is necessary to import the Lagrange function $L(D)$ for criterion (2.52) and Equation (2.30) in order to transfer the formulated task to the task solution on unconditional extreme [64]:

$$L(D) = \int_0^{x_k} \int_0^T [u(t,x) - z_j(t,x)]^2 dx\, dt + \int_0^{x_k} \int_0^T p(t,x)$$

$$\left\{ \frac{\partial u(t,x)}{\partial t} - \frac{\partial}{\partial x} D \frac{\partial u(t,x)}{\partial t} + \upsilon \frac{\partial u(t,x)}{\partial t} - F(t,x) \right\} dx\, dt, \qquad (2.54)$$

where $F(t, x)$ is the function determining the influence of external environment on the diffusion and thermo-mass-transfer processes, and $p(t, x)$ is the Lagrange function–multiplier determining the solution of the following differential equation [64]:

$$-\frac{\partial p(t,x)}{\partial t} - \frac{\partial}{\partial x} D(x) \frac{\partial p}{\partial x} - \upsilon \frac{\partial p(t,x)}{\partial x} =$$

$$= -2 \sum_{j=1}^M \frac{x_j}{|\Omega_j|} \left\{ \frac{1}{\Omega_j} \int_0^{x_k} [u(t,x) - z_j(t,x)] dz \right\}; \qquad (2.55)$$

$$\frac{\partial p(t,x)}{\partial x} = 0 \text{ on } (0, x_k)(0, T); \quad p(T, x) = 0 \text{ on } (0, x_k),$$

where x_j is the characteristic value of the spatial subregion Ω_j, where the measurement is carried out.

For determination of the coefficient of diffusion D and minimizing the criterion (2.52), it is essential to calculate the derivative of this criterion by the variable value D:

$$\frac{dL(D)}{dD} = \int_0^{x_k}\left[\int_0^T \frac{\partial u(t,x)}{\partial t}p(t,x)dt\right]dx + \int_0^{x_k}\int_0^T \frac{\partial}{\partial x}D(x)\frac{\partial p(t,x)}{\partial x}dt\,dx, \quad (2.56)$$

where $u(t, x)$ is calculated from the solution of Equations (2.30)–(2.37), but $p(t, x)$ from the conjugate system (2.55).

The correlation obtained for derivative $dL(D)/dD$ can be used as an essential condition for minimization of the criterion (2.53) and, consequently, (2.54) by the coefficient of diffusion D employing one of the gradient methods.

Therefore, the algorithm for calculating the coefficient of diffusion D can be presented as follows:

1. The initial value of the coefficient of diffusion $D^n(x)$ is introduced as a function of the spatial coordinate x on the n step of the calculating procedure.
2. The differential Equation (2.30) works out on temporal $t[0, T]$ and spatial x $[0, x_k]$ coordinates. Determination of $u(t, x)$ should be done when the solution of Equation (2.30) is found.
3. The conjugate system (2.55) is resolved on temporal $t[0, T]$ and spatial x $[0, x_k]$ coordinates. Determination of $p(t, x)$.
4. The value of $dL(D)/dD$ should be calculated by using Equation (2.56).
5. A more accurate value of the coefficient of diffusion should be calculated by using the following gradient procedure:

$$D^{n+1}(x) = D^n(x) - \iota\frac{dL(D)}{dD},$$

where ι is the positive increment step.
6. The values of both functions $J(D^n)$ and $J(D^{n+1})$ are determined on the basis of correlation (2.53). If $J(D^n) > J(D^{n+1})$, then the value of ι should be increased on the multiplier $\kappa_1 \geq 1$, the value of $J(D^{n+1})$ then can be used as the previous value, and the return to item 5 should be made.
7. Items 5 and 6 should be repeated until $J(D^{n+1}) \geq J(D^n)$. If this correlation is fulfilled, then the following condition should be checked:

$$|J(D^{n+1}) - J(D^n)| \leq \varepsilon, \quad (2.57)$$

where ε is some set error of calculation ($\varepsilon > 0$).

If condition (2.57) cannot be fulfilled, then the value of ι should be decreased by multiplication on the multiplier $\kappa_2 < 1$, and the return to item 5 should be made. In contrast, if condition (2.57) is fulfilled, then the calculation is completed.

2.6 VERIFICATION ADEQUACY OF MATHEMATICAL MODELS TO REAL GAS SENSORS

The comprehensive analysis of physical, chemical, and electrochemical processes occurring in the solid electrolyte gas sensors, allows verifying the adequacy of mathematical models to the real gas sensors. Processing the results of multiple experimental measurements of the gas sensors consists in elucidation of the type of experimental data distribution, evaluation of the parameters of the established distribution, and verification of the adequacy of the mathematical model to the real sensor.

Figure 2.16 illustrates the scheme of the algorithm of processing the results of multiple measurements. The arithmetical mean of the raw observations $\bar{R} = \check{N}[R]$ $= \sum\limits_{i=1}^{n} Ri / n$ can be used as an estimation of the mathematical expectation of precise measurements. Then the incidental errors at the ith observation can be defined as $\upsilon'_i = R_i - \bar{R}$.

The dispersion of deviation of the observation result $\check{D}[R] = \check{g}_2 (R_i) =$ $= \sum\limits_{i=1}^{n} Vi2 / (n - 1)$ can then be used as an estimation of the dispersion.

Both mathematical expectation and the dispersion of incidental value are characterized by the most important distribution features — its posture and the order of incoherence. The dissipation interval of the incidental value around its mathematical expectation characterizes the average quadratic divergence $\check{g}(R_i)$. The Bessel function for the estimation of average quadratic error of the observation result can be represented as follows:

$$\bar{g}\left(R_i\right) = \sqrt{\sum_{i=1}^{n} (Vi2 / (n - 1))} \ .$$

The moments of the highest orders [65] are used for more descriptive characterization of distribution. The asymmetry coefficient $\gamma_1 = \mu_3 \sigma^3$, where μ_3 is the central moment of the third order and $\mu_3 = (1/N) \sum\limits_{i=1}^{N} (Ri - \bar{R})^3$ is used for characterization of the symmetry or asymmetry of the extract volume N.

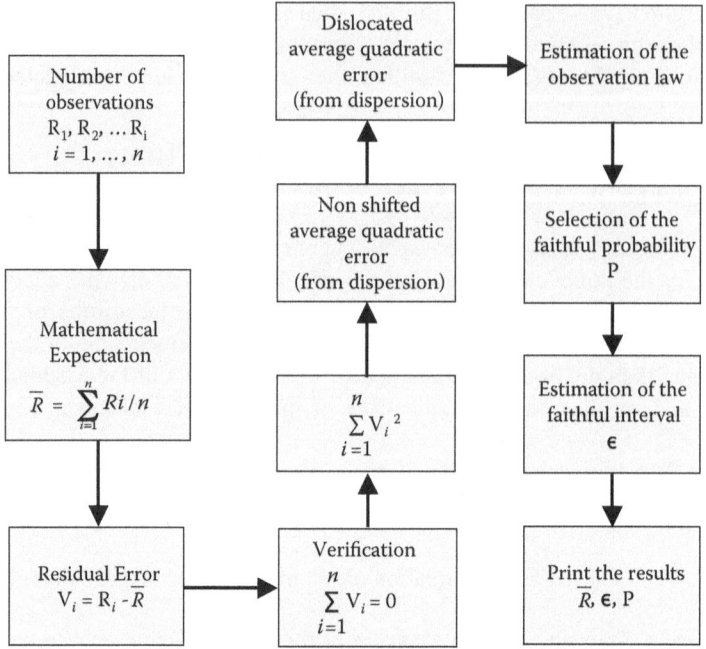

FIGURE 2.16 Scheme of the algorithm of processing the results of multiple measurements. (From Zhuiykov, S., Mathematical model of electrochemical gas sensors with distributed temporal and spatial parameters and its transformation to models of the real YSZ-based sensors, *Ionics* **12** (2006) 135–148. With kind permission of Springer Science and Business Media.)

The excess $\gamma_2 = \mu_4\sigma^4 - 3$, where μ_4 is the central moment of the fourth order and $\mu_4 = (1/N)\sum_{i=1}^{N}(Ri - \overline{R})^4$ is used for characterization of the steepness of distribution for the extract volume N.

The normal type of distribution, among all other distributions used in practice, has taken a special place because owing to the limiting theorem, the vast majority of distributions asymptotically approaches the normal distribution with the growth of the extract volume [65]. The asymmetry coefficient γ_1 and excess γ_2 are compared with the dispersions of these parameters $D(\gamma_1)$ and $D(\gamma_2)$ for verification of the correspondence of the distribution character of the incidental values R_i to the normal distribution. The dispersion functions can be expressed as follows:

$$D(\gamma_1) = \frac{6(N-1)}{(N+1)(N-3)}; \quad D(\gamma_2) = \frac{24(N-2)(N-3)}{(N+1)^2(N+3)(N+5)}.$$

In accordance with consent criterion, the asymmetry coefficient γ_1 and excess γ_2 must comply with the following inequalities:

$$\left|\gamma_1\right| \leq 3\sqrt{D(\gamma 1)}; \quad \left|\gamma_2\right| \leq 5\sqrt{D(\gamma 2)} . \tag{2.58}$$

Verification of the results, obtained by the criteria given above for the solid electrolyte gas sensors, shows that the extract with volume $N = 20$ complies with the criterion (2.58). This allows making a conclusion that the results obtained correspond to the normal distribution.

The faithful interval ε_p, in which the true value R_{true} can be found with the faithful probability p, is determined by the following equation $\varepsilon_p = \pm t_p \sigma (\bar{R})$, where $\sigma (\bar{R})$ is the average quadratic deviation, $\sigma (\bar{R}) = (\sigma (R_i)) / \sqrt{N}$, and t_p is defined by the type of distribution. For the normal distribution and the extract number $N \leq 20$, the faithful interval can be calculated by the Student's equation $\varepsilon_p = \pm t_{p,N} \sigma (\bar{R})$, where $t_{p,N}$ is the coefficient of Student's distribution.

At $p = 0.995$ and $N = 20$, $t_{p,N} = \pm 3.15$ [66]. Consequently, the true value of the measuring parameter for the solid electrolyte gas sensor with the faithful probability $p = 0.995$ is located within the faithful interval $\varepsilon_{p\ max} = \pm 5.1 \times 10^{-3}$ V by relation to the mathematical expectation $\bar{E} = \pm 1.3$ %.

During verification of adequacy of the mathematical models and the real gas sensors, statistical approaches have usually been employed, since both input and output parameters of the sensors in some respect are incidental quantities. Sometimes in practical engineering tasks, the average or maximum deviation of some characteristics, calculated by the model from the corresponding experimental values, can be used as an adequacy criterion. However, such a criterion is implemented at the practical measurement of partial gas pressures in different environments, assuming that both informative and noninformative parameters of the sensor are constant and strictly follow the scope and conditions of the laboratory experiments. This is not always the case. Furthermore, the allowed deviation between calculated and measured characteristics (10–15%) is subjective estimation.

Statistical criteria are more correct for verification of the adequacy of the model [67]. First of all, as for the question about equality, the average values of $\bar{A}_1$ and $\bar{A}_2$ (index 1 corresponds to the model, and index 2 to the experimental data) should be considered during the verification of data. It is possible to compare the average values of dispersions S_1^2 and S_2^2 only at the uniformity of complex results.

Then the average values can be evaluated by the Fisher criterion [68]. If $t_{calul} \leq t_{table}$, both $\bar{A}_1$ and $\bar{A}_2$ have very similar values. They are within the limits of the experimental error. $\bar{A}_1$ and $\bar{A}_2$ are the average values of the output emf of the sensor calculated by the presented model and measured during experiment, respectively. The uniformity check of two results can be done by characterizing the dissipation level of the output value from its mathematical expectation. If both dispersions S_1^2 and S_2^2 are homogeneous and $F_{calc.} \leq F_{table}$ for the set value of significance, then the proposed mathematical model is adequate to the real sensor. Comparison by using the above criterion is especially effective if the results of measurement are corresponded to the normal Gauss distribution.

The following equation shows calculation of the Fisher criterion:

$$t_{calul.} = \left(\bar{A}_1 - \bar{A}_2 \right) / \left(S_{1,2} \sqrt{\frac{n_1 + n_2}{n_1 n_2}} \right). \tag{2.59}$$

An average standard deviation can be expressed as follows:

$$S_{1,2} = \left(\left(\sum_{k=1}^{n_2} A_{1,k}^2 - \sum_{l=1}^{n_l} A_{1,l}^2 \right) \Big/ \left(n_1 + n_2 - 2 \right) \right)^{1/2}, \tag{2.60}$$

where n_1 is the extract volume for the model, and n_2 is the number of parallel experiments. Then, using the t-distribution table and the order of freedom for combined extract $(n_1 + n_2 - 2)$ and for selected significance level q (usually 0.005), the calculated and measured values of the criterion compared to each other. If $t_{calul} < t_{table}$, then it can be stated that the hypothesis about equality of the average values with definite probability of error was not rejected [68].

Based on the Fisher criterion, the hypothesis that the mathematical model is adequate to the real process makes sense only if the residual dispersion S_{res}^2 of the output emf $\bar{A}_1$, calculated by model in relation to the experimental data $\bar{A}_2$, is not superior to the error of experiment, determined by the reproduction of dispersion S_0^2.

The residual dispersion S_{res}^2 is calculated as follows:

$$S_{res}^2 = \left(1/f_1 \right) \sum_{n=1}^{n} \left(\bar{A}_2 - \bar{A}_1 \right)^2, \tag{2.61}$$

where $f_l = m - 1$ is the number of orders of freedom, m is the quantity of parallel experiments, n_2 is the extract volume, and

$$\bar{A}_2 = \left(1/n^2 \right) \sum_{k=1}^{n_1} A_{2k} \tag{2.62}$$

is the average emf value by the results of the parallel experiment within one group.

The reproduction dispersion S_0^2 is calculated from equation

$$S_0^2 = \left(1/m \right) \sum_{n=1}^{m} \left(1/f_2 \right) \sum_{i=1}^{n_2} \left(A_{2i} - \bar{A}_2 \right)^2, \tag{2.63}$$

where $f_2 = n_2 - 1$. Hypothesis of the adequacy of the mathematical model is not rejected if the following inequality is fulfilled:

$$F = \left(S_{res}^2 / S_0^2 \right) < F_{table} \left(f_1, f_2 \right),$$

where F_{table} is the value of the Fisher criterion, which can be found in the Fisher distribution table for the selected significance level and set f_1 and f_2. If $(S_{res}^2/S_0^2) < 1$, then the mathematical model is adequate with a significance level of $q = 0.01$.

As an example, the verification of adequacy of the mathematical model of the YSZ-based sensor, by using the statistical Fisher criterion, is shown that the hypothesis about the adequacy of the presented model to the real YSZ-based sensor with the NiO-SE sintered at 1300°C for measurement 200 ppm NO_2 at T = 800°C is not rejected by the uniformity of dispersions with a significance level of $q = 0.05$.

2.7 NOMENCLATURE

R	universal gas constant (8.3143 J/mol K)
P	partial gas pressure (Pa)
N_A	Avogadro number ($6.02 \cdot 10^{23}$ 1/mol)
D	diffusivity of gaseous component i in electrode (m²/s)
E	sensor output *emf* (mV)
Q^*	heat of gas transfer (kJ/mol)
e	charge of electron ($1.6 \cdot 10^{-19}$ Kl)
m_e	mass of electron ($9.1 \cdot 10^{-31}$ kg)
m	mass of the gas molecule (kg)
F	Faraday's constant (96487 C/mol)
μ	chemical potential of dissolved gas
Q	activation energy of the adsorbed gas
k_B	Boltzmann constant ($1.38 \cdot 10^{-23}$ J/K)
ω	probability of adsorption
υ	probability of desorption
α	kinetic coefficient
θ	gas mixture flow within the porous electrode
x	spatial coordinate in electrode cross-section direction
T	temperature
d	force of convective diffusion
N	concentration of atoms (molecules) of gas (*ppm*)
N_{eql}	concentration of oxygen atoms on the YSZ-SE boundary at the equilibrium stage
ΔG	standard Gibbs energy of formation of MeO
r	radius of oxygen atom ($6.6 \cdot 10^{-5}$ μm)
s	average grain size
ζ	width of the grain boundary
z'	number of conductivity electrons
p	porosity of SE
C	capacity of the double electrical layer
C'	capacity of the dense Helmholtz layer
C^D	capacity of the diffusive layer
C^S	capacity of the surface layer
j	density of the standard exchange current
β	transfer coefficient

η	overstrain of the electrode
φ	potential of the electrode
$\psi(x, T)$	function of coordinate x and temperature T
σ^*	integral cross-section of the dissipation of the conductivity electron on the diffusive ion
Ω	subregion
τ	minimum time for the gaseous component to be at the adsorption state
k_1, k_2	constants of speed for the direct and reverse reactions, respectively
U_0	jump of potential between SE and YSZ
φ_0	the standard electrode potential
S	an effective square of adsorbed oxygen molecules
l, n	coefficients
I	current
q	significance level
R_t	transference resistance
R_{SE}	resistance of solid electrolyte
R_{pol}	polarization resistance at the electrochemical overstrain
t	time (s)
$\check{D}[R]$	dispersion of deviation
$\check{g}(R_i)$	the average quadratic divergence
γ_1	asymmetry coefficient
N	extract volume
D	dispersion
ε_p	faithful interval
$\sigma(\bar{R})$	average quadratic deviation
$\bar{E}$	mathematical expectation
$\bar{A}$	average value
S^2	average value of dispersion
γ_1	excess
H_2O	water
O_2	oxygen
NO_x	nitrogen oxides

2.7.1 Subscripts

f	physically adsorbed
eql	equilibrium
pol	polarization
se	sensing electrode
re	reference electrode
0	initial value
res	residual
diag	diagonal
p	faithful probability

REFERENCES

1. Chebotin, V.N. and Perfiliev, M.V., *Electrochemistry of Solid Electrolytes*, Moscow, Chemistry Publishing, 1978, 312.
2. Zhuiykov, S. and Miura, N., Solid-state electrochemical gas sensors for emission control, in *Materials for Energy Conversion Devices*, Eds. S.S. Sorrel, J. Nowotny, and S. Sugihara, Cambridge, Woodhead, 2005, Chapter 12.
3. Miura, N., Nakatou, M., and Zhuiykov, S., Impedancementric gas sensor based on zirconia solid electrolyte and oxide sensing electrode for detecting total NO_x at high temperature, *Sensors and Actuators B, Chem.* **93** (2003) 221–228.
4. Miura, N. et al., Mixed potential type sensor using stabilized zirconia and $ZnFe_2O_4$ sensing electrode for NO_x detection at high temperature, *Sensors and Actuators B, Chem.* **81** (2002) 222–229.
5. Szabo, N.F. and Dutta, P.K., Correlation of sensing behaviour of mixed potential sensors with chemical and electrochemical properties of electrodes, *Solid State Ionics* **171** (2004) 183–190.
6. Martin, L.P., Pham, I.Q., and Glass, R.S., Effect of Cr_2O_3 electrode morphology on the nitric oxide response of a stabilized zirconia sensor, *Sensors and Actuators B, Chem.* **96** (2003) 53–60.
7. Zhuiykov, S. et al., High-temperature NOx sensors using zirconia and zinc-family oxide sensing electrode, *Solid State Ionics* **152** (2002) 801–807.
8. Szabo, N.F. and Dutta, P.K., Strategies for total NOx measurement with minimal CO interference utilizing a microporous zeolitic catalytic filter, *Sensors and Actuators B, Chem.* **88** (2003) 168–177.
9. Elumalai, P. et al., Sensing characteristics of YSZ-based mixed-potential-type planar NO_x sensor using NiO sensing electrodes sintered at different temperatures, *J. Electrochem. Soc.* **152** (2005) H95–H101.
10. Plashnitsa, V.V., Ueda, T., and Miura, N., Improvement of NO_2 sensing performances by an additional second component to the nano-structured NiO sensing electrode of YSZ-based mixed-potential-type sensor, *J. Applied Ceram. Tech.* **3** (2006) 127–133.
11. Brosha, E.L. et al., Development of ceramic mixed potential sensors for automotive applications, *Solid State Ionics* **148** (2002) 61–69.
12. Mochizuki, K. et al., Sensing characteristics of a zirconia-based CO sensor made by thick-film lamination, *Sensors and Actuators B, Chem.* **77** (2001) 190–195.
13. Miura, N. et al., High-temperature operating characteristics of mixed-potential-type NO_2 sensor based on stabilized-zirconia tube and NiO sensing electrode, *Sensors and Actuators B, Chem.* **114** (2006) 903–909.
14. Xiong, W. and Kale, G.M., Novel high-selectivity NO_2 sensor incorporating mixed-oxide electrode, *Sensors and Actuators B, Chem.* **114** (2006) 101–108.
15. Miura, N., Nakatou, M., and Zhuiykov, S., Development of NO_x sensing devices based on YSZ and oxide electrode aiming for monitoring car exhausts, *J. Ceram. Inter.* **30** (2004) 1135–1139.
16. Szabo, N.F. et al., Microporous zeolite modified yttria stabilized zirconia (YSZ) sensors for nitric oxide (NO) determination in harsh environments, *Sensors and Actuators B, Chem.* **82** (2002) 142–149.
17. Doquier, N. and Candel, S., Combustion control and sensors: A review, *Prog. Energy Combustion Sci.* **28** (2002) 107–150.
18. Nakamura, T. et al., NO_x decomposition mechanism on the electrodes of a zirconia-based amperometric NO_x sensor, *Sensors and Actuators B, Chem.* **93** (2003) 214–220.

19. Magori, E. et al., Thick film device for the detection of NO and oxygen in exhaust gases, *Sensors and Actuators B, Chem.* **95** (2003) 162–169.

20. Shmidt-Zhang, P. and Guth, U., A planar thick film sensor for hydrocarbon monitoring in exhaust gases, *Sensors and Actuators B, Chem.* **99** (2004) 258–263.

21. Ono, T. et al., Improving of sensing performance of zirconia-based total NO_x sensor by attachment of oxidation-catalyst electrode, *Solid State Ionics* **175** (2004) 503–506.

22. Menil, F., Coillard, V., and Lukat, C., Critical review of nitrogen monoxide sensors for exhaust gases of lean burn engines, *Sensors and Actuators B, Chem.* **67** (2000) 1–23.

23. Zosel, J. et al., Selectivity of HC-sensitive electrode materials for mixed potential gas sensors, *Solid State Ionics* **169** (2004) 115–119.

24. Menil, F., Debeda, H., and Lukat, C., Screen-printed thick films: From materials to functional devices, *J. Eur. Ceram. Soc.* **25** (2005) 2105–2113.

25. Bartolomeo, E.D. and Grilli, M.L., YSZ-based electrochemical sensors: From materials preparation to testing in the exhausts of an engine bench test, *J. Eur. Ceram. Soc.* **25** (2005) 2959–2964.

26. Garzon, F.H. et al., Solid state ionic devices for combustion gas sensing, *Solid State Ionics* **175** (2004) 487–490.

27. Wachsman, E.D. and Jayaweera, P., Selective detection of NO_x by differential electrode equilibria, in *Solid State Ionic Devices II: Ceramic Sensors*, Eds. E.D. Wachsman et al., PV 2000–32, The Electrochemical Society Proceedings Series, Pennington, NJ, 2001, 298.

28. Bartolomeo, E.D., Grilli, M.L., and Traversa, E., Sensing mechanism of potentiometric gas sensors based on stabilized zirconia with oxide electrodes: Is it always mixed potential? *J. Electrochem. Soc.* **151** (2004) H133–H139.

29. Elumalai, P. and Miura, N., Performances of NO_2 sensor using stabilized zirconia and NiO sensing electrode at high temperature, *Solid State Ionics* **176** (2005) 2517–2522.

30. Khatua, S., Held, G., and King, D.A., Model car-exhaust catalyst studied by TPD and TP-RAIRS: Surface reactions of NO on clean and O-covered Ir {100}, *Surf. Sci.* **586** (2005) 1–14.

31. Hansen, M.V. and Allen, R.G., *One Minute Millionaire*, New York, Harmony Books, 2002, 388.

32. Zhuiykov, S., Mathematical model of electrochemical gas sensors with distributed temporal and spatial parameters and its transformation to models of the real YSZ-based sensors, *Ionics* **12** (2006) 135–148.

33. Zhuiykov, S., Complete mathematical model of electrochemical gas sensors with distributed temporal and spatial parameters, in *Proc. 11th Int. Meeting on Chemical Sensors*, 11–17 July 2006, Brescia, Italy, 203.

34. Poate, J.M. et al., *Thin Films: Interdiffusion and Reactions*, New York, Wiley-Interscience, 1978, 578.

35. Belmonte, T. and Gouné, M., Numerical modelling of interstitial diffusion in binary systems: Application to iron nitriding, *Mater. Sci. Eng. A* **302** (2001) 246–257.

36. Marques, R. et al., Kinetics and mechanism of steady-state catalytic $NO + O_2$ reactions on Pt/SiO_2 and $Pt/CeZrO_2$, *J. Molecular Catal. A: Chem.* **221** (2004) 127–136.

37. Er-raki, M. et al., Soret driven thermosolutal convection in a shallow porous layer with a stress-free upper surface, *Engineering Computations: Int. J. Computer-Aided Eng.* **22** (2005) 186–205.

38. Antropov, L.I., *Theoretical Electrochemistry*, Moscow, High School Publishing, 1975, 540.

39. Zhuiykov, S., Zirconia single crystal analyser for low-temperature measurements, *Process Control and Quality* **11** (1998) 23–37.

40. Zhuiykov, S., Mathematical modelling of YSZ-based potentiometric gas sensors with oxide sensing electrodes part I: Model of interactions of measuring gas with sensor, *Sensors and Actuators B, Chem.* **119** (2006) 456–465.

41. Zhuiykov, S., Mathematical modelling of YSZ-based potentiometric gas sensors with oxide sensing electrodes part II: Complete and numerical models for analysis of sensor characteristics, *Sensors and Actuators B, Chem.* **120** (2007) 645–656.

42. Zhuiykov, S. and Nowotny, J., Zirconia-based sensors for environmental gases: A review, *Materials Forum* **24** (2000) 150–168.

43. Mukundan, R., Garson, F.H., and Brosha, E.L., *Electrodes for Solid State Gas Sensor*, U.S. Patent No. 6,605,202 (2003).

44. Miura, N. and Yamazoe, N., Approach to high-performance electrochemical NO_x sensors based on solid electrolytes, in *Sensors Update*, Eds. H. Baltes, W. Goepel and J. Hesse, Weinheim, Wiley-Vch, 2000, 191–210.

45. Xiong, W. and Kale, G.M., Microstructure, conductivity, and NO_2 sensing characteristics of α-Al_2O_3-doped (8 mol% Sc_2O_3) ZrO_2 composite solid electrolyte, *Int. J. Apply Ceram. Tech.* **3** (2006) 210–217.

46. Guth, U. and Zosel, J., Electrochemical solid electrolyte gas sensors: Hydrocarbon and NOx analysis in exhaust gases, *Ionics* **10** (2004) 366–377.

47. Miura, N. et al., Mixed potential type NO_x sensor based on stabilized zirconia and oxide electrode, *J. Electrochem. Soc.* **143** (1996) L33–L35.

48. Miura, N. et al., Stabilized zirconia-based sensor using oxide electrode for detection of NO_x in high-temperature combustion-exhausts, *Solid State Ionics* **86–88** (1996) 1069–1073.

49. Elumalai, P. and Miura, N., Influence of annealing temperature of NiO sensing electrode on sensing characteristics of YSZ-based mixed-potential-type NO_x sensor, in *Chemical Sensors VI: Chemical and Biological Sensors and Analytical Methods*, Eds. C. Bruckner-Lea et. al., PV 2004–08, The Electrochemical Soc. Proc. Series, Pennington, NJ, 2004, 80–84.

50. Zhuiykov, S. and Miura, N., Development of zirconia-based potentiometric NO_x sensors for automotive and energy industries in the early 21st century: What are the prospects for sensors? *Sensors and Actuators: B Chem.* **121** (2007) 639–651.

51. Park, C.O. and Miura, N., Absolute potential analysis of the mixed potential occurring at the oxide/YSZ electrode at high temperature in NO_x-containing air, *Sensors and Actuators B: Chem.* **113** (2006) 313–319.

52. Kuzin, B.L. and Bronin, D.I., Electrical double-layer capacitance of M, O_2/O^{2-} interfaces (M=Pt, Au, Pd, In_2O_3; O^{2-}-zirconia-based electrolyte), *Solid State Ionics* **136–137** (2000) 45–50.

53. Kennedy, I.H., Thin films solid electrolyte systems, *Thin Solid Films* **43** (1977) 41–92.

54. Skliar, M. and Tathireddy, P., Approximation of evolutional system using singular forcing, *Comput. Chem. Eng.* **26** (2002) 1013–1021.

55. Djilali, N. and Lu, D., Mathematical modelling of the transport phenomena and the chemical/electrochemical reactions on solid oxide fuel cells, *Int. J. Therm. Sci.* **41** (2002) 29–40.

56. Singh, D., Lu, D.M., and Djilali, N., A two-dimensional analysis of mass transport in proton exchange membrane fuel cells, *Int. J. Eng. Sci.* **37** (1999) 431–452.

57. Timashev, S.F. et al., Description of non-regular membrane structures: A novel phenomenological approach, *J. Membrane Sci.* **170** (2000) 191–203.

58. Belyi, A., Ovchinnikov, A.A., and Timashev, S.F., Boundary conditions of diffusive chemical kinetics, *J. Phys. Chem.* **53** (1979) 948–952.
59. Samarsky, A.A., *Theory of Differentiate Schemes*, Moscow, Science Publishing, 1983, 616.
60. Wepner, W., Interfacial processes of ion conducting ceramic materials for advanced chemical sensors, in *Proc. of 29th Int. Conf. of the American Ceramic Society*, Westerville, OH, 25–29 January 2005, **26** (2005) 15–24.
61. Martin-Molina, A., Quesada-Perez, M., and Hidalgo-Alvarez, R., Electric double layers with electrolyte mixtures: Integral equations theories and simulations, *J. Phys. Chem. B* **110** (2006) 1326–1331.
62. Zhuiykov, S., Hydrogen sensor based on a new type of proton conductive ceramic, *Int. J. Hydrogen Energy* **21** (1996) 749–759.
63. Zhuiykov, S., Development high-temperature hydrogen sensor based on pyrochlore type of proton-conductive solid electrolyte, *Ceram. Eng. Proc.* **17** (1996) 179–186.
64. Talanchuk, P., Zhuiykov, S., and Ruschenko, V.T., Computer-based design. Main principles, in *Bases of Theory and Design of Measuring Instruments*, Ed. P. Talanchuk, Kiev, High School Publishing, 1989, Chapter 8.
65. Born, M., *Einstein's Theory of Relativity*, New York, Dover, 1962, 376.
66. Lide, D.R., *CRC Handbook of Chemistry and Physics*, Boston, CRC Press, 1998, 1–30.
67. Charykin, A.K., *Mathematical Processing Results of Chemical Analysis*, Moscow, Chemistry Publishing, 1984,168.
68. Novitsky, P.V. and Zograf, I.A., *Evaluation the Errors of Measurement Results*, Moscow, Energoatomizdat Publishing, 1985, 248.

3 Metrological Characteristics of Non-Nernstian Zirconia Gas Sensors

3.1 NON-NERNSTIAN ZIRCONIA GAS SENSORS

3.1.1 Mixed-Potential NO$_x$ Sensors

So far, the zirconia-based gas sensors have been developed using trial-and-error techniques rather than following an insightful scientific approach. Many of the existing technologies utilize complex mixtures of several different materials, yet the functionality of each component is not known or well understood. Moreover, the degradation mechanisms leading to the aging behavior of zirconia sensors, which was explained in Chapter 1, are not fully understood in most cases. Fundamental knowledge and understanding of the sensing mechanisms are imperative for further development of existing and new composite materials for the SE and RE. There are several factors that determine the feasibility of a sensor technology, such as the magnitude of sensitivity to the measured gas concentration, the response rate, and the cost of the final product. A fundamental understanding of the characteristics of the materials is also important in selecting the appropriate combination of sensing elements to achieve selectivity in complex sensor array structures. Therefore, essential sensor performance parameters (e.g., stability, sensitivity, and selectivity) need to be improved, even in commercially available products. These parameters depend largely on the physical and chemical characteristics of the materials used to build the sensing devices. The specifications for sensing systems are also becoming more demanding (e.g., operation of sensors at higher temperatures or harsh environments, or lower detection thresholds for measuring pollutants).

Some of the most dangerous air pollutants, nitrogen monoxide (NO) and nitrogen dioxide (NO$_2$) (collectively referred to as NO$_x$), have been of great concern in past years owing to their adverse health effects and their abundance in the vicinity of roads, in particular in high-density urban areas. Many toxicological and epidemiological studies established adverse health effects by NO$_x$. Therefore, great efforts have been taken to reduce the emissions, and the regulations since the start of the new millennium have become stricter. For example, in the United States, the EPA on December 21, 2000, signed emission standards for model year 2007 and later heavy-duty highway engines. The first component of the regulation introduces new, very stringent NO$_x$ emission standards: 0.20 g/bhp-hr for all vehicles in use. The

TABLE 3.1
EU Emission Standards for HD Diesel
Engines

	Phase of Standard	Year	NO_x (g/kWh)
1	Euro III	2000	5.0
2	Euro IV	2005	3.5
3	Euro V	2008	2.0

Source: Data from Miura, N., Nakatou, M., and Zhuiykov, S., 2003.

NO_x and NMHC standards will be phased in for diesel engines between 2007 and 2010. In Japan, diesel emission standards require that in-use, on-road, light commercial vehicles in the specified categories should meet NO_x emissions of 0.25 g/km starting from the end of 2005 and achieve full implementation by 2011. Also, Table 3.1 illustrates some European limits for NO_x emissions applied for on-road engines. All limits are defined in mass per energy (g/kWh) over a defined test cycle. However, it has already been proved that off-road engines contribute about half of NO_x emissions from combustion engines [1]. The main emitters are construction engines and engines used in agriculture and forestry. As a lifetime of these off-road engines is longer than that of on-road engines, the NO_x reduction takes more time to become effective. Consequently, the contribution of these engines to the total NO_x emissions will relatively increase in the future. More information on NO_x limits and the corresponding test cycles in a number of countries can be found at http://www.dieselnet.com/standards.html.

Optimization of the engine combustion process has already lowered NO_x emissions significantly [2]. Specifically, an ultra-lean-burn (or direct-injection-type) engine system has been developed within the last few years to improve fuel efficiency, to reduce NO_x and CO_2 emissions from the engine during the lean cycle, and to release oxygen while oxidizing the hydrocarbons (HCs) during the rich cycle. In this new engine system, as shown in Figure 3.1, a newly developed NO_x-storage catalyst must be used in order to compensate for the low NO_x-removal ability of the conventional three-way catalyst under the lean-burn (air-rich) conditions [3]. This new engine system requires reliable high-performance NO_x sensors to be installed at the points before and after the NO_x-storage catalyst in order to improve its efficiency. Furthermore, the need for novel sensor materials and processes for reliable device operation in hostile environments is now greater than ever. The positions of different sensors in new engine systems are shown in Figure 3.2. These sensors should be able to work in the temperature range of 600–750°C for several years, ideally up to 10 years (100,000 miles), without replacement [4]. Practical implementation of these new engine systems in the future, on one hand, and strict government legislations in most of the developed countries, on the other hand, are becoming driving forces for developing NO_x sensors that work in exhaust environments. Moreover, the use of a selective catalyst reduction system for NO_x mitigation

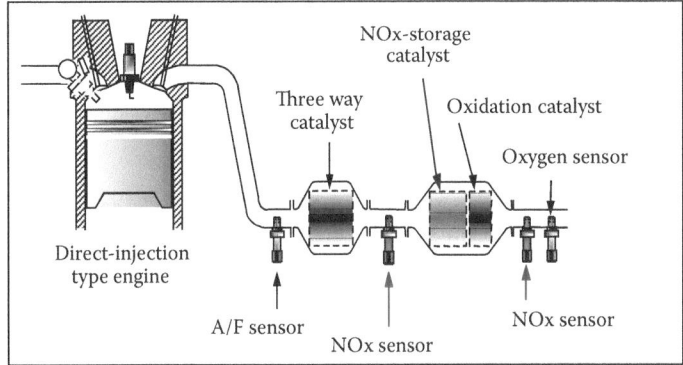

FIGURE 3.1 Catalytic converter system equipped with NO_x sensors for the exhaust gas emitted from a new-type car engine. (Reprinted from Miura, N., Nakatou M., and Zhuiykov, S., Impedancemetric gas sensor based on zirconia solid electrolyte and oxide sensing electrode for detecting total NO_x at high temperature, *Sens. Actuators B, Chem.* **93** (2003) 221–228, with permission from Elsevier Science.)

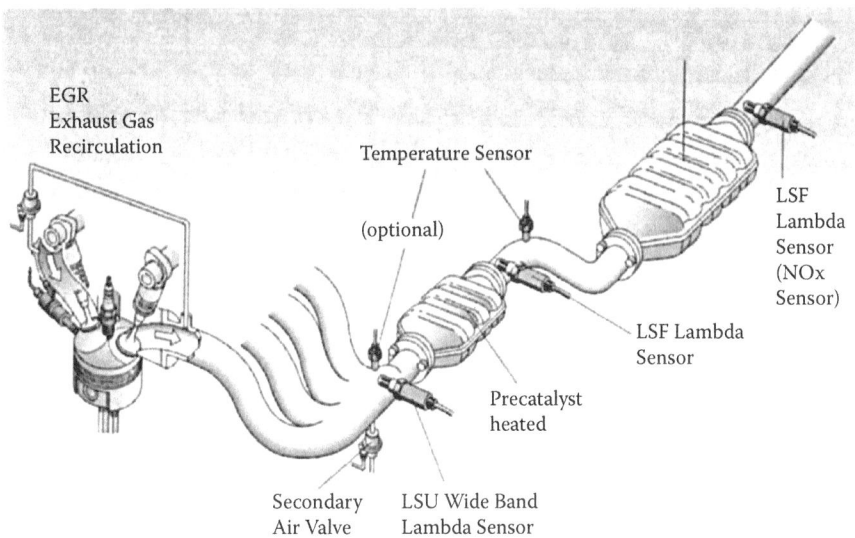

FIGURE 3.2 New advanced engine system and positions of different sensors. (Reprinted from Zhuiykov, S. and Miura, N., Development of zirconia-based potentiometric NO_x sensors for automotive and energy industries in the early 21st century: What are the prospects for sensors? *Sens. Actuators B, Chem.* 121 (2007) 639–651, with permission from Elsevier Science.)

in diesel engines has made the need for NO_x sensors for combustion environments more urgent.

One of the successfully commercialized sensors in the last century was the zirconia-based potentiometric oxygen sensor (λ-sensor), which has greatly improved

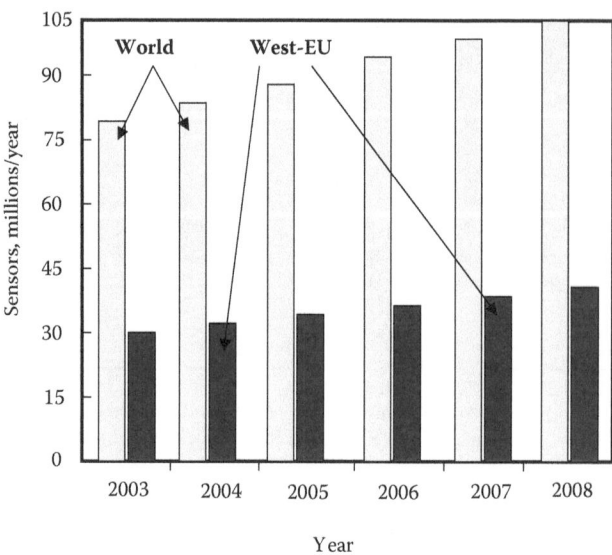

FIGURE 3.3 Demands of the automotive industry for exhaust gas sensors.

energy efficiency and reduced pollutions from vehicles [5]. Since its success in 1976, the number of oxygen sensors in vehicles has continuously increased and their worldwide production level has reached several hundred million parts over the last thirty years. Although λ-sensors are still dominating the applications of the zirconia-based gas sensors in the world, new and improved NO_x sensors are required in the twenty-first century to meet more stringent efficiency and pollution standards that will be inevitably mandated by various government agencies.

The development of new NO_x sensors has been driven by the strong demand of the automotive and combustion industries worldwide. Figure 3.3 shows market trends of the automotive exhaust gas sensors since 2003. The worldwide production of vehicles will presumably reach saturation by 2010 [5, 8], in contrast, the automotive exhaust gas sensors market is still predicted to grow to over 100 million sensors per year in the next few years. Furthermore, approximately one half of all NO_x emissions into the environment are currently due to power plants and industrial boilers [6]. Although it has been more than two decades since the commercial use of λ-sensors began, the deployment of NO_x electrochemical sensors in combustion control has been lacking. The requirements for NO_x sensors to be used in a commercial combustion application are summarized in [7, 8], and they are as follows:

1. To enable a fast response, the sensor should be placed close to the combustion zone, and it must withstand high temperatures (600–900°C) for a long time.
2. The sensor output signal should be insensitive to the moisture content.
3. The sensor should provide stable signals even in the absence of oxygen.
4. The sensor needs to be inexpensive.

Unfortunately, the use of sensors in the energy and automotive industries has been driven more by emissions legislation than by improvements in combustion efficiency. Therefore, careful control and accurate measurement of NO_x emissions by the zirconia-based sensors have received substantial scientific interest [2, 9]. Much of this work has centered on the development of a non-Nernstian type of sensor that can measure NO_x using cheaper technologies and materials than those currently being employed. Since the publication of reviews in relation to NO_x sensors for exhaust gases [9–11] at the beginning of the new millennium, the drive to improve NO_x sensors has already resulted in a number of highly successful sensors [2, 11–76]. Therefore, these types of solid-electrolyte sensors have great promise for *in-situ* monitoring of gaseous pollutants in high-temperature combustion exhausts and in other environments.

3.1.1.1 Description of Nernstian Behavior

The traditional potentiometric zirconia-based sensors widely used for automotive industry and for combustion monitoring are composed of yttria or scandia [41, 42], stabilized zirconia, and porous Pt electrodes. It has been proven that, for most automotive and combustion environments, this combination of materials is reliable, is robust, and provides accurate measurement without recalibration and refurbishment. The shape of the zirconia electrolyte in the sensor is usually designed to separate measuring and reference gases with different oxygen partial pressures. At a sufficiently high temperature, the gaseous oxygen, the mobile oxygen ions in the zirconia, and the electrons of the electrodes are in thermodynamic equilibrium. At each electrode, the following equilibrium takes place:

$$O_{2\,(gas)} + V_{\ddot{O}} + 4e^- \leftrightarrow 2O^{2-}, \qquad (3.1)$$

where $V_{\ddot{O}}$ represents a positively charged (+2) oxygen vacancy, given that it represents a site in the lattice in which the normal net charge of 2– is missing [67]. Electronic charge transfer occurs between Pt electrodes and the zirconia electrolyte as oxygen is incorporated or removed from the zirconia lattice. Moreover, the electrochemical potential of the oxygen ions has to be constant throughout whole zirconia, and especially at the TPB with both electrodes. The zirconia-based electrochemical cell, therefore, can be shown as follows:

$$O_2\,(P_{O2}\,(gas)),\ Pt\ |\ Zirconia\ (mobile\ ions\ O^{2-})\ |\ Pt,\ O_2\,(P_{O2}\,(reference)). \quad (3.2)$$

If the SE and RE of the zirconia-based sensor are exposed to different oxygen partial pressures, P_{O2} (gas) and P_{O2} (reference), this induces different chemical potentials of oxygen ions in zirconia at the interfaces with gas phases. In order that the electrochemical potential remains constant, the electrical potential has to be different. Therefore, the output *emf* of the electrochemical cell (3.2), represented as a difference between potentials on the RE and SE, obeys the well-known Nernst's law:

$$E = \bar{t_i} \cdot \frac{RT}{2nF} \cdot \ln \frac{P_{O2}\,(\text{gas})}{P_{O2}\,(\text{reference})}\,, \qquad (3.3)$$

where $\bar{t_i}$ is an average ionic transference number, R the universal gas constant, F Faraday's constant, n the valence of the measuring gas, T the absolute temperature in Kelvin, and P_{O2} (gas) and P_{O2} (reference) the oxygen partial pressures at the SE and RE, respectively. Based on knowledge of P_{O2} (reference) and the measurement of the output E, it is possible to determine the unknown partial pressure P_{O2} (gas) on the measuring side.

This equation contains only thermodynamic quantities and does not require any information about the microstructure of the system. Hence, aging effects on the microstructure of typical YSZ/Pt electrodes do not influence the sensor signals, and current λ-sensors have lifetimes of more than 100,000 miles (10 years) [51].

It is noteworthy that the surface chemistry of zirconia sensors under practical operating conditions is much more complicated than expected from the oxygen partial pressures in a simple Equation (3.3). This equation implies that only oxygen is involved in the potential forming electrode reaction, whereas in reality, a series of electrode reactions occurs together with Equation (3.1) on the SE [8]:

$$2NO + 2O^{2-} \rightarrow 2NO_2 + 4e^-, \qquad (3.4)$$

$$CO + O^{2-} \rightarrow CO_2 + 2e^-, \qquad (3.5)$$

$$CH_4 + 6O^{2-} \rightarrow CO_2 + 2H_2O + 12e^-, \qquad (3.6)$$

$$H_2 + O^{2-} \rightarrow H_2O + 2e^-, \qquad (3.7)$$

$$C_3H_8 + 10O^{2-} \rightarrow 3CO_2 + 4H_2O + 20e^-. \qquad (3.8)$$

These reactions determine the apparent potential of the zirconia-based sensor. Since the raw exhaust gas at automotive and combustion applications constitutes a non-equilibrium gas mixture, thermodynamic equilibrium has to be achieved at the SE surface of the zirconia-based sensor before monitoring the potential. Consequently, such sensors contain catalytically active materials and are operated at temperatures above 600°C, when the average ionic transference number $\bar{t_i} \geq 0.99$. For less active materials or temperatures below 600°C the apparent *emf* starts to deviate significantly from the value under equilibrium conditions due to insufficient catalytic activity. Earlier research of the YSZ-based sensors was focused on the electrode materials with high exchange currents and high catalytic activity for the desired electrode reactions. Pt electrodes were found to be the most suitable for this type of application.

3.1.1.2 Mixed-Potential Gas Sensors

When the zirconia sensor with dissimilar electrodes is exposed to the combustion exhaust, the presence of at least two independent nonequilibrium gases of different natures are able to induce the competitive oxidation and reduction reactions at the same electrode without charge exchange between electroactive chemical species. Under such circumstances, the output *emf* of the zirconia sensor departs from the theoretical performance resulting in a non-Nernstian potential. The SE potential value establishes between the Nernstian potentials of the two individual processes because the SE potential can have only one value. The system, consequently, moves to reach a saturation in which the potentials of the two electrochemical reactions are the same [23]. This value is called *mixed potential*, and it depends on the current density of two processes in the open circuit conditions [52, 53].

The mixed potential developed is a function of various electrode parameters including, morphology, adsorption, catalytic, and electrocatalytic properties [54]. To get a measurable potential difference between two electrodes, there must be asymmetry between them. Therefore, in most of the mixed-potential sensors the RE is usually Pt and the SE is oxide and/or an oxide mixture [13]. As a result, depending on the nature of the SE, it is possible that both reducible and oxidizable gases can be analyzed by the single sensor having a simple design.

An example would be the case of the oxide-SE on a zirconia sensor exposed to an atmosphere containing oxygen and a NO + NO_2 mixture at high temperatures. The oxidation of NO to NO_2 takes place depending on oxygen partial pressure at temperatures lower than 600°C. The equilibrium shifts with decreasing temperature and rising P_{O_2} in the sample gas toward the side of NO_2. The oxidation of NO is a suitable reaction for mixed-potential electrodes on zirconia because oxygen is involved, and the reaction takes place in a temperature range in which the thermodynamic equilibrium is not established [77]. It is clear that under such circumstances, a change in the concentration of one of the gases contributing to the current balance will alter the potential of null current flow to a different degree on different electrocatalytic surfaces. Thus, the measuring potential should be very strongly influenced by the temperature and the NO concentration. In the sample gas is not playing a decisive role since oxygen reaction is slow at temperatures lower than 600°C. Consequently, the two following competitive reactions take place on the SE [77]:

$$1/2O_2 + V_{ad} \rightarrow O_{ad}$$

$$\frac{O_{ad} + V_{ad} + 2e^-(Me) \rightarrow O_0^x + V_{ad}}{1/2O_2(g) + V_0'' + 2e^-(Me) \rightarrow O_0^x} \quad , \tag{3.9}$$

for the oxygen reaction, and

$$NO + O_{ad} \leftrightarrow NO_2 + V_{ad}. \tag{3.10}$$

for the NO oxidation with adsorbed oxygen, and

$$NO(g) + V_{ad} \rightarrow NO_{ad}$$

$$NO_{ad} + O_0^x \rightarrow NO_{2,ad} + V_0'' + 2e^- \qquad . \qquad\qquad (3.11)$$

$$\frac{NO_{2,ad} \leftrightarrow NO_2(g) + V_{ad}}{NO(g) + O_0^x \rightarrow NO_2(g) + V_0'' + 2e^-}$$

for the NO oxidation with oxygen from the zirconia. The fastest reaction has the strongest influence on establishing the mixed potential. In fact, the mixed-potential NO_x sensors can be represented as single-chamber gas sensors [28, 31, 57] or two-chamber gas sensors [2, 12, 14, 17, 23]. For the single-chamber sensors, both the SE and RE are exposed to the same gas mixture, and for the two-chamber sensors, the SE is exposed to the measuring gas and the RE is exposed to the reference gas with a fixed O_2 concentration. Air has been used as a reference gas for the most of the two-chamber NO_x sensors. Characteristically, the direction of the *emf* response was positive and negative to NO_2 and NO, respectively, as was seen in the previous publications [2, 55, 56]. Figure 3.4 shows typical dependence of *emf* on the logarithm of NO and NO_2 concentration for the YSZ-based sensor using the $NiCr_2O_4$-SE [17].

The vast majority of researchers investigating mixed-potential-type zirconia-based sensors in recent years discovered that the sensing properties of these devices

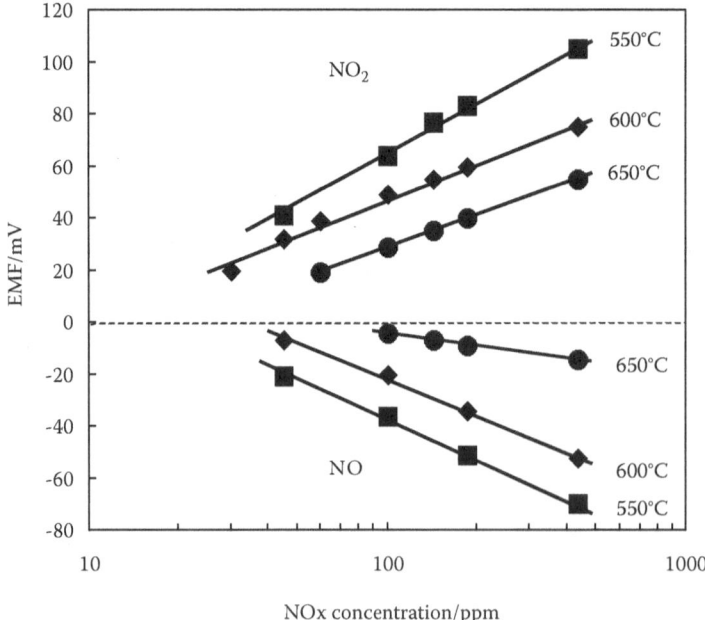

FIGURE 3.4 Typical dependence of *emf* on the logarithm of NO and NO_2 concentration for the YSZ-based sensor using a $NiCr_2O_4$-SE. (Reprinted from Zhuiykov, S. et al., Potentiometric NO_x sensor based on the stabilised zirconia and $NiCr_2O_4$ sensing electrode operating at high temperatures, *Electrochem. Comm.* **3** (2001) 97–101, with permission from Elsevier Science.)

can be improved substantially by replacing the Pt-SE with a suitable semiconductor oxide electrode. Electron-conducting metal oxides offer many possibilities in the design of mixed-potential sensors [52]. As far as practical implications are concerned, this replacement is in favor of the cheaper oxide-SE. In relation to the NO_x mixed-potential sensors, single and spinel oxides can adopt relatively high operating temperatures and, therefore, could be incorporated into practical sensors capable of detecting NO_x in vehicle exhausts and in combustion applications. Our search for oxide electrode materials since the new millennium revealed that among 12 spinel-type oxides derived from trivalent transition metals (Co, Fe, Mn, and Cr) and divalent transition metals (Cu, Zn, and Cd) tested as SE materials, ZnO and zinc-family oxide-SE ($ZnCr_2O_4$ and $ZnFe_2O_4$) provided the largest *emf* response to both NO and NO_2 in the temperature range of 550–700°C [2, 13, 15, 20, 23]. Independently, several research groups found that some of the oxides, such as Cr_2O_3 [12, 24, 35, 50, 59, 70, 75, 76], WO_3 [39, 45, 71], $LaFeO_3$ [32, 45], as well as mixed oxides ($CuO+CuCr_2O_4$ [42] and ITO [41]), could provide promising NO_x-sensing characteristics in the same temperature range, if the sintering conditions of the SE were appropriately selected. This suggested the importance of a further search of both modified sintering conditions and electrode materials capable of working at temperatures as high as 700–900°C. The temperature of engines can occasionally reach about 900°C during vehicles' acceleration. Therefore, it is vital to have an SE material which can withstand such high temperatures for a long time. Recently, we found that the NiO-SE was capable of working at temperatures up to 900°C [19, 31, 57, 60–62, 68, 76]. Figure 3.5 illustrates that among many investigated single-metal oxides, NiO provides the highest NO_2 sensitivity at 850°C [60]. Furthermore, we also found that the NiO-SE is relatively stable and gives the highest sensitivity to NO_x in a temperature range of 700–900°C among all the oxides tested and reported so far [19, 31, 60]. Table 3.2 summarizes the sensing characteristics to NO_x for the mixed-potential zirconia-based sensors using the nonmetal SE reported by different authors to date.

3.1.1.3   Concepts of the Total-NO_x Sensor Based on Mixed Potential

Based on information about various oxides used for the SE in NO_x sensors given above, the new concept of potentiometric measurement of total NO_x regardless of the NO_2 (NO) ratio in the exhaust gases at high temperatures has been developed recently [24, 34, 35]. In 1999, the ability to measure total NO_x in exhaust gases by a new zirconia-based laminated-type sensor was reported by Riken Corporation, Japan [34]. Since then, this sensor structure has been modified, and the new total-NO_x detection system is shown conceptually in Figure 3.6 [35]. The main functions in this system are as follows:

1.   High-temperature exhaust gas enters the first half cavity through the gas inlet attached with a newly developed oxidation catalyst layer and an electrode for oxidation of all HCs and CO in the nonequilibrium gas

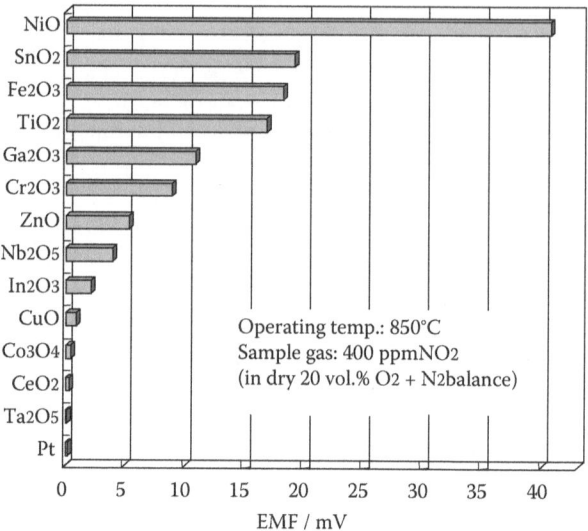

FIGURE 3.5 A histogram showing the sensing response (at 850°C) of various oxide-SEs to 400 ppm NO_2. (Reprinted from Zhuiykov, S. and Miura, N., Development of zirconia-based potentiometric NO_x sensors for automotive and energy industries in early 21st century: What are the prospects for sensors? *Sens. Actuators B, Chem.* 121 (2007) 639–651, with permission from Elsevier Science.)

mixture by excess of oxygen pumping from atmospheric air. Oxygen concentration can be monitored and controlled by an O_2 sensor.

2. Gases (NO and NO_2) in the nonequilibrium NO_x mixture react in the first half cavity, so that the equilibrium concentration of NO and NO_2 can be achieved under the fixed conditions of the sensor. Consequently, an almost constant NO_2-NO ratio exists thereafter independently of the fluctuating exhaust conditions.

3. NO in the equilibrium NO_x mixture at a constantly controlled high temperature diffuses toward the second half of the cavity, where it is oxidized electrochemically to NO_2 by the use of an NO conversion electrode (Pt-Rh). Thus, almost pure NO_2 can be obtained if the NO_x conversion efficiency is high.

4. Finally, the NO_x converted to NO_2 can be measured by the mixed-potential-type NO_x sensor using an oxide-sensing electrode.

From a practical point of view, the simple zirconia-based NO_x sensor of the mixed-potential type, comprising only a SE, a RE, and a zirconia electrolyte, shows large outputs at low NO_x concentrations. However, the sensor *emf* to NO_2 and NO traditionally shows different directions and will compensate each other in real vehicle exhausts [23]. Hence the simplest design of NO_x sensors, compatible with the design of λ-sensors, cannot be successfully employed for the total NO_x measurement either in vehicle exhausts or in combustion applications. It has to

TABLE 3.2

Typical Examples of the SE Materials for Potentiometric Mixed-Potential-Type NO_x Sensors Reported by Different Authors

	Materials of the Oxide-Sensing Electrode	Operating Temp. °C	Measuring Conc. ppm	Year(s) of Publication	Reference Number
	WO_3	500–700	5–200	2000	8, 52
	WO_3	450–700	20–1000	2004, 2005	39, 45
	$NiCr_2O_4$	550–650	15–500	2001–2004	17, 56
	$ZnCr_2O_4$	550–650	20–500	2001–2005	2, 13, 21
	$ZnFe_2O_4$	550–700	20–500	2002–2005	13, 20, 23
	ZnO	550–700	50–450	2004, 2005	15, 48
	Cr_2O_3	500–600	100–800	1996, 2003, 2004	12, 50, 56, 75
NO_x	Cr_2O_3+oxidation catalyst	500–600	20–1000	2000–2006	24, 34, 35, 59
	NiO	700–900	50–400	2004–2006	19, 31, 57, 60–63, 64, 68, 76
	$LaFeO_3$	450–700	20–1000	2001, 2004	32, 45
	$La_{0.8}Sr_{0.2}FeO_3$	450–700	20–1000	2004	45
	$La_{0.85}Sr_{0.15}CrO_3/Pt$	600–700	20–1500	2005	29, 48
	Tin-doped indium (ITO)	613	100–450	2005	41
	$La_{0.6}Sr_{0.4}Fe_{0.8}Co_{0.2}O_3$	500	100–600	2004	8
	$CuO + CuCr_2O_4$	518–659	10–500	2005	42

Source: Data from Miura, N., Nakatou, M., and Zhuiykov, S., 2003.

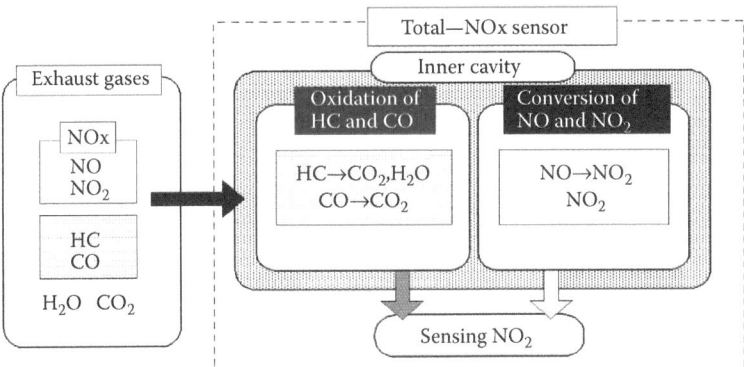

FIGURE 3.6 Sensing system for the total-NO_x detection. (Reprinted from Zhuiykov, S. and Miura, N., Development of zirconia-based potentiometric NO_x sensors for automotive and energy industries in early 21st century: What are the prospects for sensors? *Sens. Actuators B, Chem.* 121 (2007) 639–651, with permission from Elsevier Science.)

be modified in order to address the requirements of the automotive and combustion applications to NO_x sensors. Therefore, the laminated structure of the total NO_x zirconia-based sensor, consisting of the oxidation-catalyst electrode and the space for complete conversion from equilibrated NO to NO_2, has been designed [24, 34, 35]. The design indeed provides a constant oxygen level at the RE by pumping O_2 from the outside atmosphere into the sensor structure. This sensor structure also allowed solving another problem associated with nonequilibrium vehicle exhausts—interference output *emf* by the presence of reducing gases. It has been confirmed that the NO_x-sensing cell based on mixed potentials is not interfered with by H_2O and CO_2. Thus, all reducing gases can be converted to CO_2 and H_2O by the oxidation catalyst.

3.1.1.4 Development of the NO_x Sensor's Design

One factor having a tremendous impact on the sensor's characteristics is the processing route followed to prepare the material. Sensing devices often bulk structures. The configuration that is most commonly employed involves either sintered powders in the form of a dense pressed pellet or porous thick films deposited on a tube or a planar substrate. Recently, there has been growing interest in thin-film-processing routes (by means of sputtering techniques, physical vapor deposition, chemical vapor deposition, metal-organic chemical vapor deposition, spin coating, etc.). There is no doubt that planar, layered designs will impact sensor development. Different fabrication methods are known to result in a variety of microstructures and in varying response characteristics for a particular sensor; there have been no systematic studies to identify an optimal processing technique.

Success or failure of the accurate and reliable measurement of total NO_x in vehicle exhausts very much depends on the sensor design. The general trend toward the planar structure of the sensor has been confirmed by different research groups all around the world since the mixed-potential-type NO_x sensor based on the stabilized zirconia and oxide-SE was proposed in 1996 [55, 56]. Different manufacturing technologies have been employed for this process, but the trend still goes on. In thin- and thick-film oxide deposition technology, the gas-sensing effect predominantly has the surface nature, because the change in concentration of conduction electrons in metal oxides is a result of surface chemical reactions such as chemisorption, reduction/reoxidation, and/or catalysis taking place during interaction with the surrounding gas. The results of experimental and theoretical studies demonstrate that different surface sites participate in combustion, electrical response, interaction with eater vapors, and adsorbed species. Different test gases interact differently to the same degree with different surface sites as well [16]. Therefore, the surface and bulk modifications of the oxide-SE of the zirconia-based NO_x sensor have to be optimized in the planar sensor structure to provide the maximum sensing outcome. One of the examples of the planar sensor structure for the total NO_x measurement is presented in Figure 3.7. The sensor is fabricated by sintering YSZ green sheets on which electrodes and a heater are screen-printed. The NO_x-SE is made of Cr_2O_3, which has activity to both NO_x and oxygen. The SE and RE are formed in the inner cavity for introduction of exhaust gas. An oxidation-catalyst electrode and a NO_x

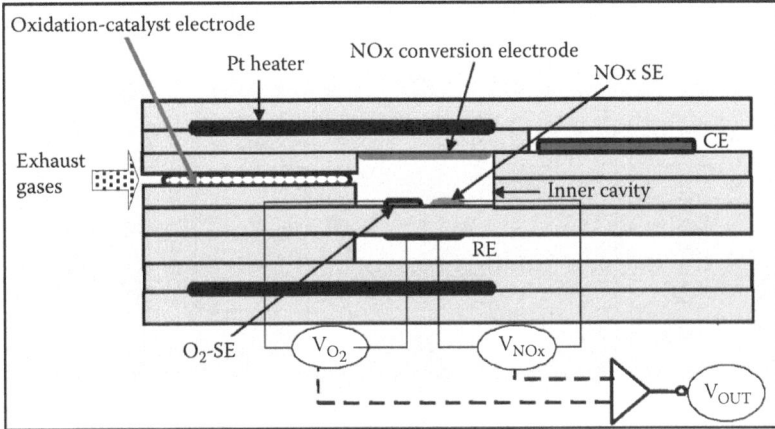

FIGURE 3.7 Cross-sectional view of the laminated-type total-NO_x sensor attached with oxidation catalyst. (Reprinted from Zhuiykov, S. and Miura, N., Development of zirconia-based potentiometric NO_x sensors for automotive and energy industries in the early 21st century: What are the prospects for sensors? *Sens. Actuators B, Chem.* 121 (2007) 639–651, with permission from Elsevier Science.)

conversion electrode are also formed in the inner cavity for the parallel position. Owing to the Pt activity only to O_2, it has been used as a material of the RE and it was arranged beside the SE in order to cancel out the dependence of the sensor output on the O_2 concentration. The YSZ heater plate with printed Pt is also attached to the sensing element. The sensor temperature is controlled to be kept at 600°C by the control unit.

The sensing performance of the zirconia-based total-NO_x sensor is shown in Figure 3.8. Figure 3.8, *a*, shows the dependence of the sensor output on the NO concentration at 600°C. In fact, NO has been changed from 100 to 1000 ppm, but the sensor output is still linear to the logarithmic NO concentration. From the results obtained, it was confirmed that this sensor was capable of measuring a wide range of NO_x concentration from 20 to 1000 ppm regardless of the NO_2-NO ratio in gas mixtures. The influence of O_2 on the sensor output was measured when O_2 was changed from 0 to 20 vol. % while the NO concentration was fixed at 100 ppm. The sensor *emf* slightly decreased with increasing the O_2 concentration in the vicinity of 5%. It is suspected that the difference in O_2 activity between the NO_x-SE and RE may cause the output change. The influence of the propane on the sensor *emf* is shown in Figure 3.8, *b*. In this case, the C_3H_8 concentration was changed from 0 to 6000 ppm during the test, while the NO concentration was fixed at 100 ppm in the absence of oxygen and remained the same. This figure also shows the sensor output without an oxidation catalyst for comparison. It is clear from this figure that in the absence of an oxidation catalyst, the sensor *emf* has strong interference at C_3H_8 concentrations over 3000 ppm. The influence of CO on the sensor output was also checked. The output *emf* of the sensor without an oxidation catalyst was interfered substantially by the presence of CO in the concentration range above 1 vol. %, while

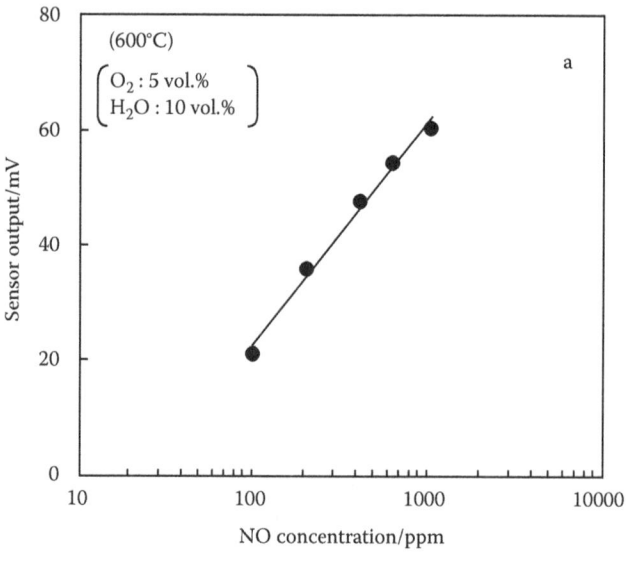

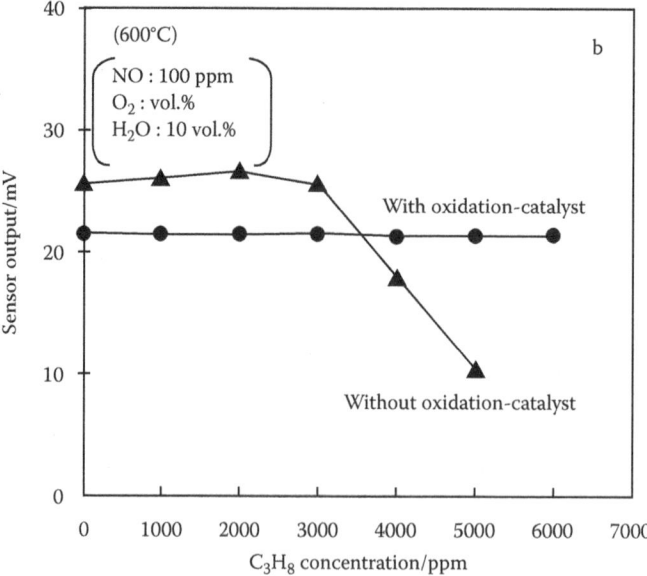

FIGURE 3.8 Dependence of the zirconia-based sensor output at 600°C on (*a*) NO$_x$ concentrations and (*b*) C$_3$H$_8$ concentrations. (Reprinted from Zhuiykov, S. and Miura, N., Development of zirconia-based potentiometric NO$_x$ sensors for automotive and energy industries in the early 21st century: What are the prospects for sensors? *Sens. Actuators B, Chem.* 12 (2007) 639–651, with permission from Elsevier Science.)

the output *emf* of the sensor with an oxidation catalyst did not change by the presence of CO up to 5 vol. %. It is evident from these results that the use of the oxidation-catalyst layer in the planar sensor structure is a very effective way to overcome the interference from reducing gases on the sensor output signal.

In the tests using lean-burn gasoline engines, the sensor *emf* correlated fairly well to the NO_x concentration measured by an independent NO_x analyzer in the range of engine rotation between 1000 and 2600 rpm. An accuracy of the sensor output in the engine rotation was ±4 ppm at 45 ppm, NO_x ±11 ppm at 101 ppm, and ±11 ppm at 306 ppm, respectively, when calculated as NO_x concentrations. It means that the total error of measurement was no more than 10%. In another test it was confirmed that the sensor had enough durability for fuel-rich gas (A/F = 12) [35].

So far, the Cr_2O_3-SE has shown promising NO_x-sensing characteristics since it was reported in 1996 [56]. A few years later, Cr_2O_3 was successfully employed as a NO_2-SE in a total-NO_x sensor [59, 75]. Another research group that employed this material as a SE also reported its sensing properties [12]. However, it is very hard to compare these results with the previous results. Although the details are unknown, it is our strong belief that the sintering conditions for Cr_2O_3 were different between these two particular cases, and, consequently, the NO_x-sensing characteristics differed significantly. It was concluded that the primary factor for determining the NO_x sensitivity was the electrochemical activity of the SE for the O_2 reaction [12]. Furthermore, the authors also admitted that there were some variations in the absolute signal magnitude for NO or NO_2 of the different sensors, presumably due to discrepancy in the fabrication process. This fact indirectly confirmed that apart from the properties of the material of the SE, the applied technology played a crucial role in achieving high NO_x-sensing characteristics.

Another research group reported the NO_x-sensing properties of the nanosized (about 50 *nm*) $LaFeO_3$-SE applied on a zirconia substrate by thermal decomposition of a heteronuclear complex $La[Fe(CN)_6] \cdot 5H_2O$ at 700°C [32]. In fact, the sintered thick-film SEs consisted of a porous structure with large grains (2–10 μm), which were made of homogeneous nanosized particles free from intragranular pores. The sensor showed a high *emf* response to NO_2—about 80 mV at 400°C. Unfortunately, slow response and recovery time of 4 and 15 minutes, respectively, were far from the industrial demand for this type of NO_x sensor. The sluggish response of the sensor can be ascribed to the fact that the time to reach equilibrium for the SE is larger at a lower temperature. This publication also shows that an increase in working temperature of the sensor enhances the sensor response and recovery time with the simultaneous sacrifice of sensitivity.

Further development of the nanostructured perovskite type of oxide-SE for the planar zirconia-based NO_x sensor [28, 43] indicated that both the sensing characteristics and the working temperature can be improved. Figure 3.9 shows *emf* responses of the zirconia-based sensors with the $La_{0.8}Sr_{0.2}FeO_3$-SE and $LaFeO_3$-SE, respectively, to different NO_2 concentrations at different temperatures [26]. The grain sizes of the SEs varied from 50 to 150 *nm*. Interestingly, the *emf* responses of the sensors were found to be in opposite directions with respect to other oxide-based sensor responses: negative *emf* values upon exposure to NO_2. The results obtained from the *emf* measurements (different sign for *p*- and *n*-type oxide-based SEs of the

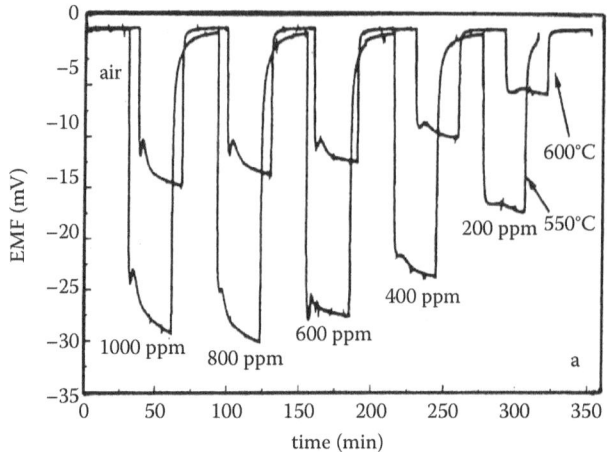

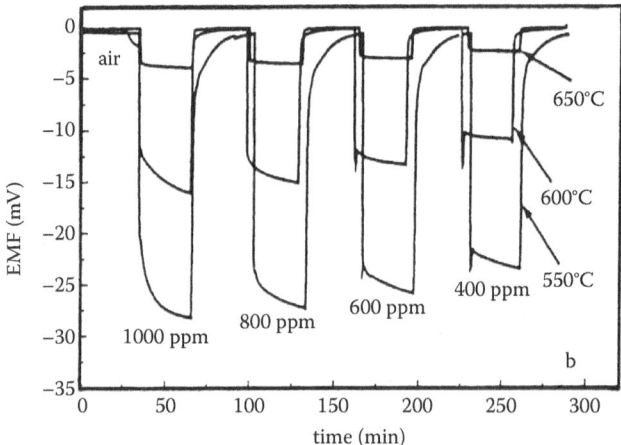

FIGURE 3.9 EMF responses of (*a*) $La_{0.8}Sr_{0.2}FeO_3$-based and (*b*) $LaFeO_3$-based sensors to different concentrations of NO_2 at different operating temperatures. (Reprinted from Zhuiykov, S. and Miura, N., Development of zirconia-based potentiometric NO_x sensors for automotive and energy industries in the early 21st century: What are the prospects for sensors? *Sens. Actuators B, Chem.* 121 (2007) 639–651, with permission from Elsevier Science.)

sensors) cannot be explained only by the occurrence of the electrochemical reactions at the TPB. Based on the sensor design, the electrochemical reactions took place at the same rate at both the Pt-RE and oxide-SE, and the authors therefore concluded that the difference in chemical sorption-desorption behavior at two dissimilar electrodes was the predominant factor for the gas-sensing mechanism [28]. In addition, it was found that the sensor response time was strongly affected by the grain size of the powder used for preparation of the SE: the sensor response was much larger when the grain size was reduced down to nanosized dimensions. Upon increasing the oxide grain size, the response became smaller, slower, and more unstable.

Other interesting results were obtained when a mixed-oxide $CuO+CuCr_2O_4$-SE was applied on the scandia-stabilized zirconia [42]. The particle size of both phases of the SE was approximately between 200 nm and 1 μm. This sensor was capable of measuring NO_2 concentrations from 5 to 500 ppm at 659°C. The authors concluded that this mixed-oxide-SE had negligible interference from CO and CH_4 in the concentration range of 100–500 ppm. The *emf* was also found to be insensitive to the change in O_2 concentration from 4.2 vol. % to 16.2 vol. % at 663°C. However, one of the most distinguishing features of this sensor was fast response. The 90% of response and recovery times at 659°C were 8 and 10 seconds, respectively [42]. The authors claimed that the rapid and stable response and recovery characteristics of the NO_2 sensor with the attached $CuO+CuCr_2O_4$-SE were mainly attributed to the phase composition and microstructure of the novel SE and RE.

Analysis of the data in Table 3.2 about oxide-SEs used for NO_x sensors shows that so far, NiO was found to give the highest NO_2 sensitivity among various spinel- and single-oxide groups reported to date at temperatures over 700°C. Certainly, the highest NO_2 sensitivity could be achieved at the lowest operational temperature. An increment in operating temperature leads to faster kinetics (higher catalytic activity, especially to the anodic reaction of oxygen) at the TPB, as well as to the higher conversion of NO_2 to NO on the surface of NiO grains. Both these factors decrease the NO_2 sensitivity substantially at higher operating temperatures, accompanied by quick recovery.

It was reported for the first time in 2004 that the NiO-SE could successfully work at a temperature as high as 850°C. Other publications followed in the following years [31, 57, 60–62, 76], where both the properties and the applied technological features of this oxide as a potential candidate for the practical SE of NO_x sensors were comprehensively investigated. Recent review of materials for high-temperature electrochemical NO_x gas sensors [76] revealed the comparative assessment of the tubular and planar design of the mixed-potential NO_x sensors with different materials of SEs working at different environments. It was concluded that the character of responses to NO_x concentrations for two design configurations is similar. A linear response would increase with increasing temperature, so the decrease in slope is owing to changes in the electrode kinetics, which consequently depend on the sintering conditions. Furthermore, the comparison between the results for the WO_3- and NiO-SEs was made. The response for the tubular design employing the WO_3-SE is significantly higher than that for the planar design with the same SE. However, this discrepancy decreases at higher temperatures. The decreasing difference indicated that the RE in the planar design of the sensor is less affected by the presence of NO_x, which is expected as the system approaches equilibrium at higher temperatures [76].

Recently, similar results were independently obtained for the ZnO-SE and NiO-SE by other researchers [29, 48] investigating NO_x-sensing properties of ZnO/Pt, NiO/Pt, and $La_{0.85}Sr_{0.15}CrO_3$/Pt-SEs. Although it is very difficult to compare these results with our results for the NiO-SE [19, 31, 55, 60–62] and ZnO-SE [15] due to the variations in sintering and measuring conditions, it is evident that the results are similar to each other for the measurements at a temperature of 700°C. Furthermore, our sensors attached with the NiO-SE exhibited very high NO_2 sensitivity at a high temperature of 850°C.

The major difference between recently published results and those achieved by another research group [31, 50] is owing to the sintering conditions and, as a result, to the created morphology of the SE. For examples, researchers [48] reported NiO-SE sintering conditions of 1100°C for 1 hour, whereas we already investigated and reported how the different sintering temperatures from 1100°C to 1400°C affected the microstructure and morphology of the NiO-SE and, consequently, the NO_x-sensing characteristics of the sensor [57]. To compare the NO_2-sensing behavior at a fixed operating temperature, the *emf* responses to 200 ppm NO_2 at 800°C were recorded for the planar sensors using NiO-SEs sintered at various temperatures. It has been reported that the *emf* values to 200 ppm NO_2 substantially increased from 3 to 55 mV when the sintering temperature of the NiO-SE increased from 1100 to 1400°C, though the recovery rate decreased. Thus, in order to achieve reasonable balance between relatively high NO_2 sensitivity and an acceptable recovery rate, a compromise would be to select a sintering temperature of 1300°C for the NiO-SE [31].

Other small discrepancies are probably due to the different testing conditions. We investigated the NO_x-sensing performance at a fixed concentration of 5 vol. % O_2 in nitrogen balance during all experiments, while 7 vol. % O_2 was reported in [48]. However, overall conclusions are very similar: the results obtained suggest that operating temperature has a strong effect on the sensing performance, with higher temperatures leading to a diminished response magnitude but a shorter recovery from NO_x exposure. The difference in the morphology of the SE, which is predominantly due to the different sintering conditions, may lead to additional enhanced NO_2 sensitivity at high temperatures. This will enable more accurate NO_x measurements of the planar sensor.

The research results, published in 2006, showed that among the various precious metals tested, Rh gave an improvement in NO_2 sensitivity [63]. In addition, when the Rh content varied from 1 to 12 wt. %, the maximum NO_2 sensitivity was obtained at 3 wt. %. The sensitivities to 50 ppm of NO_2 of the unloaded and Rh-loaded NiO-SEs were 37 mV and 65 mV, respectively, in the presence of 5 vol. % O_2 and 5 vol. % water vapor at 800°C.

Figure 3.10 illustrates the correlation between the *emf* response and the concentration of NO_2 for the sensor using the unloaded and Rh-loaded NiO-SEs at 800°C. It is seen that the *emf* value varies linearly with the logarithmic NO_2 concentration in both cases. In addition, at each NO_2 concentration, a large increase in *emf* value was observed in the case of the Rh-loaded SE. It is noteworthy that the sensitivity to 50 ppm of NO_2 (Δemf) was as high as about 80 mV even at 800°C in the case of the Rh-loaded NiO-SE. In order to clarify the reason for the enhanced sensitivity of the Rh-loaded SE, polarization-curve measurements were carried out. As a result, the polarization curve for the cathodic reaction of NO_2 was found to make a large shift to the positive direction, while the anodic polarization curve in the base gas was hardly changed.

It should also be noted that the NO and NO_2 reactions can be enhanced or suppressed, respectively, if the SE potential is polarized anodically (positively) or

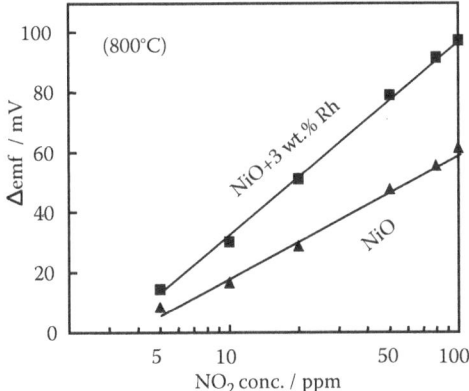

FIGURE 3.10 Dependence of *emf* on the NO_2 concentration at 800°C for the sensor using the unloaded and the Rh (3 wt. %)-loaded NiO-SE. (Base gas: 5 vol. % O_2 + 5 vol. % H_2O + N_2 balance.) (Reprinted from Zhuiykov, S. and Miura, N., Development of zirconia-based potentiometric NO_x sensors for automotive and energy industries in the early 21st century: What are the prospects for sensors? *Sens. Actuators B, Chem.* 121 (2007) 639–651, with permission from Elsevier Science.)

if the reverse effect occurs with cathodic polarization [2, 17]. This indicates that if the NO_x sensor has an additional Pt counterelectrode (CE), the selectivity to NO or NO_2 can be controlled by polarizing the SE. Such a sensor may be considered to be a certain type of sensor, where a positive or negative potential is applied to the SE in order to provide exclusive sensitivity to the measuring component of a gas mixture. In order to obtain selected selectivity to the measuring gas, the tubular device using the $NiCr_2O_4$-SE was biased by a voltage (V_{s-c}) relative to the Pt-CE, and then the SE potential (E_s) relative to the Pt-RE was measured under exposure to various atmospheres. The shift of E_s from the value in the base air to those in air NO or NO_2 (the sensitivity V_s) was dependent on V_{s-c}. Figure 3.11 shows the influence of V_{s-c} on E_s to NO (in air), NO_2 (in air), and base air at 550°C. Analysis of data in this figure suggests that, if polarization is selected adequately, it is possible to single out the sensitivity increment due to NO while suppressing that of NO_2, or vice versa. This leads to the selective detection of NO (NO_2) over NO_2 (NO) using a three-electrode sensor design. Thermodynamically, NO is dominating in the NO_x mixture at high temperature; the equilibrium composition at 600°C is 90 vol. % NO and 10 vol. % NO_2 [56]. Therefore, from the practical point of view, it is interesting to investigate the possibility of the selective detection of NO or total NO_x, if the working temperature of the NO_x sensor is more than 600°C. Figure 3.12 represents the results of this investigation when the bias under anodic polarization was about +175 mV. It was observed that the sensor using the $NiCr_2O_4$-SE is mostly selective to NO over NO_2 under these conditions. The semilog plot of the V_s as a function of NO or NO_2 concentration is essentially linear for both gases. Based on the linear relationship between V_s and NO concentration, it expected that the lowest possible detection limit would be around 14–15 ppm NO [17]. It appears to indicate that the change of the bias voltage in the three-electrode (oxide-SE, Pt-CE, and Pt-RE)

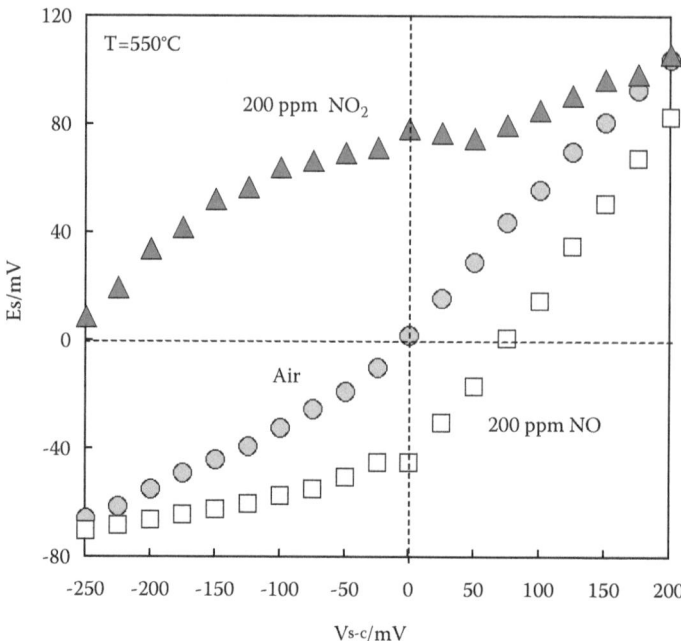

FIGURE 3.11 E_s versus V_{s-c} correlations for the device using a three-electrode structure in air, 200 ppm NO and 200 ppm NO_2, at 550°C. (Reprinted from Zhuiykov, S. et al., Potentiometric NO_x sensor based on the stabilised zirconia and $NiCr_2O_4$ sensing electrode operating at high temperatures, *Electrochem. Comm.* **3** (2001) 97–101, with permission from Elsevier Science.)

tubular sensor could bring about the significant improvement of NO selectivity. Unfortunately, this sensor appears to be insufficiently sensitive to NO_x at temperatures higher than 650°C.

One of the interesting results published in 2006 focused on the improvement of NO_2 sensitivity by adding WO_3 to the NiO-based SE in different proportions [68]. Unfortunately, the authors failed to explain how an addition of the second phase to the NiO-SE has changed the NO_2 sensitivity of the sensor. Nevertheless, this effect is very interesting and worth considering in detail because there is a possibility that further improvements in the development of the NO_2-sensitive electrodes may be based on very similar adjustments. It was reported that 10 wt. % of WO_3 added to NiO before sintering improved the NO_2 sensitivity of the zirconia-based mixed-potential sensor up to ~40% at 800°C [68]. The response and recovery rates of the sensor were almost the same as for the pure NiO-SE. In contrast, then, the WO_3 content has been increased from 10 to 50 wt. %, and the NO_2 sensitivity has been decreased by about ~30% at the same testing conditions. For the zirconia-based mixed-potential NO_x sensors, differences among electrodes depend primarily on the reaction kinetics within the bulk of the SE and on the TPB rather than on the conduction properties of the electrolyte. Consequently, the geometrical changes within the bulk of the SE can affect whether or not a reaction is diffusion limited.

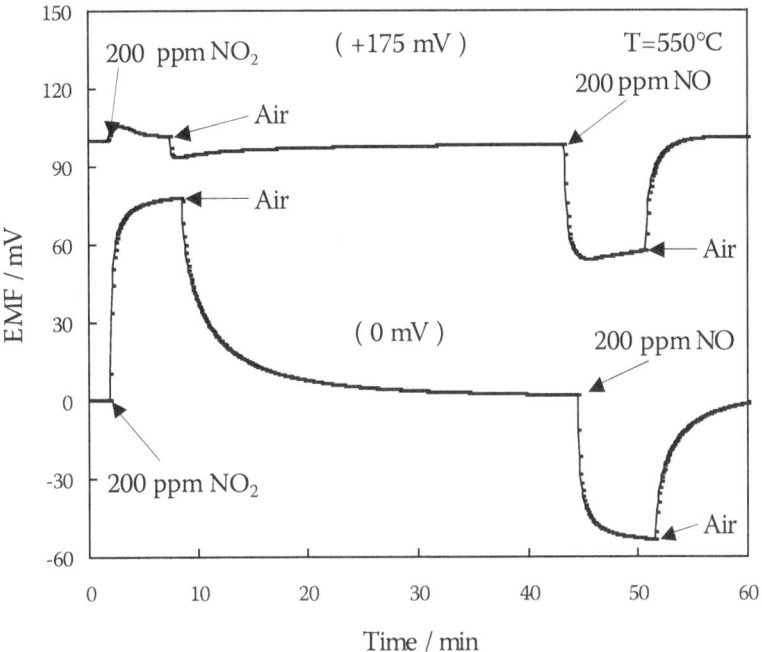

FIGURE 3.12 Response transients of the YSZ sensor using the NiCr$_2$O$_4$-SE to 200 ppm NO$_2$ and 200 ppm NO at 0 mV and +175 mV bias, at 550°C. (Reprinted from Zhuiykov, S. et al., Potentiometric NO$_x$ sensor based on the stabilised zirconia and NiCr$_2$O$_4$ sensing electrode operating at high temperatures, *Electrochem. Comm.* **3** (2001) 97–101, with permission from Elsevier Science.)

Apparently, the search for an additional oxide added to an existing SE, which would enhance the overall NO$_2$ sensitivity of the zirconia-based sensors, was the purpose of this exercise. This idea was not new. A similar approach was employed a few years ago for the mixed-potential NO$_2$ sensor with a combination of ZnFe$_2$O$_4$-ZnCr$_2$O$_4$ powders sintered together as a SE on the zirconia surface [13]. As a result, the NO$_2$ sensitivity was increased by about 50% at 700°C compared with that for the single oxide-SE. However, no improvement was obtained in the response rate for the sensor [13]. Another purpose to sinter a NO$_2$-sensitive SE as a mixture or combination of two single oxides with different properties is based on the predicted assumption that one of the oxides will evaporate during sintering and subsequently would increase the bulk porosity of the SE. Changes in the bulk porosity of the SE would challenge the magnitude of kinetics within the SE. Therefore, in order to control these changes, the appropriate selection of oxides, their mixtures, and, more importantly, the sintering conditions should be done.

It has been explained in the previous chapter that the SE matrix affects the catalytic activity to the gas-phase decomposition reaction of NO$_2$ (see Figure 2.6). Thus, a more porous structure of the SE would ultimately lead to the higher NO$_2$ sensitivity. However, in relation to the NiO-WO$_3$ system, it was not as simple as has

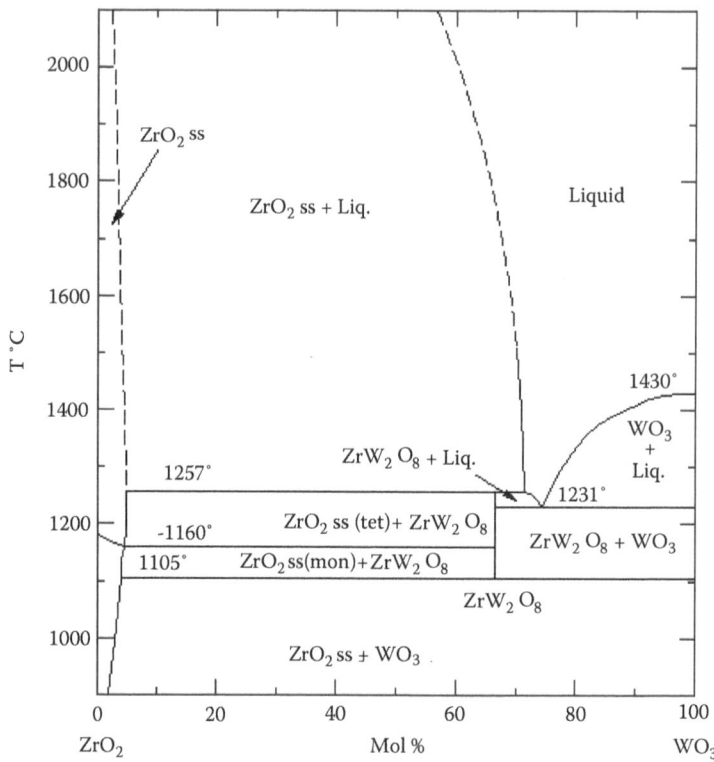

FIGURE 3.13 Phase diagram for the ZrO_2-WO_3 system.

been explained [68]. A mixture of NiO-WO_3 powders was heated to 1400°C and then was sintered for 2 hours at this temperature in air. During sintering at such high temperature, WO_3 did not evaporate, as has been claimed. Instead, two new phases, $NiWO_3$ and γ-ZrW_2O_8, were prepared on the surface of zirconia during heating between 800 and 1000°C and between 1150 and 1250°C, respectively, as is clearly shown in the ZrO_2-WO_3 phase diagram illustrated in Figure 3.13 [78]. ZrW_2O_8 has been known since the end of the twentieth century as a unique ceramic material with negative thermal expansion over the wide range of temperatures. ZrW_2O_8 is stable at the temperature range of 1105–1257°C [79]. It has also been reported that ZrW_2O_8 is a relatively dense ceramic, and its density changes from 82% to 89% in the temperature range of 1155–1200°C, respectively [80]. Analysis of the presented ZrO_2-WO_3 phase diagram suggests that if, during the sintering of the NiO-SE, the NiO-WO_3 mixture was heated to 1400°C and then was sintered for 2 hours afterwards, it is likely that γ-ZrW_2O_8 will be formed in the case of a 50 wt. %/50 wt. % mixture of NiO-WO_3 powders. These dense particles of ZrW_2O_8, formed on the zirconia surface, would decrease the TPB and, consequently, the number of active sites for the NO_2 reaction, despite the overall increase in the bulk porosity of SE.

This would lead to the substantial drop in NO_2 sensitivity at the same temperature range, as has been reported [68]. Having said that, it should also be admitted that for the successful sintering of NiO-WO_3 powders into γ-ZrW_2O_8, 6 hours is generally required at working temperatures of 1155–1200°C [80] or hot isostatic pressing. Thus, it is unlikely that the large number of dense γ-ZrW_2O_8 particles would appear after sintering NiO-SE, if only 10 wt. % of WO_3 was added to NiO before sintering. Therefore, in the case of a 10 wt. % WO_3 addition to NiO, only a trace of $NiWO_3$ solid solution was formed on the zirconia surface with simultaneous increase in the bulk porosity of the NiO-SE, which ultimately enhanced NO_2 sensitivity of the sensor. Traces of $NiWO_3$ increased the number of active centers for the cathodic reaction of NO_2, improving the TPB because even the well-developed bulk porosity of the SE cannot be solely responsible for ~40% enhancement in NO_2 sensitivity.

Consequently, it can be concluded that the enhancement of the NO_2 sensitivity and selectivity for the zirconia mixed-potential sensors with the oxide-SE will be focused on the technological improvements in fabrication of the SE. Further development of the sintering technology for the SE will be based on predicted calculation of how the initial binary oxide mixtures will be transferred into solid solutions with domination of one oxide phase and trace of another. In these solid solutions, the electrode kinetic of the bulk reactions within the SE will be responsible for sensitivity and selectivity of the sensor.

3.1.2 MIXED-POTENTIAL HYDROCARBON SENSORS

In addition to NO_x, hydrocarbons (HCs) are typical pollutants causing photochemical smog and the greenhouse effect. In the last few years, the reduction of HC emissions from vehicle exhausts has been addressed by the development of three-way catalytic converters capable of simultaneously oxidizing HCs and carbon monoxide [2, 27, 53]. This is because the current generation of low-emission vehicles is being replaced with ultra-low-emission vehicles as required, in order to comply with increasingly stringent world emission standards [81]. However, the level of HC emissions from automotive engines at so-called cold-starts is one of the biggest trade secrets, which companies involved in the development of automobiles do not want to reveal in scientific media. To deal with automotive emissions appropriately, apart from the NO_x sensors, the sensitive *on-board* HC sensors should be installed in the vehicles. The electrochemical mixed-potential HC sensors based on zirconia seem to be advantageous over other potential candidates in respect to their mechanical, thermal, and chemical stabilities [82]. A hydrocarbon gas sensor has to satisfy the following requirements: must display high sensitivity to HCs in atmosphere containing various inflammable gases, be independent on the oxygen concentration, and detect HCs on the *ppm* level; and it should have a fast response and be stable over the extended period of time in the exhaust gases. Thus, the development of a potentiometric zirconia-based HC sensor with appropriate HC-sensitive electrode materials has been advanced during the last decade, and several research groups have already reported progress in the development of HC sensors [83–94].

At the end of the twentieth century, a number of HC-sensitive electrode materials for solid electrolyte sensors have been investigated. These materials include terbium

(Tb)-doped zirconia [85, 90], gold, gold alloys and gold-oxide composites [87, 88], and pure and mixed oxides [89].

The exhaust environment is particularly harsh for mixed-potential sensors. The sensor in the exhaust of an internal combustion engine must be able to withstand the fluctuation of oxygen concentration and temperature, and provide sufficient mechanical and chemical stability in order to operate at least 100,000–150,000 miles. Similar to the mixed-potential NO_x sensors, the sensitivity of a zirconia-based sensor with a HC-sensitive electrode is influenced by the electrochemical and chemical reactions on the SE as well as on the TPB, autocatalytic reactions in the gas phase, and adsorption-desorption behavior on and within the SE [2]. If the material of a SE would recrystallize over time, this would change the TPB and consequently the characteristics of the sensor. In an effort to address these problems, new materials for the SE of the *on-board* HC sensor — Pr_6O_{11} [82], perovskite electrode materials $LaCoO_3$ [88], $La_{0.8}Sr_{0.2}CoO_{3-\delta}$ [92], $La_{0.8}Sr_{0.2}CrO_3$ [27], $LaMnO_3$ [53], $Y_{1.6}Tb_{0.3}Zr_{0.54}O_{\partial}$ [94], and composite electrodes Au-oxide/YSZ with Ga_2O_3, In_2O_3, and Nb_2O_5 [89] — have been investigated within the last few years.

Figure 3.14 shows a cross-section of the novel HC sensor that consists of an oxygen pump cell and a gas-detection cell [82]. In this sensor, Pt is used for the active electrode of the gas-detection cell and Pr_6O_{11} is used for the inactive electrode owing to its low catalytic activity and relatively high electric conductivity. The sensing characteristics to 500 ppm of C_3H_8 of the proposed sensor at a temperature of 800°C are shown in Figure 3.15. The published results [82] allowed identifying

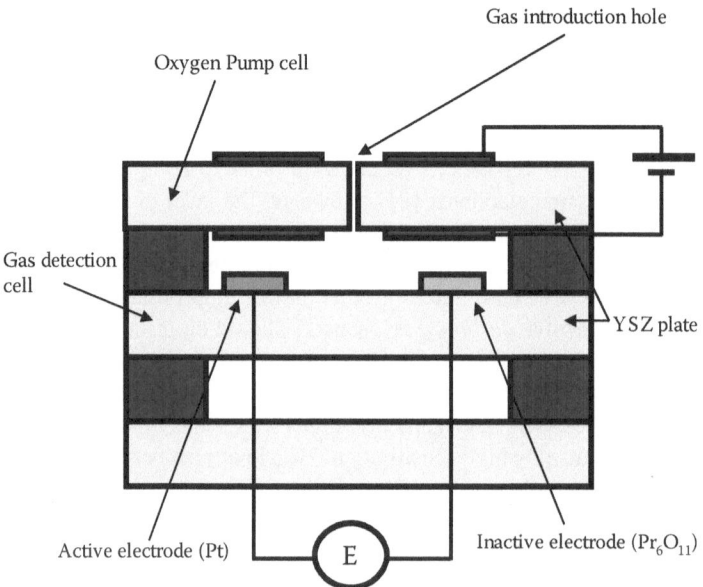

FIGURE 3.14 A cross-sectional view output of a HC sensor attached with a Pr_6O_{11}-SE. (Reprinted from Inaba, T., Saji, K., and Sakata, J., Characteristics of an HC sensor using a Pr_6O_{11} electrode, *Sens. Actuators B: Chem.* **108** (2005) 374–378, with permission from Elsevier Science.)

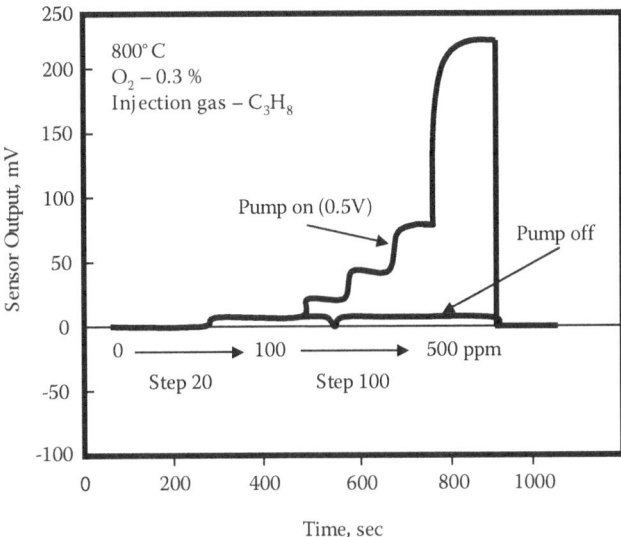

FIGURE 3.15 Output of a HC sensor attached with a Pr_6O_{11}-SE at 800°C with and without the pump being driven. (Reprinted from Inaba, T., Saji, K., and Sakata, J., Characteristics of an HC sensor using a Pr_6O_{11} electrode, *Sens. Actuators B: Chem.* **108** (2005) 374–378, with permission from Elsevier Science.)

the advantages of controlling the O_2 concentration in the gas-detection cell using the pump cell. As is clearly shown in this figure, the sensor output corresponded to the C_3H_8 concentration when the pump was off. The output *emf* was smaller than the Nernst *emf*, which was caused by a decrease in the C_3H_8 concentration according to a part of the combustion of C_3H_8 at the pump cell electrode (Pt). However, when the pump was on, the sensor output increased substantially. Although the O_2 concentration in the gas-detection space decreases owing to the pump cell being on, the O_2 consumed by C_3H_8 combustion at the active electrode essentially does not change. Consequently, the reason for increasing the sensor output is a shift in the ratio of the O_2 concentration at the inactive electrode with respect to the active electrode. These results indicate that the regulation of the O_2 concentration in the detection cell of the sensor is an effective way to enhance the sensor sensitivity to the selected gas. It was also revealed that in O_2-H_2 and O_2-CO atmospheres, the electric potential generated on the Pr_6O_{11} electrode is based on mixed potential and was almost the same as the potential at the Pt electrode. Thus, the sensitivity of the gas-detection cell to H_2 and CO was reduced. Using this method and the sensor design given in Figure 3.14, the traces of HC can be detected with high sensitivity even in a high-oxygen-concentration atmosphere at high temperatures.

The comparison of mixed-potential *emf* from perovskite, fluorite, and spinel metal-oxide electrodes used in the mixed-potential HC sensors was presented [95] for justification of using the precatalyst to mitigate cross-reference. The thermodynamic, chemical and mechanical stability in the exhaust gases, sufficient electron conductivity to control device impedance, as well as the ability to generate stable

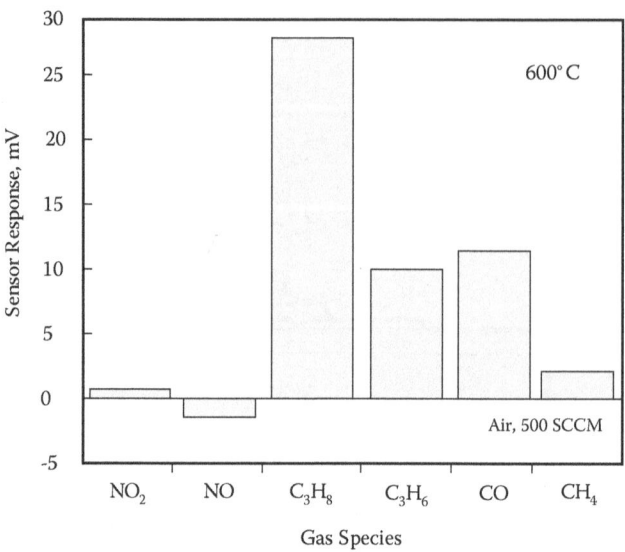

FIGURE 3.16 Sensitivity to the different gases of the Tb-doped YSZ-SE deposited on a sapphire substrate versus a Pt counterelectrode at 600°C in flowing air. (Reprinted from Brosha, E.L. et al., Mixed potential NO_x sensors using thin film electrodes and electrolytes for stationary reciprocating engine type applications, *Sens. Actuators B: Chem.* **119** (2006) 398–408, with permission from Elsevier Science.)

non-Nernstian potentials at elevated temperatures are the essential prerequisite criteria for using metal oxides as a SE of the HC sensors. Figure 3.16 illustrates sensitivity to the different gases for the Tb-doped YSZ-SE deposited on the sapphire substrate versus Pt counter electrode at 600°C in flowing air. The Tb-doped YSZ possesses fluorite structure with p-type electronic conductivity at elevated temperatures and is the latest electrode material studied the last few years. Figure 3.16 shows negligible sensitivity to NO and NO_2 with moderate sensitivity to hydrocarbons and CO. There was very small *emf* response to any gases at the temperatures below 600°C. However, characterization at 700°C exhibited a little gain to NO and NO_2 sensitivity, with the conclusion that this SE material would be well suited as a high-temperature HC/CO sensor [95]. So far, mixed-potential sensors with a $La_{0.8}Sr_{0.2}CrO_3$–SE showed the largest sensitivity to NO_x, preferential to nonmethane hydrocarbons and CO [27]. Sensors with Mg-doped $LaCrO_3$ and $Y_{1.6}Tb_{0.3}Zr_{0.54}O_{2-\partial}$ exhibited intermediate preferential sensitivity levels toward CO and HCs with minimal sensitivity to NO_x [95]. The level of the output *emf* was heavily dependent on the sensor temperature. Therefore, to obtain an essential sensitivity to HCs, the sensor temperature must be kept high and constant at all times with a high level of accuracy. It can be achieved by using a planar multielectrode thin-film sensor structure [91]. With the proper selection of SE combinations, it is also possible to control gas selectivity to NO_x, CHs, and CO in the control environment by an array of the mixed-potential sensors installed into the exhaust line.

In contrast to the zirconia-based mixed-potential NO_x sensors, the higher the sintering temperatures, which decrease porosity, the better the sensor response. This is contrary to the results for the WO_3 electrode described [83], for which the largest response was obtained for the highest surface area. Therefore, the trend in the magnitude of the response time cannot always be improved with enlargement of the surface-to-volume ratio because in the mixed-potential-type sensors, the enhanced reaction rate is not always desirable for a larger response. The mixed potential itself can shift in either direction depending on which reaction is reduced in rate. Consequently, depending on the practical applications, the changing rate of one reaction can increase the sensor *emf*, which may offset any detrimental effects of a decrease in response time [83].

3.1.3 Impedance-Based Zirconia Gas Sensors

The mixed-potential type of potentiometric sensors, described in the previous sections, appears to be advantageous for combustion applications and for *on-board* NO_x, CO, and HC sensors owing to their high sensitivities, especially in the measurement of a lower concentration (less than 100 ppm).

It was first reported in 2001 [32] that the evaluation by impedance spectroscopy of a biased zirconia sensor with an attached $LaFeO_3$-SE has shown that only the electrode resistance is a function of NO_2 content. Consequently, impedance spectroscopy offers a method for directly probing the electrode reactions that are the basis for mixed-potential-type gas sensors [67]. As a result of further development by using impedance spectroscopy in zirconia-based gas sensors with oxide-SEs, a new type of YSZ-based sensor for detecting total NO_x and HCs at high temperatures has been proposed recently [2, 14, 21, 62, 74, 96–100]. In this case, the change in the complex impedance of the device attached with a specific oxide-SE was measured as a sensing signal.

For NO_x measurement, SEs consisting of spinel-type oxides $CrMn_2O_4$, $NiCr_2O_4$, $NiFe_2O_4$, and $ZnCr_2O_4$ have been investigated in measuring gas consisting of 200 ppm NO_2 + 200 ppm NO in base air at 700°C. For the first three spinel oxides, a large, flat, semicircular arc was observed in each Nyquist plot in the examined frequency range (0.1 Hz–100 kHz); the shape of the semiarc for each sensor was almost similar for the gases. These results indicated that the impedance values of these devices are not affected by the coexistence of NO_x in the sample gas under these conditions, and, consequently, these sensors are insensitive to NO_x gas at high temperatures. On the other hand, for the sensor employing a $ZnCr_2O_4$-SE, the impedance behavior was entirely different from the above-mentioned results. As shown in Figure 3.17, the resistance value (Z', the intercept) at the intersection of the large semiarc with the real axis at low frequencies (around 0.1 Hz) varied with the concentration of both NO and NO_2 in the sample gas [74]. In addition, the resistance value decreased with an increase in the concentration of both NO and NO_2. Such a trend is completely different from that for the mixed-potential-type NO_x sensor, whose response direction to NO is opposite to that of NO_2 (see Figure 3.4). Meanwhile, the Z' value (the intercept, about 2000 ohm) at the intersection of

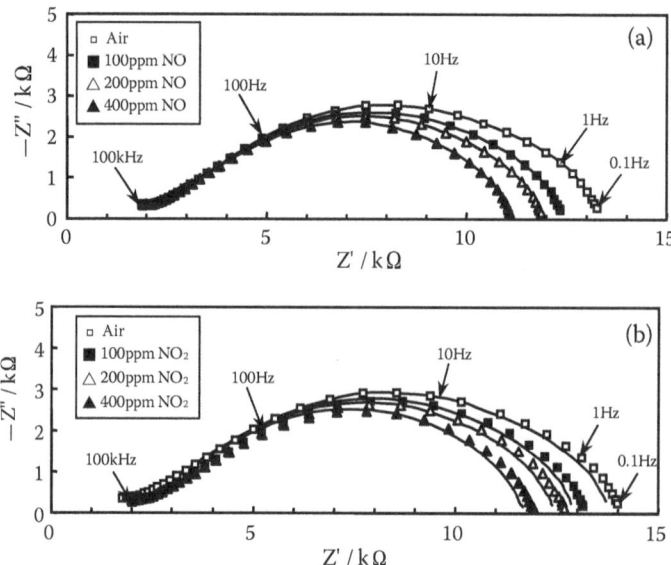

FIGURE 3.17 Complex impedance plots in base air and the sample gas with each of the various concentrations of (*a*) NO and(*b*) NO_2 at 700°C for the YSZ-based sensor attached with a $ZnCr_2O_4$-SE. (Reprinted from Miura, N., Nakatou M., and Zhuiykov, S., Impedance-metric gas sensor based on zirconia solid electrolyte and oxide sensing electrode for detecting total NO_x at high temperature, *Sens. Actuators B, Chem.* **93** (2003) 221–228, with permission from Elsevier Science.)

the large semiarc at high frequencies (~50 kHz) was kept almost constant even if the concentration of NO_x was changed from 0 ppm to 400 ppm.

When the sensor used only Pt electrodes, the impedance was independent of the frequency and had a fixed Z' value of ~50 ohm at 700°C [2]. This value was not affected by the existence of NO_x and was much smaller than the Z' at higher frequency for the sensor with a $ZnCr_2O_4$-SE. It appears that the frequency-independent value represents the YSZ bulk resistance. In order to verify this assumption, additional Nyquist plots for the three electrodes ($ZnCr_2O_4$-SE, Pt counterelectrode, and Pt-RE) attached to a YSZ-based sensor were obtained in the frequency range of 0.1 Hz–1 MHz in base air and different NO_x concentrations at 700°C. At higher frequencies, each plot contained a second, smaller semiarc, the shape of which was unaffected by the coexistence of NO_x in sample gas. This suggests that the measured resistance (Z') at higher frequencies (10 kHz – 1 MHz) for the $ZnCr_2O_4$-SE basically depends on the $ZnCr_2O_4$ bulk resistance, including the small YSZ bulk resistance [14, 21, 74]. It was also concluded that the Z' at lower frequencies (~0.1 Hz) is due to the impedance of the electrochemical reaction occurring at the oxide-SE/YSZ interface. The apparent equivalent circuit for the device attached with the $ZnCr_2O_4$-SE is based on the above-mentioned results, shown in Figure 3.18, where R_b means resistance of YSZ bulk, R_o is resistance of the oxide-SE, and C_o is capacitance of the oxide electrode, respectively. R_i represents resistance of the electrode at the

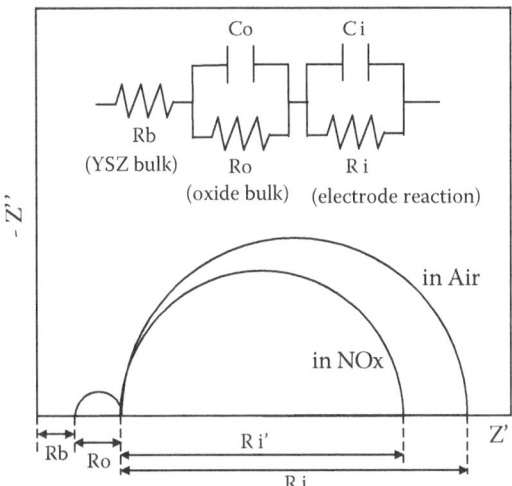

FIGURE 3.18 Apparent equivalent circuit for the YSZ-based sensor using a ZnCr$_2$O$_4$-SE. (Reprinted from Miura, N., Nakatou M., and Zhuiykov, S., Impedancemetric gas sensor based on zirconia solid electrolyte and oxide sensing electrode for detecting total NO$_x$ at high temperature, *Sens. Actuators B, Chem.* **93** (2003) 221–228, with permission from Elsevier Science.)

oxide-SE/YSZ interface, and C$_i$ is capacitance of the electrode reaction at the SE/YSZ interface. It is evident that only the electrode-reaction resistance R$_i$ is affected by the interaction between the interface and NO$_x$ (e.g., adsorption and reactions), which indicates that the R$_i$ can be used as the output-sensing signal for NO$_x$ detection.

The difference between the impedance in base air ($|Z|_{air}$) and the impedance in a sample gas ($|Z|_{gas}$) containing NO$_x$ concentration at the fixed frequency of 1 Hz has been defined as "gas sensitivity" of the device [14]. Of interest is the shift of so-called gas sensitivity due to the rising working temperature of an YSZ-based sensor using a ZnCr$_2$O$_4$-SE. Such measurements were performed using various NO$_x$ concentrations at the fixed temperatures of 650°C and 700°C. Figure 3.19 shows the dependence of the gas sensitivity on the concentration of NO and NO$_2$ for this sensor. A near linear correlation between gas sensitivity and the NO and NO$_2$ was found over the range 50–400 ppm at 700°C. Better linear correlation has been observed when the impedance was measured at the lower frequency of 0.1 Hz at the sacrifice of sampling time. Moreover, the most interesting result obtained is the near identical sensitivity to NO and NO$_2$ at 700°C, which shows that this sensor is capable of detecting the total NO$_x$ (NO and NO$_2$) in the gas mixture at 700°C, regardless of the NO-NO$_2$ ratio. Although this is a very important facet to the development of practical NO$_x$ sensors for vehicles, it is clear that the linearity and NO:NO$_2$ ratio are slightly less advantageous at 650°C.

Experience with the mixed-potential type of sensors employing spinel SEs suggests that the change in O$_2$ concentration may affect the performance of imped-ance-based sensors. In order to check this point, the O$_2$ concentration in the sample

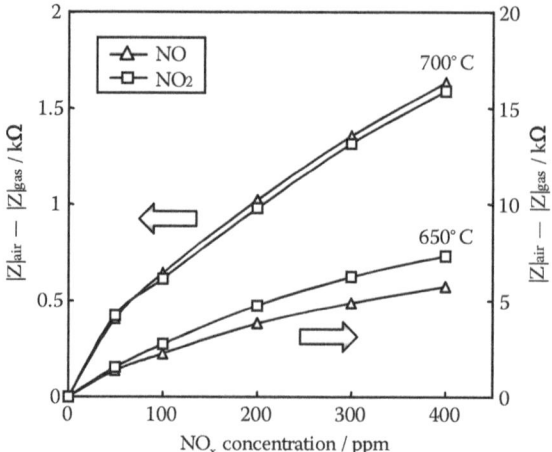

FIGURE 3.19 Dependence of gas sensitivity ($|Z|air - |Z|gas$) on concentration of NO (or NO$_2$) at 650°C and 700°C for the YSZ-based device using a ZnCr$_2$O$_4$-SE. (Reprinted from Miura, N., Nakatou M. and Zhuiykov, S., Impedancemetric gas sensor based on zirconia solid electrolyte and oxide sensing electrode for detecting total NO$_x$ at high temperature, *Sens. Actuators B, Chem.* **93** (2003) 221–228, with permission from Elsevier Science.)

gas was changed from 5 vol. % to 80 vol. % at 700°C whilst gas sensitivity to 100 ppm NO and NO$_2$ was recorded. The impedance value $|Z|$, as shown in Figure 3.20, exhibited very strong linear correlation to the logarithm of oxygen concentration at 1 Hz. Meanwhile, the gas sensitivities to NO and to NO$_2$ were almost equal at all O$_2$ concentrations examined. This result indicates that the oxygen concentration in the gas existing in the close vicinity of the SE should be monitored and should be held constant at all times. Thus, an oxygen sensor and oxygen pump based on YSZ

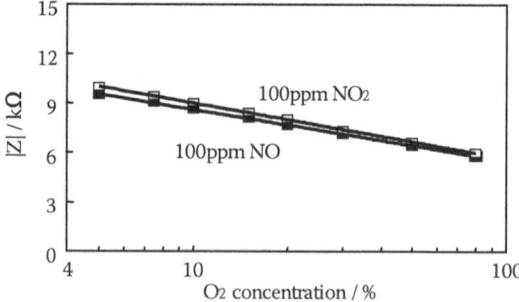

FIGURE 3.20 Dependence of the impedance value $|Z|$ on oxygen concentration in the sample gas at 700°C for the YSZ-based device using a ZnCr$_2$O$_4$-SE. The concentration of NO (or NO$_2$) was adjusted at 100 ppm. (Reprinted from Miura, N., Nakatou M., and Zhuiykov, S., Impedancemetric gas sensor based on zirconia solid electrolyte and oxide sensing electrode for detecting total NO$_x$ at high temperature, *Sens. Actuators B, Chem.* **93** (2003) 221–228, with permission from Elsevier Science.)

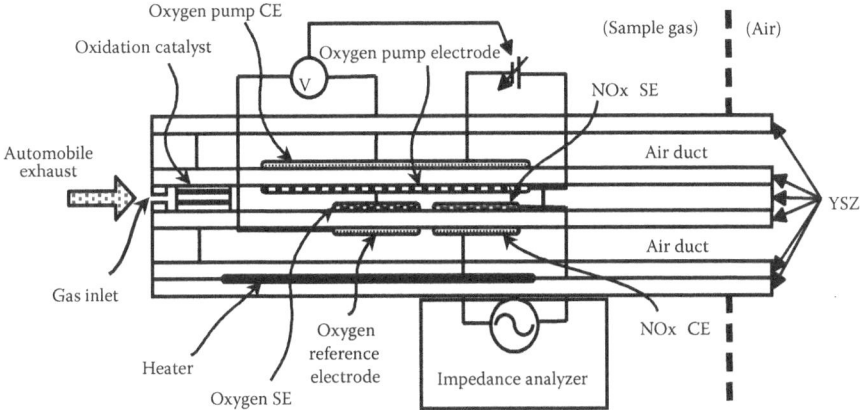

FIGURE 3.21 A cross-sectional view of the proposed laminated-type complex impedance-based NO_x sensor. (Reprinted from Miura, N., Nakatou, M., and Zhuiykov, S., Development of NO_x sensing devices based on YSZ and oxide electrode aiming for monitoring car exhausts, *J. Ceramics Intern.* **30** (2004) 1135–1139, with permission from Elsevier Science.)

should be used for monitoring and controlling, respectively, the oxygen concentration. These functioning devices could be installed together in the laminated-type YSZ-based sensor, in the same way as the modified mixed-potential-type sensor [35, 59]. A cross-sectional view of one of the proposed laminated types of the complex impedance-based NO_x sensor is presented in Figure 3.21. The change in O_2 concentration in the sample gas can be compensated by means of the O_2-sensing electrode (Pt), which is installed near the oxide-SE. An oxidation catalyst has also been included in the proposed design of the complex impedance-based NO_x sensor in order to oxidize the combustible gases in the vehicle exhaust to CO_2 and H_2O before reaching the NO_x SE. Therefore, in this laminated-type sensor structure, combustible gases have no interference to NO_x measurement, and the change in O_2 concentration around the SE is controlled by the oxygen sensor and is regulated by the oxygen pump. The main advantage of this design is that the total NO_x content can be measured and the NO_x sensitivity can be protected from the influence of coexisting combustible gases as well as from the variation of O_2 concentration in exhaust gas.

The complex impedance-based sensor with appropriate SE material can be used for measurement of CO [95], H_2O [97], and dry hydrogen-containing gases at high temperatures [98]. Au-doped Ga_2O_3 has been employed as an SE in a YSZ-based complex impedance CO sensor at 550°C. The impedance value of 0.42 Hz was strongly correlated to the CO concentration range of 100–800 ppm. The response and recovery time to 800 ppm CO were around 10 seconds [96]. An In_2O_3–SE has been used in a YSZ-based impedancemetric sensor for measurement of H_2O concentrations from 70 ppm up to 3% at 900°C [97]. The sensing performances of all complex impedance-based sensors reported so far are unique and differ significantly from those potentiometric Nernstian and non-Nernstian mixed-potential YSZ sensors. The change of complex impedance (gas sensitivity) has been attributed to the change in the resistance of the electrode reactions at the oxide-SE/YSZ interface

[74]. Further development of the complex impedance-based YSZ gas sensors in 2005–2007 was attributed to the selection of SE material sensitive to HCs at high temperatures [99], to improvement of NO_x sensing by use of YSZ/Cr_2O_3 composite electrodes [100] and to the adoption of the modified impedance method for the periodic *in-situ* diagnostics of the solid electrolyte/liquid-metal electrode interface during the life span of YSZ-based sensors measuring oxygen partial pressure in melts [101, 102].

Several single-metal oxides have been investigated in order to select the best SE material for HC detection at 600°C in the presence of 1 vol. % H_2O. The summary of the relative sensitivity of the sensors using each of the investigated oxides to propene and methane (400 ppm each) is shown in Figure 3.22 [99]. SnO_2 has shown the highest sensitivity to both HCs among oxides examined. However, since methane has not been regulated as an air pollutant, ZnO was chosen as the SE material providing rather high sensitivity to propene as well as negligible sensitivity to methane. It was also found that ZnO possesses low sensitivity to NO_2 but is still sensitive to NO. In order to compensate for NO sensitivity and convert NO to NO_2 at high temperatures, 1.5 wt. % Pt was imbedded into a ZnO-SE to improve its oxidation catalytic activity. Figure 3.23 depicts a SEM image of the cross-section of the ZnO-SE/YSZ interface. It has been confirmed that the structure of the ZnO-SE

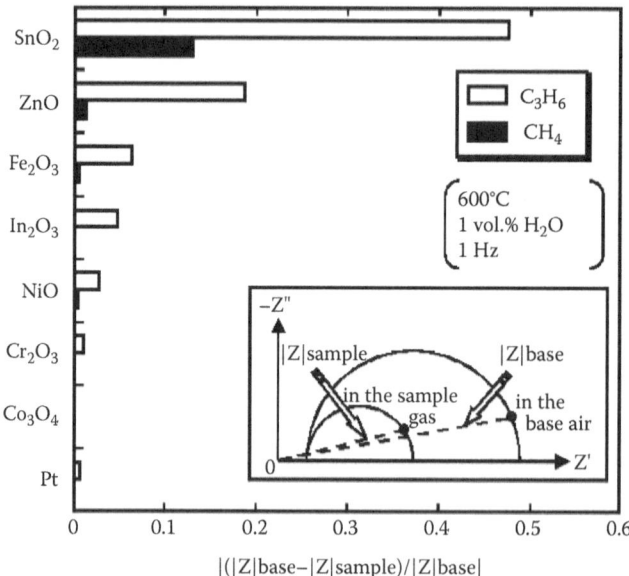

|(|Z|base−|Z|sample)/|Z|base|

FIGURE 3.22 Comparison of the absolute values of relative sensitivity to C_3H_6 and CH_4 (400 ppm each) of the sensors using each of various single-oxide-SEs at 600°C in the presence of 1 vol.% H_2O at 1 Hz. The inset shows modeling complex impedance plots. The points on dashed lines correspond to 1 Hz. (Reprinted from Nakatou, M. and Miura, N., Detection of propene by using new-type impedancemetric zirconia-based sensor attached with oxide sensing-electrode, *Sens. Actuators B, Chem.* 120 (2006) 57–62, with permission from Elsevier Science.)

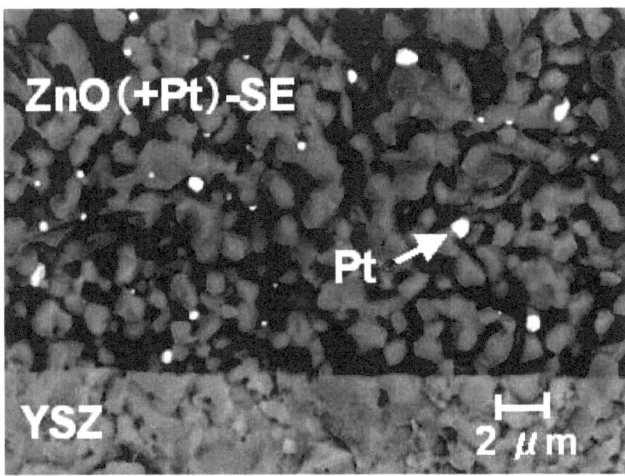

FIGURE 3.23 A SEM (back-scattering) image of a cross-sectional view of a ZnO (+1.5 wt. % Pt)-SE. (Reprinted from Nakatou, M. and Miura, N., Detection of propene by using new-type impedancemetric zirconia-based sensor attached with oxide sensing-electrode, *Sens. Actuators B, Chem.* 120 (2006) 57–62, with permission from Elsevier Science.)

is rather porous, and Pt particles (white dots) have been clearly identified in the bulk of the ZnO-SE. The sample gas can diffuse through the porous structure of the SE, and NO will oxidize to NO_2 at the presence of the dispersed Pt in the SE.

Further tests confirmed that the sensitivity to NO has been decreased. However, the sensitivity to propene was still low, compared to that of the sensor using the SnO_2-SE. In order to enhance C_3H_6 sensitivity, the applied potential of + 50 mV was set during impedance measurements. From the practical point of view, it is very important to examine the influence of coexisting combustible gases at the changing oxygen level on the sensor performance in humidified atmosphere. The complex impedance-based YSZ sensor attached with a ZnO-SE has shown that it is insensitive to CO_2 (10–20 vol. %) and H_2O (0.5–2.0 vol. %) at the changing of the oxygen level from 5 to 10 vol. % under polarization of +50 mV. The sensitivity to other gases at 600°C in the presence of 1 vol. % of H_2O is presented in Figure 3.24 [99]. The C_3H_6 sensitivity was also found to be independent of the total flow rate of the sample gas. Consequently, the results of this investigation confirmed that the complex impedance-based YSZ sensor with a ZnO-SE (+1.5 wt. % Pt) is capable of selectively detecting the C_3H_6 concentrations at high temperatures under harsh operating conditions.

3.2 FUTURE TRENDS

The present chapter has surveyed the latest results in the development of solid-state zirconia gas sensors for monitoring such important gaseous pollutants as NO_x and C_xH_y. Recent research and development of oxide-SEs as well as the improvement in design of the total-NO_x sensors in the early twenty-first century revealed that the

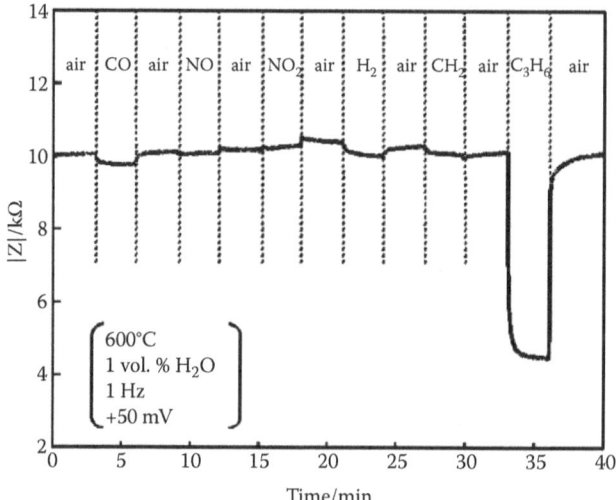

FIGURE 3.24 Response and recovery transients to various gases (NO and NO_2, 200 ppm; others, 400 ppm each) of the sensor using a ZnO (+1.5 wt.% Pt)-SE at 1 Hz. (Reprinted from Nakatou, M. and Miura, N., Detection of propene by using new-type impedancemetric zirconia-based sensor attached with oxide sensing-electrode, *Sens. Actuators B, Chem.* 120 (2006) 57–62, with permission from Elsevier Science.)

trend toward implementing thin- and thick-film technology for the planar sensor structure continues. Monitoring of exhaust gases, particularly in vehicles, and environmentally important pollutants are two main areas of application for these sensors. Although the vast majority of the zirconia-based gas sensors show excellent sensing performances for emissions control and environmental monitoring, these sensors generally are prototypes used in laboratory studies to explore new possibilities in gas sensing. Despite the promising performance in the controlled conditions of the laboratory, and initial promising results in the engine exhaust environment [39, 59], these sensors have yet to be tested for long-term stability, which is an essential prerequisite to industry acceptance. In early sensors, signal degradation of as much as 35% after only 300 hours of use has been reported [76]. This decrease was caused by deterioration of the catalytic activity of the SE, and in the modified sensor this deterioration was reduced to 5% after 1000 hours of operation [103]. Furthermore, the recent tests for long-term stability have shown that the decrease of 5–15% in the sensor output can be achieved after 6000 hours of continuous operation [104]. However, the improvement of long-term stability is still in progress because it needs to be extended substantially to make the zirconia-based NO_x sensors commercially valuable.

Potentiometric, non-Nernstian, zirconia-based, mixed-potential sensors offer several advantages. The recent shift from random to carefully selected properties for SEs of both single oxides and spinels has increased the working temperature of these sensors to 700°C, which is compatible with the working temperature at the vehicle exhausts. These devices are comparatively simple in design, and they exhibit good sensitivity and selectivity to the measuring gas. The fabrication of several electrodes on the single solid electrolyte is a realistic goal, which gives rise to the possibility of sensors for NO_x, CO, and C_xH_y. Use of planar construction can also overcome the need for a seal between two electrodes, which simplifies the sensor design. However, the reaction kinetics can be affected by such factors as processing conditions and microstructural changes during fabrication technology, which consequently lead to variations between sensors within one batch. Furthermore, it will change the sensor response over time. Since the requirements for catalytic activity at high temperatures of SEs are relatively easy to meet, there are many possibilities for SEs for non-Nernstian potentiometric gas sensors. However, from the practical point of view, the main problems are related to the nonideal selectivity of the single electrodes and lack of long-term stability of their interfaces. One of the ways to decrease cross-sensitivity from other gases, including oxygen, is the improvement of the sensor's design [59, 75]. However, even in the modified design of the mixed-potential NO_x sensors, chemical reactions at high temperatures tend toward equilibrium. Thus, the sensor's response tends to decrease with increasing temperature, which places an upper limit on the working temperature of these sensors [76]. The recent research results have shown that the upper temperature limits have been expanded by careful selection of new SE materials.

It is well-known that a sensor's response to the measuring gas involves a complex interaction of the several factors involving both the YSZ-SE-gas TPB and the methods to determine their effects on the overall sensor output signal. It is clear that optimization of the processing methods, microstructure, and properties of SEs will enhance significantly the sensing properties of mixed-potential sensors. The improvement of sensitivity and selectivity for the zirconia mixed-potential sensors with oxide-SEs will be concentrated on the fabrication technologies of the SE. A better understanding of the whole complex of physical, chemical, and electrochemical reactions is necessary for selection oxides and oxide mixtures for SEs of the mixed-potential sensors. The development of this understanding will be based on predicted and careful experimental verification of properties of the binary oxide mixtures or solid solutions with domination of one oxide phase and trace of another. Therefore, most of the future improvement will be focused on the technological improvements of new and existing materials of the SE. Hopefully, this will provide a comprehensive understanding of the interaction between the different mechanisms that determine the sensitivity to the measuring gas.

Further development of new impedance-based gas sensors is likely to allow their introduction to niche applications in the field of zirconia-based, solid-state gas sensors. Impedance spectroscopy has the potential to measure changes not detectable with simple current or voltage measurements. Although it is not practical to implement complete impedance spectroscopy in an operating sensor, an optimized frequency suitable for a specific gas sensor may be used. So far, these sensors are still more complex and expensive as compared to the potentiometric zirconia-based gas sensors. However, they have two important advantages: (1) measurement of total NO_x concentration, regardless of the NO/NO_2 ratio in exhausts; and (2) near equal sensitivity to NO and NO_2 at 700°C. These are essential prerequisites to their practical implementation in vehicle exhausts. Therefore, further investigation is required in order to obtain better understanding of the sensing mechanism of these sensors and to establish long-term stability. These sensors have to operate for long periods of time without substantial deterioration of the output signal.

Finally, planar, thick-film, YSZ-based sensors for O_2, C_xH_y, and NO_x are expected to reinforce their place in the market owing to their rapid response and potential for implementation as multicomponent gas sensors in vehicle exhausts. The performance of recently developed ultra-lean-burn engines and NO_x storage catalysts depends significantly on the performance of such sensors. Thus, solid-state electrochemical sensors must reach even higher levels of performance and reliability, so continued development of these sensors is required in order to address more stringent requirements.

REFERENCES

1. Burtscher, H., Physical characterization of particulate emissions from diesel engines: A review, *J. Aerosol Sci.* **36** (2005) 896–932.
2. Zhuiykov, S. and Miura, N., Solid-state electrochemical gas sensors for emission control, in *Materials for Energy Conversion Devices*, Eds. C.C. Sorrell, J. Nowotny, and S. Sugihara, Cambridge, Woodhead Publishing, 2005, Chapter 12.
3. Khair, M., Lemare J., and Fisher, S., Achieving heavy-duty diesel NO_x/PM levels below the EPA 2002 standards: An integral solution, *SAE paper* (2000) 2000-01-0187.
4. Mukundan, R. and Garson, F., Electrochemical sensors for energy and transportation, *The Electrochem. Soc. Interface* **Summer** (2004) 30–35.
5. Docquier, N. and Candel, S., Combustion control and sensors: A review, *Prog. Ener. Comb. Sci.* **28** (2002) 107–150.
6. Marquis, B.T. and Vetelino, J.F., A semiconductor metal oxide sensor array for the detection of NO_x and NH_3, *Sens. Actuators B: Chem.* **77** (2001) 100–110.
7. Eskilsson, D. et al., Optimisation efficiency and emissions in pellet burners, *Biomass Bioenergy* **27** (2004) 541–546.
8. Zhuiykov, S. and Miura, N., Development of zirconia-based potentiometric NO_x sensors for automotive and energy industries in the early 21st century: What are the prospects for sensors? *Sens. Actuators B, Chem.* **121** (2007) 639–651.
9. Ménil, F., Coillard, V., and Lucat, C., Critical review of nitrogen monoxide sensors for exhaust gases of lean burn engines, *Sens. Actuators B: Chem.* **67** (2000) 1–23.
10. Zhuiykov, S. and Nowotny, J., Zirconia-based sensors for environmental gases: A review, *Mater. Forum* **24** (2000) 150–168.
11. Miura, N., Lu, G., and Yamazoe, N., Progress in mixed-potential type devices based on solid electrolyte for sensing redox gases, *Solid State Ionics* **136–137** (2000) 533–542.
12. Szabo, N.F. and Dutta, P.K., Correlation of sensing behaviour of mixed potential sensors with chemical and electrochemical properties of electrodes, *Solid State Ionics* **171** (2004) 183–190.
13. Zhuiykov, S. et al., High-temperature NO_x sensors using zirconia and zinc-family oxide sensing electrode, *Solid State Ionics* **152** (2002) 801–807.
14. Miura, N., Nakatou, M., and Zhuiykov, S., Impedance-based total-NO_x sensor using stabilized zirconia and $ZnCr_2O_4$ sensing electrode operating at high temperature, *Electrochem. Comm.* **4** (2002) 284–287.
15. Miura, N. et al., Mixed-potential-type NO_x sensor based on YSZ and zinc oxide sensing electrode, *Ionics* **10** (2004) 1–9.

16. Korotcenkov, G., Gas response control through structural and chemical modification of metal oxide films: State of the art and approaches, *Sens. Actuators B: Chem.* **107** (2005) 209–232.

17. Zhuiykov, S. et al., Potentiometric NO_x sensor based on the stabilised zirconia and $NiCr_2O_4$ sensing electrode operating at high temperatures, *Electrochem. Comm.* **3** (2001) 97–101.

18. Bartolomeo, E.D. et al., Zirconia-based electrochemical NO_x sensors with semiconducting oxide electrodes, *J. Amer. Ceram. Soc.* **87** (2004) 1883–1889.

19. Elumalai, P. et al., Dependence of NO_2 sensitivity on thickness of oxide-sensing electrodes for mixed-potential-type sensors using stabilized zirconia, *Ionics* 12 (2006) 331–337.

20. Zhuiykov, S. et al., Stabilised zirconia–based NO_x sensor using $ZnFe_2O_4$ sensing electrode, *Electrochem. Solid-State Lett.* **4** (2001) H19–H21.

21. Miura, N., Nakatou, M., and Zhuiykov, S., Development of NO_x sensing devices based on YSZ and oxide electrode aiming for monitoring car exhausts, *J. Ceramics Intern.* **30** (2004) 1135–1139.

22. Szabo, N.F. et al., Microporous zeolite modified yttria stabilized zirconia (YSZ) sensors for nitric oxide (NO) determination in harsh environments, *Sens. Actuators B, Chem.* **82** (2002) 142–149.

23. Miura, N. et al., Mixed potential type sensor using stabilized zirconia and $ZnFe_2O_4$ sensing electrode for NO_x detection at high temperature, *Sens. Actuators B, Chem.* **81** (2002) 222–229.

24. Ono, T. et al., Improving of sensing performance of zirconia-based total NO_x sensor by attachment of oxidation-catalyst electrode, *Solid State Ionics* **175** (2004) 503–506.

25. Menil, F., Debeda, H., and Lukat, C., Screen-printed thick films: From materials to functional devices, *J. Euro. Ceram. Soc.* **25** (2005) 2105–2113.

26. Bartolomeo, E.D. and Grilli, M.L., YSZ-based electrochemical sensors: From materials preparation to testing in the exhausts of an engine bench test, *J. Euro. Ceram. Soc.* **25** (2005) 2959–2964.

27. Garzon, F.H. et al., Solid state ionic devices for combustion gas sensing, *Solid State Ionics* **175** (2004) 487–490.

28. Bartolomeo, E.D., Grilli, M.L., and Traversa, E., Sensing mechanism of potentiometric gas sensors based on stabilized zirconia with oxide electrodes: Is it always mixed potential? *J. Electrochem. Soc.* **151** (2004) H133–H139.

29. West, D.L., Montgomery, F.C., and Armstrong, T.R., Use $La_{0.85}Sr_{0.15}CrO_3$ in high-temperature NO_x sensing elements, *Sens. Actuators B: Chem.* **196** (2005) 758–765.

30. Garson, G.H., Mukundan, R., and Brosha, E.L., Solid-state mixed potential gas sensors: Theory, experiments and challenges, *Solid State Ionics* **136–137** (2000) 633–638.

31. Elumalai, P. and Miura, N., Performances of planar NO_2 sensor using stabilized zirconia and NiO sensing electrode at high temperature, *Solid State Ionics* **31–34** (2005) 2517–2522.

32. Yoon, J.W. et al., The NO_2 response of solid electrolyte sensors made using nano-sized $LaFeO_3$ electrodes, *Sens. Actuators B: Chem.* **76** (2001) 483–488.

33. Guillet, N. et al., Development of a gas sensor by thick film technology for automotive applications: Choice of materials—realization of a prototype, *Mater. Sci. Eng.* C **21** (2002) 97–103.

34. Kunimoto, A. et al., New total-NO_x sensor based on mixed potential for automobiles, *SAE paper* (1999) 1999-01-1280.

35. Ono, T. et al., Total NO_x sensor based on mixed-potential for detecting of low NO_x concentrations, *SAE paper* (2005) 2005-01-0451.

36. Skelton, D.C. et al., A surface-science-based model for the selectivity of platinum-gold alloy electrodes in zirconia-based NO_x sensors, *Sens. Actuators B: Chem.* **96** (2003) 46–52.

37. Martin, L.P., Pham, R.Q., and Glass, R.S., Effect of Cr_2O_3 electrode morphology on the nitric oxide response of a stabilized zirconia sensor, *Sens. Actuators B: Chem.* **96** (2003) 53–60.

38. Szabo, N.F. and Dutta, P.K., Strategies for total NO_x measurement with minimal CO interference utilizing a microporous zeolitic catalytic filter, *Sens. Actuators B: Chem.* **88** (2003) 168–177.

39. Grilli, M.L. et al., Planar non-Nerstian electrochemical sensors: Field test in the exhaust of a spark ignition engine, *Sens. Actuators B: Chem.* **108** (2005) 319–325.

40. Figueroa, O.L. et al., Temperature-controlled CO, CO_2 and NO_x sensing in a diesel engine exhaust stream, *Sens. Actuators B: Chem.* **107** (2005) 839–848.

41. Li, X., Xiong, W., and Kale, G.M., Novel nanosized ITO electrode for mixed potential gas sensor, *Electrochem. Solid-State Lett.* **8** (2005) H27–H30.

42. Xiong, W. and Kale, G.M., Novel high-selectivity NO_2 sensor incorporating mixed-oxide electrode, *Sens. Actuators B: Chem.* **114** (2006) 101–108.

43. Bartolomeo, E.D. et al., Nano-structured perovskite oxide electrodes for planar electrochemical sensors using tape casted YSZ layers, *J. Euro. Ceram. Soc.* **24** (2004) 1187–1190.

44. Tanaka, S. and Esaka, T., High NO_x sensitivity of oxide thin films prepared by RF sputtering, *Mater. Res. Bull.* **35** (2000) 2491–2502.

45. Bartolomeo, E.D. et al., Planar electrochemical sensors based on tape-cast YSZ layers and oxide electrodes, *Solid State Ionics* **171** (2004) 173–181.

46. Szabo, N.F. et al., Microporous zeolite modified yttria stabilized zirconia (YSZ) sensors for nitric oxide (NO) determination in harsh environments, *Sens. Actuators B: Chem.* **82** (2002) 142–149.

47. Riegel, J., Neumann, N., and Wiedenmann, H.-M., Exhaust gas sensors for automotive emission control, *Solid State Ionics* **152–153** (2002) 783–800.

48. West, D.L., Montgomery, F.C., and Armstrong, T.R., "NO-selective" NO_x sensing elements for combustion exhausts, *Sens. Actuators B: Chem.* **111–112** (2005) 84–90.

49. Mukundan, R., Garson, F.H., and Brosha, E.L., *Electrodes for Solid State Gas Sensor*, U.S. Patent No. 6,605,202 (2003).

50. Martin, L.P., Pham, I.Q., and Glass, R.S., Effect of Cr_2O_3 morphology on the nitric oxide response of a stabilized zirconia sensor, *Sens. Actuators B, Chem.* **96** (2003) 53–60.

51. Goepel, W., Reinhardt, G., and Rosch, M., Trend in the development of solid state amperometric and potentiometric high temperature sensors, *Solid State Ionics* **136–137** (2000) 519–531.

52. Miura, N. and Yamazoe, N., Approach to high-performance electrochemical NO_x sensors based on solid electrolytes, in *Sensors Update*, Eds. H. Baltes, W. Goepel, and J. Hesse, Weinheim, Wiley-Vch, 2000, 191–210.

53. Brosha, E.L. et al., Development of ceramic mixed potential sensors for automotive applications, *Solid State Ionics* **148** (2002) 61–69.

54. Guth, U. and Zosel, J., Electrochemical solid electrolyte gas sensors: Hydrocarbon and NO_x analysis in exhaust gases, *Ionics* **10** (2004) 366–377.

55. Miura, N. et al., Mixed potential type NO_x sensor based on stabilized zirconia and oxide electrode, *J. Electrochem. Soc.* **143** (1996) L33–L35.

56. Miura, N. et al., Stabilized zirconia-based sensor using oxide electrode for detection of NO_x in high-temperature combustion-exhausts, *Solid State Ionics* **86–88** (1996) 1069–1073.

57. Elumalai, P. et al., Sensing characteristics of YSZ-based mixed-potential-type planar NO_x sensor using NiO sensing electrodes sintered at different temperatures, *J. Electrochem. Soc.* **152** (2005) H95–H101.

58. Park, C.O. and Miura, N., Absolute potential analysis of the mixed potential occurring at the oxide/YSZ electrode at high temperature in NO_x-containing air, *Sens. Actuators B: Chem.* **113** (2006) 316–319.

59. Ono, T. et al., Sensing performances of mixed potential type NO_x sensor attached with oxidation-catalyst electrode, *Electrochemistry* **71** (2003) 405–407.

60. Miura, N. et al., NO_x sensing characteristics of mixed-potential-type zirconia sensor using NiO sensing electrode at high temperatures, *Electrochem. Solid-State Lett.* **8** (2005) H9–H11.

61. Miura, N. et al., High-temperature operating characteristics of mixed-potential-type NO_2 sensor based on stabilized-zirconia tube and NiO sensing electrode, *Sens. Actuators B: Chem.* **114** (2006) 903–909.

62. Miura, N. et al., Zirconia-based gas sensors using oxide sensing electrode for monitoring NO_x in car exhaust, *Proc. of 29th Int. Conf. Adv. Ceram. & Comp., Advances in Ceramic Materials*, American Ceramic Society, **26** (2005) 3–13.

63. Wang, J. et al., Mixed-potential-type zirconia-based NO_x sensor using Rh-loaded NiO sensing electrode operating at high temperatures, *Solid State Ionics* **177** (2006) 2305–2311.

64. Figueroa, O.L. et al., Temperature-controlled CO, CO_2 and NO_x sensing in a diesel engine exhaust stream, *Sens. Actuators, B: Chem.* **107** (2005) 839–848.

65. Park, C.O., Akbar, A.S., and Wepner, W., Ceramic electrolytes and electrochemical sensors, *J. Materials Sci.* **38** (2003) 4639–4660.

66. West P.L., Montgomery, F.C., and Armstrong, T.R., "Total NOx" sensing elements with compositionally identical oxide electrodes, *J. Electrochem. Soc.* **153** (2006) H23–H28.

67. Tuller, H.L., Defect engineering: Design tools for solid state electrochemical devices, *Electrochem. Acta* **48** (2003) 2879–2887.

68. Wang, J. et al., Improvement of NO_2 sensitivity of mixed-potential-type zirconia-based sensor by the addition of WO_3 to NiO sensing-electrode, in *Proc. 11th Int. Meeting on Chemical Sensors*, Brescia, Italy, 11–17 July 2006, 32.

69. Xiong, W. and Kale, G.M., Microstructure, conductivity, and NO_2 sensing characteristics of α-Al_2O_3-doped (8 mol % Sc_2O_3) ZrO_2 composite solid electrolyte, *Int. J. Apply Ceram. Tech.* **3** (2006) 210–217.

70. Ramamoorthy, R., Akbar, S.A., and Dutta, P.K., Dependence of potentiometric oxygen sensing characteristics on the nature of electrodes, *Sens, Actuators B: Chem.* **113** (2006) 162–168.

71. Grilli, M.L. et. al., Planar electrochemical sensors based on YSZ with WO_3 electrode prepared by different chemical routes, *Sens. Actuators B: Chem*, **111–112** (2005) 91–95.

72. Hasegawa, I., Tamura, S., and Imanaka, N., Solid electrolyte type nitrogen monoxide gas sensor operating at intermediate temperature region, *Sens. Actuators B: Chem.* **108** (2005) 314–318.

73. Zhuiykov, S., Mathematical modelling of YSZ-based potentiometric gas sensors with oxide sensing electrodes—Part II: Complete and numerical models for analysis of sensor characteristics, *Sens. Actuators B, Chem.* **120** (2007) 645–656.

74. Miura, N., Nakatou M., and Zhuiykov, S., Impedancemetric gas sensor based on zirconia solid electrolyte and oxide sensing electrode for detecting total NO_x at high temperature, *Sens. Actuators B, Chem.* **93** (2003) 221–228.

75. Elumalai, P., Hasei, M., and Miura, N., Influence of thickness of Cr_2O_3 sensing-electrode on sensing characteristics of mixed-potential-type NO_2 sensor based on stabilized zirconia, *Electrochemistry* **2** (2006) 197–201.

76. Fergus, J.W., Materials for high temperature electrochemical NO_x gas sensors, *Sens. Actuators B: Chem.* **121** (2007) 652–663.

77. Kading, S. et al., YSZ-cells for potentiometric nitric oxide sensors, *Ionics* **9** (2003) 151–155.

78. Chang, L.L.Y., Scroger, M.G., and Phillips, B., System ZrO_2-WO_3, *J. Am. Ceram. Soc.* **50** (1967) 211–215.

79. Evans, J.S.O. et al., Compressibility, phase transitions, and oxygen migration in zirconium tungstate, ZrW_2O_8, *Science* **275** (1997) 61–65.

80. Chen, J.C. et al., Synthesis of negative-thermal-expansion ZrW_2O_8 substrates, *Scripta Materialia* **49** (2003) 261–266.

81. *EURO 5 and 6 Emissions Standards for Cars and Vans*, Europ. Federation for Transport & Environment, September 2006, 26.

82. Inaba, T., Saji, K., and Sakata, J., Characteristics of an HC sensor using a Pr_6O_{11} electrode, *Sens. Actuators B: Chem.* **108** (2005) 374–378.

83. Fergus, J.W., Solid electrolyte based sensors for the measurement of CO and hydrocarbon gases, *Sens. Actuators B: Chem.* (in press).

84. Miura, N., et al., Mixed-potential-type propylene sensor based on stabilized zirconia and oxide electrode, *Electrochem. Commun.* **2** (2000) 77–80.

85. Schmidt-Zhang, P. and Guth, U., A planar thick film sensor for hydrocarbon monitoring in exhaust gases, *Sens. Actuators B: Chem.* **99** (2004) 258–263.

86. Zosel, J. et al., Au–oxide composites as HC-sensitive electrode material for mixed potential gas sensors, *Solid State Ionics*, **152–153** (2002) 525–529.

87. Zosel, J. et al., Response behaviour of perovskites and Au/oxide composites as HC-electrodes in different combustibles, *Solid State Ionics* **175** (2004) 531–533.

88. Muller, R. et al., Investigations on selected gallium and strontium doped lanthanum chromites as electrode materials for HC detection, *Ionics* **8** (2002) 262–266.

89. Zosel, J. et al., Selectivity of HC-sensitive electrode materials for mixed potential gas sensors, *Solid State Ionics* **169** (2004) 115–119.

90. Brosha, E.L. et al., Mixed potential sensors using lanthanum manganate and terbium yttrium zirconium oxide electrodes, *Sens. Actuators B: Chem.* **87** (2002) 47–57.

91. Mukundan, R., Brosha, E.L., and Garzon, F.H., *Method for Forming a Potential Hydrocarbon Sensor with Low Sensitivity to Methanol and CO*, U.S. Patent No. 6,656,336.

92. Hashimoto, A. et al., High-temperature hydrocarbon sensors based on a stabilized zirconia electrolyte and proton conductor-containing platinum electrode, *Sens. Actuators B: Chem.* **81** (2001) 55–63.

93. Brosha, E.L. et al., CO/HC sensors based on thin films of $LaCoO_3$ and $La_{0.8}Sr_{0.2}CoO_{3-\delta}$ metal oxides, *Sens. Actuators B: Chem.* **69** (2000) 171–182.

94. Hibino, T. and Wang, S., A novel sensor for C1–C8 hydrocarbons using two zirconia-based electrochemical cells, *Sens. Actuators B: Chem.* **61** (1999) 12–18.

95. Brosha, E.L. et al., Mixed potential NO_x sensors using thin film electrodes and electrolytes for stationary reciprocating engine type applications, *Sens. Actuators B: Chem.* **119** (2006) 398–408.

96. Wu, N. et al., Impedance-metric Pt/YSZ/Au–Ga$_2$O$_3$ sensor for CO detection at high temperature, *Sens. Actuators B: Chem.* **110** (2005) 49–53.

97. Nakatou, M. and Miura, N., Impedancemetric sensor based on YSZ and In$_2$O$_3$ for detection of low concentrations of water vapour at high temperature, *Electrochem. Comm.* **6** (2004) 995–998.

98. Nakatou, M. and Miura, N., Detection combustible hydrogen-containing gases by using impedancemetric zirconia-based water-vapour sensor, *Solid State Ionics* **176** (2005) 2511–2515.

99. Nakatou, M. and Miura, N., Detection of propene by using new-type impedancemetric zirconia-based sensor attached with oxide sensing-electrode, *Sens. Actuators B, Chem.* **120** (2006) 57–62.

100. Martin, L.P. et al., Impedance metric NO$_x$ sensing using YSZ electrolyte and YSZ/Cr$_2$O$_3$ composite electrodes, *J. Electrochem. Soc.* **154** (2007) J94–J104.

101. Zhuiykov, S., Sensors measuring oxygen activity in melts: Development of impedance method for "*in-situ*" diagnostics and control electrolyte/liquid-metal electrode interface, *Ionics* **11** (2005) 352–361.

102. Zhuiykov, S., "*In-situ*" diagnostics of solid electrolyte sensors measuring oxygen activity in melts by developed impedance method, *Meas. Sci. Technol.* **17** (2006) 1570–1578.

103. Kato, N. and Kokune, N., Long-term stable NO$_x$ sensor with integrated in-connector control electronics, *SAE paper* (1999) 1999-01-020.

104. Orban, J.E. et al., Long-term ageing of NO$_x$ sensors in heavy-duty engine exhausts, *SAE paper* (2005) 2005-01-3793.

4 Zirconia Sensors for Measurement of Gas Concentration in Molten Metals

4.1 ZIRCONIA SENSORS FOR THE MEASUREMENT OF OXYGEN ACTIVITY IN MELTS

4.1.1 POLYCRYSTALLINE ZIRCONIA SENSORS

In the processing of molten metals, the chemical composition of the molten metal has to be controlled because interactions between a molten metal and the atmosphere can change the composition of the metal. In some cases, such undesired elements as oxygen or hydrogen can be incorporated into the melts. Therefore, to improve control of the various metallurgical processes, electrochemical sensors have been employed by industry [1–18]. The most successful and widely used example of such sensors is the zirconia oxygen sensor used in steelmaking [1, 3, 6, 10, 11]. Sensors based on zirconia solid electrolytes have several advantages in the processing of melts. The conductivity of zirconia increases with increasing temperature, so the high operating temperature required during the processing of molten metals is well suited to the zirconia-based gas sensors. The supporting electronics are relatively simple, since the output of an electrochemical sensor is an *emf*. In addition, zirconia-based electrolytes are generally stable compounds which can withstand the harsh chemical environment in melts. Furthermore, an additional phase (referred to as an *auxiliary SE*) can be added to provide sensitivity to a species that is not present in the electrolyte. Thus, zirconia sensors can be designed for detecting a wide variety of gases through judicious selection of the electrolyte and SE materials. The use of electrochemical sensors in metallurgical processing will continue to expand as the performance, reliability, and cost-effectiveness of current sensors are improved and as new sensors are developed.

Oxygen from reaction with the atmosphere is removed from molten steel by adding aluminum or silicon alloys, which react with the oxygen to form oxides. Determining the optimal amounts of these alloys to be added requires precise measurement of oxygen in the steel, which is provided by an oxygen sensor [3, 10]. This sensor is based on polycrystalline zirconia. The RE is a metal–metal oxide mixture (most commonly, Cr/Cr_2O_3), the equilibrium of which establishes a reference oxygen partial pressure. Although oxygen sensors have been used for many years, there are still areas for improvement. Current sensors are used for one measurement and then discarded. Replacing these disposable sensors with extended-life sensors would both improve the quality of data obtained (i.e., continuous measurements could be made) and reduce the costs. One approach for extending the lifetime of current oxygen sensors has been alternative fabrication techniques, which improve the seal between the reference electrode and the molten metal. Another approach has been made on the design modification of a nonisothermal sensor, in which the RE is outside the molten steel [15, 16]. This reduced temperature lessens the requirements on the RE seal, but introduces an additional voltage due to the temperature difference between the two electrodes. Another approach to extending the life of oxygen sensors is to use an applied voltage to electrochemically reverse the degradation of the reference.

During molten aluminum (Al) processing, the most important dissolved gas is hydrogen, which is produced when Al reacts with moisture to form aluminum oxide and hydrogen. The solubility of hydrogen in the liquid Al is much higher than that in solid Al, so the dissolved hydrogen from the reaction of molten Al with water vapor in the atmosphere can lead to porosity in the cast products. Therefore, the melt must be degassed prior to casting. Systems for measuring the hydrogen content in molten Al are currently commercially available. In response to the need for lower cost hydrogen sensors, research has been performed on developing low-cost zirconia electrochemical sensors [5]. In the most common systems, an inert gas (usually nitrogen) flows through a probe and over the molten metal, such that chemical equilibrium between the hydrogen partial pressure in the gas and the concentration of hydrogen in the molten Al can be established. Zirconia-based sensors can also be used to improve control of the alloy composition, which can affect the resulting microstructure. This is particularly important in cases where a small amount of a reactive or volatile alloying element is added. In such cases, the alloying element may be preferentially lost, which can cause significant changes to a small initial concentration.

Research into new ionic conductors based on the stabilized zirconia for molten metal applications has reached a level where most studies on such materials concentrate mainly on obtaining incremental improvements in conductivity by better processing control and refinement of the microstructure. Further increases in the conductivity are important in terms of enhancing the efficiency of the sensors as valuable tools for improving the quality control in the processing of molten metals. However, much less attention has been given to the investigations of the hafnia- (HfO_2) and HfO_2-ZrO_2-based solid electrolytes. This could be explained by the fact that HfO_2 is considerably more expensive than ZrO_2, which makes it less attractive for practical applications.

However, in some molten metals where the temperature can reach 1200–1600°C, HfO_2-based or HfO_2-ZrO_2-Y_2O_3 sensors can be successful alternatives to the traditional zirconia-based sensors due to their high ionic and low electronic conductivities at such extreme temperatures [18, 19]. The limits of applicability of ZrO_2 electrolytes, especially at temperatures higher than 1100°C, are determined by partial electronic conductivity, which increases as oxygen partial pressure (P_{O_2}) decreases to the level of $10^{-23} - 10^{-28}$ Pa. Partial electronic conductivity permits an oxygen-ion

TABLE 4.1
Properties of Selected Pure Oxides

Property	ZrO_2	HfO_2	Y_2O_3
Structure	Monoclinic	Monoclinic	Cubic
Transition temperature (°C)			
Monoclinic $\leftrightarrow$ tetragonal	1205	1700	—
Tetragonal $\leftrightarrow$ cubic	2370	2700	—
Cubic $\leftrightarrow$ hexagonal	—	—	2350
Melting point (°C)	2677	2900	2420
Boiling point (°C)	4300	—	—
Density at 20°C (g/cm³)	6.1	9.68	5.03
Coefficient of the thermal expansion, α, at 1000°C ($\times$ 10^{-6} K^{-1})	7.01	5.3	8.3
Specific electrical conductivity, k, at 1000°C (($\Omega\bullet$cm)$^{-1}$)	10^{-2}	—	—
Thermal conductivity, λ, at 100°C (W $\bullet$ (m$\bullet$K)$^{-1}$)	1.67–2.09	—	14

Source: Data from Zhuiykov, S., 2000.

flux across the electrolyte and causes *emf* mismeasurement owing to the interfacial polarization [18]. ZrO_2 also may be chemically reduced at low Po_2 levels. One possible way to minimize this short-circuit electronic conductivity is to use fully stabilized zirconia [20]. However, the fully stabilized zirconia possesses a substantially lower level of ionic conductivity than partially stabilized zirconia. Thus, the development of alternative ion-conducting materials, such as HfO_2-ZrO_2-based electrolytes for thermodynamic control of oxygen impurity in high-melting metals and alloys in nonferrous metallurgy, for the semiconductor industry, and for the copper refineries, is required. Such application environments are characterized by both very high temperatures and extremely low oxygen potential. The physical, chemical, and thermomechanical properties of selected pure oxides are summarized in Table 4.1.

The presence of alumina in ZrO_2 electrolytes leads to higher resistance against thermal stresses [4, 21]. α-alumina exhibits the absence of chemical reactions in ZrO_2-Y_2O_3 systems, lower coefficient of thermal expansion, and higher heat conductivity than zirconia. Consequently, alumina could not only improve the thermomechanical properties of stabilized HfO_2-ZrO_2 composites but also play a decisive role in the heterojunction between the HfO_2-ZrO_2 electrolyte and an alumina that is also an insulator in practical designs of the oxygen sensors. A further complicating factor that has been observed in the traditional zirconia systems is the effect of high temperature (~1000°C) annealing on the deterioration of conductivity of the ZrO_2 electrolyte [22, 23]. This aging process could certainly be reduced by the addition of Y_2O_3. However, the electrolyte material, in this case, will suffer a consistent decrease in the overall conductivity with increasing Y_2O_3 content. Different HfO_2-ZrO_2 electrolyte systems have been investigated in an attempt to overcome this problem of aging with very promising results [19]. For samples of the HfO_2-ZrO_2-Y_2O_3-Al_2O_3 system, where Al_2O_3

FIGURE 4.1 Plug-type zirconia sensor design for different applications. (Reprinted from Zhuiykov, S., Investigation of conductivity, microstructure and stability of HfO_2-ZrO_2-Y_2O_3-Al_2O_3 electrolyte compositions for high-temperature oxygen measurement, *J. Europ. Ceram. Soc.* **20** (2000) 967–976, with permission from Elsevier Science.)

was 50 mol %, the presintered pellets of HfO_2-ZrO_2-Y_2O_3 were crushed in an agate mortar, a certain amount of Al_2O_3 was added and mixed, and then powders obtained were also rubber-pressed into pellets and sintered at 1700°C for 5 hours in air. The sintered pellets were cut into pieces of about 4 mm in diameter and less than 5 mm in length, and then were welded into Al_2O_3 insulating tubes, as shown in Figure 4.1. The plug-type sensor design is a robust form of construction and avoids any possible leaks through the junction of the solid electrolyte and insulating tube. The amount of Al_2O_3 was calculated to avoid significant anisotropy in the coefficients of thermal expansion of both solid electrolyte pellets and alumina tube. The sensors were checked for leakage at room temperature before they were sealed. Measurements were made using both an increasing and decreasing temperature cycle to ensure the results were consistent. A built-in R-type thermocouple was used.

4.1.2 PE′ PARAMETER MEASUREMENTS AND SENSING PROPERTIES

A known thermodynamic *emf* method was applied to determine the parameter *pe'*. This parameter *pe'* describes the relation between the partial ionic and *n*-type electronic conductivity of the electrolyte, and is defined as the Po_2 at which the ionic conductivity and the *n*-type electronic conductivity of the electrolyte are equal. Basically, it is accepted that this parameter must be consistent and known to a high degree of accuracy if precise and meaningful measurements are to be made. This method has already been described comprehensively in previous publications [8, 24]. When zirconia-based electrolytes are exposed to the high temperatures (T > 1100°C) and low oxygen partial pressures ($Po_2 < 10^{-20}$ Pa), usually encountered in metal melts, they exhibit mixed ionic and n-type electronic conductivities. Under these conditions, the solid electrolyte sensor generates emf that is influenced by the electrical properties of the solid electrolyte. Schmalzried [25] has analyzed the contribution of electronic conductivity in the zirconia electrolytes to the measured emf of an electrochemical cell in the Po_2 region less than 10^5 Pa and has shown that, in the presence of n-type electronic conductivity, the emf of the sensor can be expressed as

$$E = \frac{RT}{4F} \int_{Po_2(I)}^{Po_2(II)} t_{ion} d\left(\ln Po_2\right) = \frac{RT}{F} \ln \frac{pe'^{1/4} + Po_2(II)^{1/4}}{pe'^{1/4} + Po_2(I)^{1/4}}, \qquad (4.1)$$

where Po_2 (II) and Po_2 (I) are the respective oxygen partial pressures at two electrolyte-electrode interfaces, R is the gas constant, F is the Faraday constant, and T is the absolute temperature. To satisfy the condition Po_2 (II) $>> pe' > Po_2$ (I) [24], an Al melt contained in an alumina crucible was used to represent the extremely low oxygen partial pressure Po_2 (I). Aluminium was selected from the group of high-melting metals because it has the lowest oxygen potential at a temperature of 700°C and higher [19]. Several investigators [11, 24] have analyzed and emphasized further the importance of an accurate knowledge of the pe' value specifically for measurement of low oxygen partial pressure by the ZrO_2 sensors.

The emf measurements were made using sensors (Figure 4.1) consisting of the HfO_2-ZrO_2-Y_2O_3-Al_2O_3 electrolyte system and an inner Cr-Cr_2O_3 RE. Similar sensors based on the ZrO_2-Y_2O_3 electrolyte with 10 mol % of Y_2O_3 were used for comparison [19]. Both types of sensors were immersed into the melt at a temperature range of 1000–1400°C. Stable emf recordings were obtained after 7–9 minutes. These emf values were then used to calculate the pe' values by rearranging Equation (4.1) as follows:

$$pe' = \frac{\left\{\left[\exp\left(\frac{EF}{RT}\right)\right]Po_2(I)^{1/4}\right\} - Po_2(II)^{1/4}}{1 - \exp\left(\frac{EF}{RT}\right)}. \qquad (4.2)$$

Low values of the parameter pe' (ionic transference number, t_{ion}, is 0.5) indicate low contributions of partial electronic conductivity to the total conductivity, which result in higher ionic conductivities (k) according to the equation

$$k_{ion} = Kpe'^{-1/4}, \qquad (4.3)$$

where K is a constant and $Po_2 = pe'$ and $k\,e' = k_{ion}$ at $t_{ion} = 0.5$. By knowing pe', it is possible to calculate t_{ion} as a function of Po_2 using the expression

$$t_{ion} = \left[1 + \left(\frac{Po_2}{pe'} \right)^{-1/4} \right]^{-1}. \qquad (4.4)$$

Figure 4.2 shows the XRD trace of the sample of HfO_2-ZrO_2 electrolyte doped with 15 mol % Y_2O_3, from which the predominance of the cubic hafnia-zirconia phase was identified. This solid electrolyte system was selected owing to its lowest level of the pe' parameter at high temperatures, as has been reported [19]. The lattice parameter of the cubic phase was calculated (a = 5.138 Å), and it was in good agreement with those lattice parameters of the same solid electrolyte system which were reported by others (a = 5.137 Å) [11] and (a = 5.138 Å) [26]. When the intensity scale of the XRD peaks was enlarged, a shoulder was found at the peak of 2θ = 62.3° and a fine splitting of the peak at 2θ = 73.4° was also observed. It is suggested that the two peaks may correspond to the (004) and (400, 040) peaks for the tetragonal zirconia phase. However, the intensity ratio of these two peaks is a reversal of that for a single tetragonal phase. Therefore, this result appears to indicate a mixture of cubic hafnia and zirconia and tetragonal zirconia phases with the cubic form as the dominant one. A tetragonal zirconia cell was subsequently calculated, based on the (400) split peaks, giving the result of a = 5.132 Å and c = 5.140 Å. In addition to the hafnia and zirconia phases, a few peaks with very weak intensities were also observed in the XRD trace (e.g., d = 3.31 Å and d = 2.02 Å in Figure 4.2). Due to the low intensity of these peaks, they have not been positively identified.

Investigation of HfO_2-ZrO_2-Y_2O_3-Al_2O_3 solid electrolyte microstructure revealed that the radiuses of both Al_2O_3 and HfO_2-ZrO_2 grains in this structure vary from 5 to 15 μm. However, in average, the hafnia-zirconia grains are bigger than the alumina grains. Apparently, it indicates that this structure may be the most preferred for optimization of the conductivity and in terms of the microhardness values and fracture toughness of the composite systems. Investigations of the phase assemblage of the HfO_2-ZrO_2-Y_2O_3-Al_2O_3 solid solutions have shown the presence mainly of two phases, one of which is the phase with the cubic fluorite-type structure of the HfO_2-ZrO_2-Y_2O_3 composition.

The values of the parameter pe' were measured as a function of temperature for both HfO_2-ZrO_2-Y_2O_3-Al_2O_3 and HfO_2-ZrO_2-Y_2O_3. The thermodynamic *emf* method with plug-type sensors has been used in these measurements. The results of these measurements were shown in Figure 4.3. Additional data, reported by other investigations [18], were presented for comparison. As demonstrated earlier [19], a minimum parameter pe' value was observed at ~15 mol % Y_2O_3 in the entire temperature range, which was in good agreement with the previously published data for the high temperature range of 1200–1600°C [18]. Some discrepancies are obvious at lower temperatures due to a distortion of the microstructure caused by the sluggish reactions in ceramic systems. It is possible that the additional sintering effect occurred during the immersion of the sensors into the melt. Although the shapes of the curves

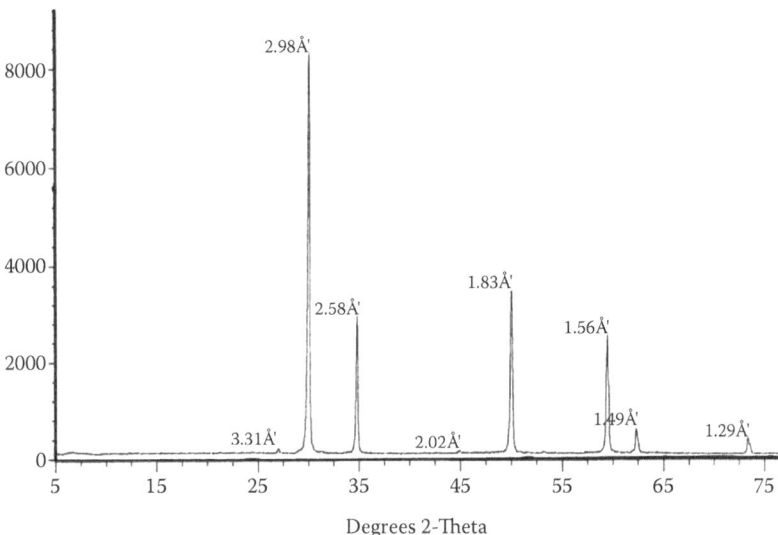

FIGURE 4.2 X-ray powder diffraction patterns of a HfO_2-ZrO_2-Y_2O_3 solid electrolyte system showing the presence of mainly a cubic phase. (Reprinted from Zhuiykov, S., Investigation of conductivity, microstructure and stability of HfO_2-ZrO_2-Y_2O_3-Al_2O_3 electrolyte compositions for high-temperature oxygen measurement, *J. Europ. Ceram. Soc.* **20** (2000) 967–976, with permission from Elsevier Science.)

for the HfO_2-ZrO_2-Y_2O_3-Al_2O_3 and HfO_2-ZrO_2-Y_2O_3 electrolytes were similar to the shapes of the curves for the ZrO_2-Y_2O_3 electrolyte, the measured values of the parameter *pe'* for the ZrO_2-Y_2O_3 system were considerably lower than for the HfO_2-ZrO_2-Y_2O_3. Therefore, the HfO_2-based electrolytes exhibit considerably lower parameter *pe'* (lower partial electronic conductivity) than the zirconia-based electrolytes. Generally, the ionic conductivity is governed by the diffusion of oxygen ions across oxygen-ion vacancies. Up to a certain doping content, the mobility of charge carriers is due to O^{2-} ion vacancies. However, the mobility of the vacancies will be diminished if the doping content increased. At high concentrations of the stabilized oxide, there is a large probability that a given anionic site becomes the neighbor of two or more dopant cations. These sites then act as deep traps for oxygen vacancies. Deep trapping would certainly contribute to the rapid decrease in conductivity beyond its maximum. Therefore, an additional increase of the Y_2O_3 content in the solid electrolyte system did not shift toward a lower partial electronic conductivity. It is also interesting to note that the maximum conductivity of the HfO_2-Y_2O_3 system was obtained at 10 mol % of Y_2O_3 [26]. In addition, the minimum *pe'* parameter for the HfO_2-ZrO_2-Y_2O_3 system was found at 15 mol % of Y_2O_3. It is therefore apparent that the optimum of Y_2O_3 content should be within the range of 10–15 mol % for practical sensors.

The investigated HfO_2-ZrO_2-Y_2O_3 electrolyte systems exhibited linear Arrhenius plots of the lattice conductivity as a function of temperature. The difference in ionic conductivities between HfO_2-ZrO_2-Y_2O_3 and ZrO_2-Y_2O_3 solid electrolytes decreases

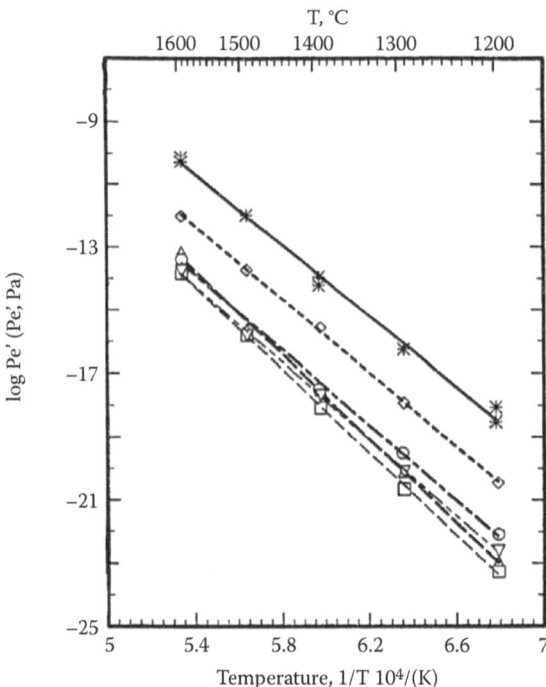

FIGURE 4.3 Plot of log Pe' versus $1/T$ functions for selected binary and ternary HfO_2- and ZrO_2- solid electrolytes in plug-type sensors: (*) ZrO_2-15 mol% CaO [18]; ($\Diamond$) ZrO_2-10 mol% Y_2O_3; (Δ) HfO_2-15 mol% Y_2O_3 [18]; ($\bigcirc$) $HfO_2-ZrO_2-Al_2O_3-15$ mol% Y_2O_3; and ($\square$) $HfO_2-ZrO_2 - 15$ mol% Y_2O_3. (Reprinted from Zhuiykov, S., Investigation of conductivity, microstructure and stability of $HfO_2-ZrO_2-Y_2O_3-Al_2O_3$ electrolyte compositions for high-temperature oxygen measurement, *J. Europ. Ceram. Soc.* **20** (2000) 967–976, with permission from Elsevier Science.)

as the temperature rises and at temperatures higher than 1300°C. The HfO_2-based solid solutions have shown higher ionic conductivities than the corresponding ZrO_2- based solid solutions [19]. The experimental evaluations of the parameter pe' for $HfO_2-ZrO_2-Y_2O_3$ exhibited that the HfO_2-based solid solutions possess considerably lower parameter pe' values (by one to two orders of magnitude) than ZrO_2-based solid solutions at the high temperature range of 1200–1600°C. Therefore, one of the advantages of the HfO_2- or HfO_2-ZrO_2-based electrolytes is seen for the sensor applications in the environments where the high temperature (> 1100°C) and extremely low oxygen partial pressures ($Po_2 = 10^{-12}–10^{-25}$ Pa) have been combined. These applications are as follows: thermodynamic control of oxygen impurity in high-melting metals, alloys, and fully killed steel melts, as well as in copper refineries and in carburazing industries. Under the above-mentioned working conditions, the HfO_2-based electrolytes have much less electronic conductivity than the ZrO_2-based electrolytes.

The results obtained [11, 18] also indicated that the sensors based on $HfO_2-ZrO_2-Y_2O_3$ electrolyte systems have a higher chemical resistivity and thermal shock stability than ordinary ZrO_2-based electrolytes. Consequently, they may be used as alternative solid electrolytes for sensors measuring oxygen partial pressure in fully killed steel melts. Application of the HfO_2-ZrO_2-Y_2O_3-Al_2O_3 solid solution for sensor purposes allows the development of alternative inexpensive sensor designs with lower consumption of oxide electrolyte materials. In this respect, progress could be made by the plug-type sensor structures combining an alumina insulating tube with the minimized size of the HfO_2-ZrO_2-Y_2O_3-Al_2O_3 pellet.

4.1.3 SINGLE-CRYSTAL ZIRCONIA SENSORS

Although polycrystalline zirconia sensors are the most commonly used *in-situ* sensors for oxygen measurement in high-temperature molten melts, the lowest working temperature of such sensors is restricted by impurities in stabilized zirconia which cause electronic conductivity. The electronic conductivity will consequently cause a loss of energy due to short-circuiting. Therefore, in most sensor applications, the electronic conductivity should be lower than the ionic conductivity by more than three orders of magnitude [8]. Despite the demand for accurate oxygen sensors with an operating temperature lower than 600°C, the electronic conductivity has a significant effect on the polycrystalline zirconia sensors at temperatures below 600°C. Several investigations [4, 27–31] provided detailed information on various aspects of decreasing the influence of electronic conductivity of polycrystalline zirconia at low temperatures. The previous analytical work [29] supplied a great deal of valuable information on different aspects of using the tetragonal phase of zirconia for low temperature sensors. Additionally, the surfaces of zirconia in these sensors have a hydrofluoric acid treatment before applying Pt electrodes. This treatment reduced polarization of the electrodes and allowed accurate oxygen measurement at the comparatively lower temperature of 400°C. Although the chemistry of the treatment is fairly simple, a detailed knowledge of total conductivity and stability of this zirconia phase was not available at the time this analysis was performed, drastically limiting the accuracy of the results obtained.

One of the factors which has influenced the total conductivity of polycrystalline zirconia sensors at low temperatures is the electrode material. In order to measure an oxygen concentration, the material of electrodes should be able to adsorb and dissociate a molecule of O_2. Pt is normally used at temperatures of 600°C and above, but other materials are of interest for lower temperature application. Several investigations of zirconia oxygen sensors have been carried out in which various metal–metal oxide combinations are utilized as electrodes $(U,Pr)O_{2+x}$ + Pt; $(U,Sc)O_{2+x}$ + Pt [28, 30]. These materials allowed the reduction of the lower temperature limit for combustion zirconia gas sensors down to 450°C. However, these electrode materials have some level of radioactivity that causes difficulties for wide industrial application. Although these investigations have expanded knowledge about electrochemical properties of the zirconia-oxide-SE interface considerably, a number of open questions exist concerning reduction of the working temperature of zirconia sensors. More recently, a zirconia single crystal [32] in combination with metal–metal oxide electrodes [33] has been investigated. It has been found that the chemical resistance of the zirconia single crystal is three to five times higher than that of the conventional polycrystalline zirconia electrolytes [34–36]. It has also been found that the zirconia sensor with liquid metal–metal oxide electrodes has much less electrode polarization at low temperatures than the polycrystalline zirconia sensor with porous electrodes based on Pt, Au, and their alloys. Therefore, a zirconia single crystal could be used as an alternative solid electrolyte for low-temperature oxygen measurements. The main difference between polycrystalline and single-crystal solid electrolytes is that the single crystal has no grain boundaries and, consequently, a minimum of segregated impurities. In this situation, the negative influence of defects at the grain boundaries on the electrophysical properties of the solid electrolyte is removed [37]. Consequently, the contribution of electronic conductivity relative to the total conductivity of the solid electrolyte could be decreased. Therefore, the sensors based on zirconia single crystals can be used for determination of oxygen potential in low-melting metals (T ~ 550°C) such as Zn, Pb, Sn, Ga, and their alloys.

A zirconia single-crystal oxygen probe and oxygen sensor elements are presented in Figure 4.4 and Figure 4.5. The two-phase electrodes containing $Bi+Bi_2O_3$ and $In+In_2O_3$ were prepared from an intimate mixture of the metal and its oxide in a 1:1 molar ratio. The X-ray pattern of the mixture after equilibration indicated no change in the phase combinations. YSZ single crystals, containing 10 mol % of yttria, as shown in Figure 4.6, were grown in the Institute of Physics and Power Engineering, Russia, by an inductive high-frequency (HF) melting technique [38]. This technique is based on the gradual melting mixture of initial ultra-high-purity components when the energy of the HF field is totally absorbed by the heating electrolyte. The equipment employed for producing zirconia single crystals includes all basic components of modern plants for growing high-melting crystals: a sealed working chamber, a HF generator (power 60 kW, frequency 5.28 MHz), a water-cooling system, an

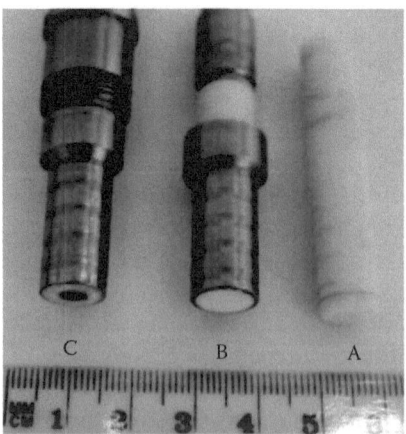

FIGURE 4.4 Oxygen sensor elements and oxygen probe based on a YSZ single crystal: *a*: assembly of YSZ into insulating tube; *b*: assembly of the insulating tube into the stainless steel sensor cover; and *c*: oxygen probe. (From Zhuiykov, S., Zirconia single crystal analyser for low-temperature measurements, *Proc. Control and Quality* **11** (1998) 23–37. With permission.)

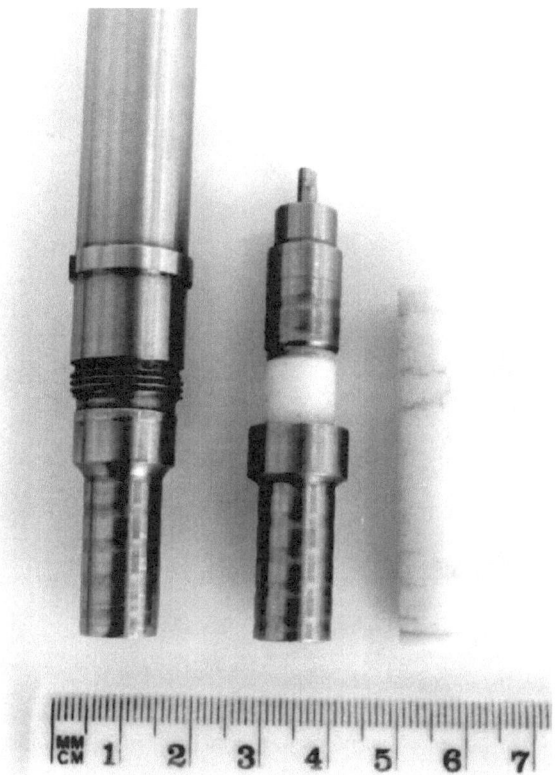

FIGURE 4.5 Sideway view of zirconia single-crystal sensor elements and oxygen probe.

automatic process control system, a power supply unit, a system for evacuating the working chamber and for filling it with an inert gas, and control and measuring equipment. YSZ single-crystal samples were oriented by the Laue X-ray back-reflection technique, cut, polished, and finished with diamond paste. Finally, single-crystal cylinders 4 mm in diameter and less than 5 mm in length were tightly assembled into a ceramic insulating tube (Figure 4.4, *a*) by isostatic pressing at 1700°C. Solid-state diffusion at the cylinder-tube interface was observed. The cylinder-tube design is a robust form of construction and avoids any possible leaks through the junction of the single crystal and insulating tube. Typically, the ceramic insulating tube is made of the mixture 58.8–69.2 mol % $MgAl_2O_4$ and 30–40 mol % MgO. The dimensions of this tube are 9 mm outer diameter and 50 mm long. Special grades of the stainless steel and Mo were selected for the sensor cover and the current conductor, respectively. The sensor cover was also tightly assembled on the outer surface of the insulating tube (Figure 4.4, *b*). All materials (single crystal, ceramic insulating tube, and metal cover) exhibited compatible thermal expansion coefficients. The value of these coefficients was carefully studied in a quartz dilatometer by heating samples in air. No significant anisotropy in coefficients of the thermal expansion was observed. $Bi-Bi_2O_3$ REs were placed inside the ceramic insulating tubes for contact with single crystals and were insulated carefully from ambient air.

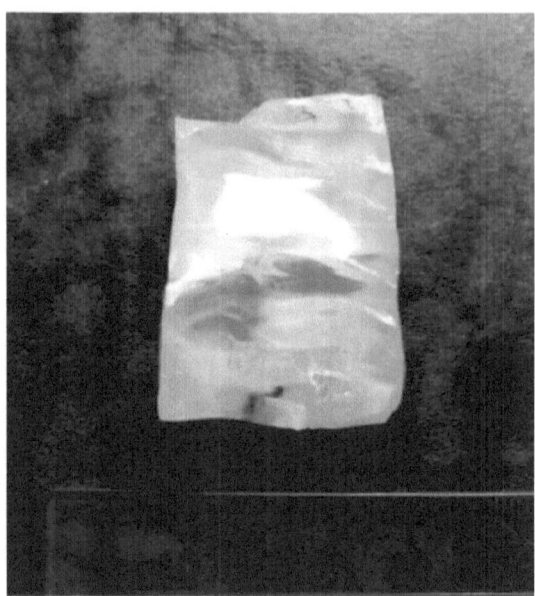

FIGURE 4.6 Sample of a zirconia single crystal made by an inductive high-frequency melting technique. (From Zhuiykov, S., Zirconia single crystal analyser for low-temperature measurements, *Proc. Control and Quality* **11** (1998) 23–37. With permission.)

The material of the current conductor wire must not react with or be soluble in the molten RE; otherwise, the oxygen potential of the RE will vary and its *emf* will be affected, impairing measurement accuracy and stability. If the current conductor wire will be alloyed with the RE and/or molten metal, a new *emf* is generated between the different alloys formed at the two ends of the current conductor wire, in addition to the *emf* generated by the oxygen potential difference. Thus, the correct *emf* generated by the oxygen potential cannot be determined. Since molybdenum neither reacts with nor dissolves in Bi-Bi$_2$O$_3$ REs, it was therefore used as a current conductor material. The YSZ-based sensors for control of oxygen activity in molten metals were mounted into high-temperature probes, as shown in Figure 4.4, *c*. The open-circuit *emf* of these oxygen sensors,

$$Mo/Bi\text{-}Bi_2O_3//(ZrO_2)_{0.9} - (Y_2O_3)_{0.1}//[O]\ Mo, \qquad (4.5)$$

where O is oxygen dissolved in the melt and Mo is molybdenum a current conductor, was measured for a temperature range of 280–1000°C. The liquid-metal RE made of bismuth saturated by oxygen has a stable oxidation potential and possesses the minimum electric resistance at the boundary of the melt and YSZ. Sensors were checked for leakage at room temperature before they were sealed. The temperature of the sensors was measured to ±0.5°C by a built-in K-type thermocouple placed adjacent to the RE. The *emf* was measured with a high-input impedance (>10 MΩ) digital multimeter 179 TRMS accurate to ±0.1 mV.

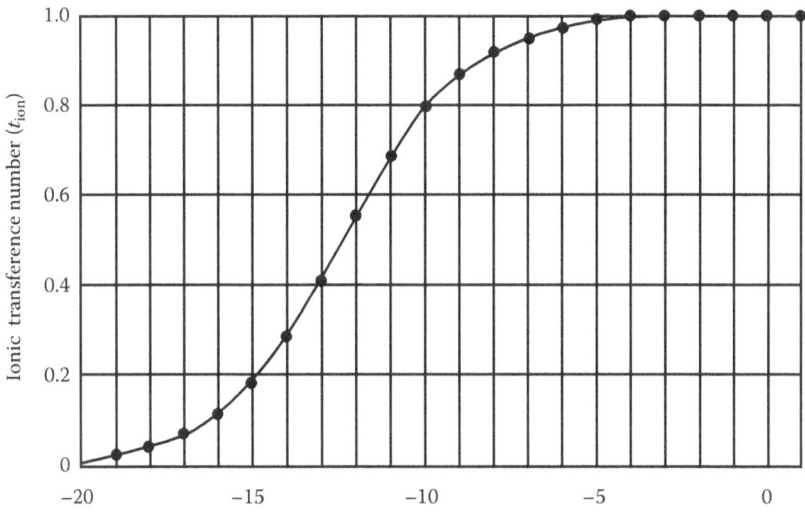

Oxygen concentration log O_2 (O_2, *ppm*)

FIGURE 4.7 Ionic transference number (t_{ion}) as a function of oxygen concentration for the sensor based on a zirconia single crystal at 400°C. (From Zhuiykov, S., Zirconia single crystal analyser for low-temperature measurements, *Proc. Control and Quality* **11** (1998) 23–37. With permission.)

Figure 4.7 illustrates the ionic transference number (t_{ion}) as a function of oxygen concentration in molten Ga in the range 10^{-19}–100 ppm for a sensor, shown in Figure 4.4, at a temperature of 400°C. A measuring electrode was not required for determining the oxygen potentials in molten gallium because molten metal is the ideal electrode for such measurements. These results appear to indicate that the sensors based on a zirconia single crystal are preferable for use at low temperatures and low oxygen potentials. Sensors based on polycrystalline zirconia cannot work precisely at temperatures lower than 450°C due to a high electronic conductivity level. At lower temperatures, some discrepancies are obvious for polycrystalline zirconia because of a distortion of the microstructure caused by sluggish reactions in ceramic systems. This is owing to the impurity phase segregation, dislocations, presence of secondary inclusions, lattice mismatching, and imperfect contact between the zirconia grains in the polycrystalline ceramics. Polycrystalline ZrO_2 is therefore liable to be affected by grain boundary effects and a limited amount of porosity. These features of the polycrystalline zirconia also have a detrimental effect on the ion-transport properties of the solid electrolyte at low temperatures.

Table 4.2 shows the concentration of chemical impurities in a zirconia single crystal [8]. The concentration of impurities shown here is low and therefore is unlikely to have a significant influence on the ionic conductivity of the zirconia single crystal at low temperatures. The level of impurities concentration shown in Table 4.2 indirectly confirmed that the partial electronic conductivity of the single-crystal zirconia is negligible.

TABLE 4.2
Concentration of Chemical Elements as Impurities in Zirconia Single Crystal

Impurity Element	Concentration (%)	Impurity Element	Concentration (%)
Si	1.3×10^{-3}	Mg, Mn	1×10^{-4}
W	1.0×10^{-3}	Sb	3×10^{-4}
Ca	2×10^{-3}	Cr	6×10^{-4}
Fe, Mo, Ba, Bi	3×10^{-3}	Sn	3×10^{-4}
Al, Ti, Nb, Pb	1×10^{-4}	Ni	4.3×10^{-4}
Co	6×10^{-4}	Cu, Zn	2×10^{-4}

Source: Data from Zhuiykov, S., 2000.

In order to examine how impurities affect the resistance of both single-crystal and polycrystalline zirconia, samples were prepared and resistance was measured. Samples of polycrystalline zirconia and the zirconia single crystal with the same concentration of Y_2O_3 were used at temperatures of 400–800°C. The results of testing are shown in Figure 4.8. Based on the fact that the higher the sintering conditions of polycrystalline zirconia, the less the resistance and porosity of solid electrolyte [8], polycrystalline zirconia samples were sintered for 6 hours at a temperature of 1700°C. The curves are approximately parallel, that is, they have similar activation energies. Furthermore, the impurities in the polycrystalline zirconia increased the resistance of the ceramic insignificantly at temperatures from 500°C to 800°C.

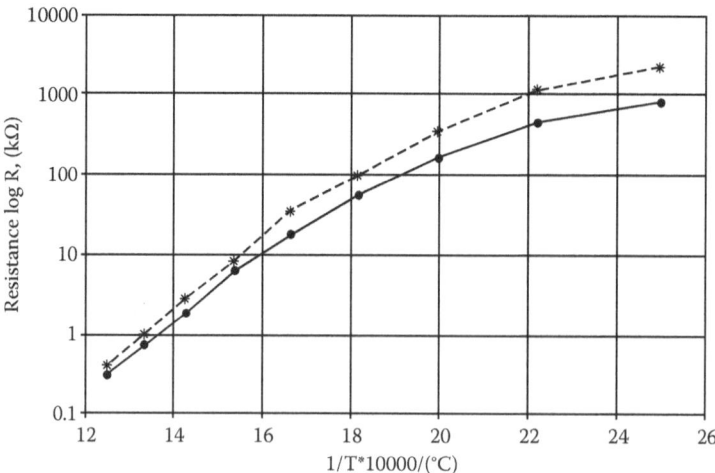

FIGURE 4.8 Temperature dependence of the logarithm of resistance of the zirconia: (*) polycrystalline ceramic; and (●) single crystal. (From Zhuiykov, S., Zirconia single crystal analyser for low-temperature measurements, *Proc. Control and Quality* **11** (1998) 23–37. With permission.)

However, the difference in resistance of the single-crystal and polycrystalline zirconia at temperatures lower than 500°C is still significant.

The oxygen-sensing properties of the sensors based on the zirconia single crystal with an $In+In_2O_3$ and $Bi+Bi_2O_3$ RE were also reported [8]. The *emf* between the SE and RE in these oxygen sensors followed the Nernst Equation (3.3). For the $In+In_2O_3$ RE, the following equilibrium exists:

$$2\ In + 3/2\ O_2 \leftrightarrow In_2O_3. \tag{4.6}$$

The reference oxygen pressure P_{O2} (II) of the $In+In_2O_3$ mixture is fixed by equation [39]:

$$ln\ P_{O2}\ (II) = \Delta G_{In2O3}/RT, \tag{4.7}$$

where ΔG_{In2O3} is the standard Gibbs energy of the In_2O_3 formation [39, 40]. The same equation can be written for the Bi_2O_3 RE. The temperature dependencies of the thermodynamic oxygen potential for the $In+In_2O_3$ and $Bi+Bi_2O_3$ REs are defined by the following equations [8]:

$$\Delta G_{In2O3} = -\ 620,535 + 225.29T, \tag{4.8}$$

$$\Delta G_{Bi2O3} = -\ 585,420 + 289.9T. \tag{4.9}$$

The modified thermodynamic *emf* for the zirconia oxygen sensors with $In+In_2O_3$ and $Bi+Bi_2O_3$ REs can be calculated from Equations (3.3) and (4.7):

$$E = (\Delta G_{In2O3} - RT\ ln\ P_{O2}\ (I))/4F, \tag{4.10}$$

$$E = (\Delta G_{Bi2O3} - RT\ ln\ P_{O2}\ (I))/4F. \tag{4.11}$$

Figure 4.9 shows the reversible *emf* responses as a temperature function for the above sensors in ambient air (P_{O2} (I) = 20.9%). The microporous thin-film Pt-ZrO_2-Y_2O_3 SE with a large inner surface allows a higher oxygen reaction rate and thus enhances oxygen ionic conductivity at relatively low temperatures. The measured *emf* values were within ±1 mV of the theoretical values given by Equations (4.10) and (4.11). The measurements were made in the temperature range of 300–700°C for the $In+In_2O_3$ RE and 360–1100°C for the $Bi+Bi_2O_3$ RE. Highly dispersed Pt particles in the thin-film Pt-ZrO_2-Y_2O_3 SE have enhanced the oxygen exchange rate at the electrode-zirconia interface by influencing the oxygen adsorption equilibrium, promoting the overall charge transfer at the TPB:

$$1/_2\ O_2\ (gas) + 2e^-\ (electrode) \leftrightarrow O^{2-}\ (electrolyte). \tag{4.12}$$

The test results, presented in Figure 4.9, show that using liquid $In+In_2O_3$ and $Bi+Bi_2O_3$ REs and a thin-film Pt-ZrO_2-Y_2O_3 SE support the view that the zirconia single crystal behaves as a superior ionic conductor even at a temperature as low as 360°C. Furthermore, additional tests of the zirconia single-crystal sensor with the

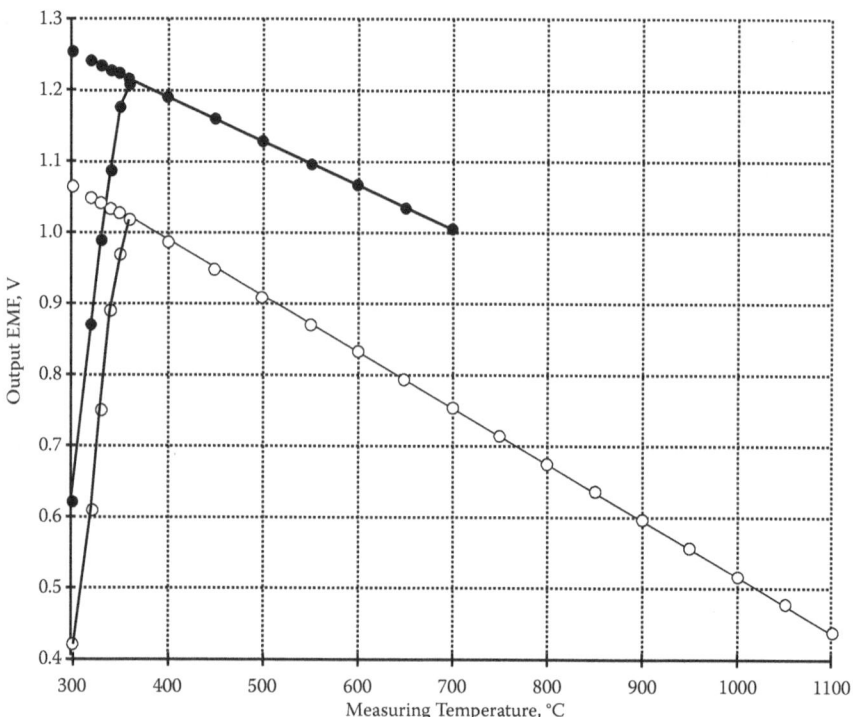

FIGURE 4.9 Temperature dependence of *emf* for zirconia single-crystal sensors with a molten metal–metal oxide RE. Measuring gas is air. (●) $In+In_2O_3$ RE; and (○) $Bi-Bi_2O_3$ RE. Symbols with full line represent data taken during heating and cooling. (From Zhuiykov, S., Zirconia single crystal analyser for low-temperature measurements, *Proc. Control and Quality* **11** (1998) 23–37. With permission.)

$(U,Pr)O_{2+x} + Pt$; $(U,Sc)O_{2+x} + Pt$ SE revealed that this sensor exhibits the Nernstian behavior only at temperatures higher than 480°C. Therefore, the microporous Pt-ZrO_2-Y_2O_3 SE appears more effective than the porous Pt in promoting oxygen transfer at low temperatures. A Pt-ZrO_2-Y_2O_3 SE is also affected less by higher temperature exposure. The influence of changes in oxygen partial pressure and of temperature on electrode resistance is consistent with a mechanism of oxygen exchange involving the uptake of oxygen at the gas-electrode interface, its diffusion through the crystal lattice of the electrode material to the TPB, and its transfer to the electrolyte [28].

The *emf* of the zirconia single-crystal sensor with the $Bi+Bi_2O_3$ RE as a function of measuring temperature at the different oxygen concentrations is shown in Figure 4.10. Several important features can be observed from this figure. First of all, there was a strong correlation between Nernstian and measured *emf* in both high-temperature and low-temperature regions. This implies that the zirconia single-crystal sensor can measure O_2 concentration from 0.1 ppm up to 20,000 ppm not only at low temperatures but also at high temperatures. Secondly, the impurity segregation would not obviously change the surface exchange processes on the electrolyte-liquid

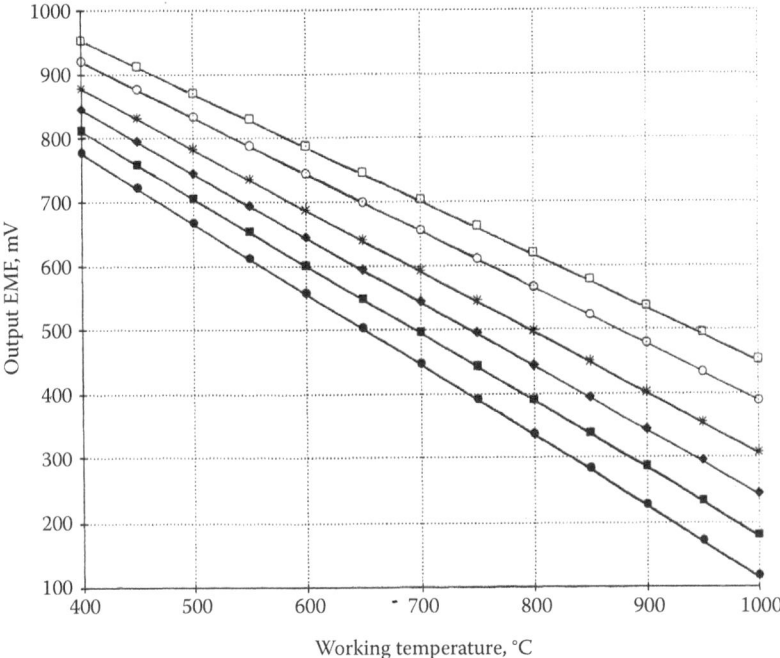

FIGURE 4.10 Output *emf* for the sensor based on a zirconia single crystal with a Bi-Bi$_2$O$_3$ RE at different O$_2$ concentrations: (●) 0.1 ppm; (■) 1 ppm; (♦) 10 ppm; (*) 100 ppm; (○) 2000 ppm; and (□) 20,000 ppm. (From Zhuiykov, S., Zirconia single crystal analyser for low-temperature measurements, *Proc. Control and Quality* **11** (1998) 23–37. With permission.)

RE interface. Consequently, the accuracy of the oxygen measurements at low and high temperatures differs insignificantly.

Further investigation of the zirconia single-crystal oxygen sensor revealed that 90% of the response and recovery time at 400°C was fast for this temperature and was about 15–20 seconds. Complete recovery takes a little more time and basically depends upon the SE properties. However, such a delay does not present a serious problem, even at low measuring temperatures. Measurements were repeated for other O$_2$ concentrations and temperatures. The reproducibility of the sensor response time was reasonably good. However, the changes in the gas flow rate substantially affected the sensor's behavior at temperatures lower than 400°C. Then the gas flow rate was higher than 110 cm^3/min, and the temperature of the RE was higher than the temperature of the thin-film SE, resulting in temperature gradient, which is responsible for inaccuracy in oxygen measurement [40]. This problem is not apparent at temperatures higher than 550°C. The response time for the sensor at temperatures above 700°C is limited only by the inherent speed of the apparatus. True response times are probably shorter than those shown. It has been observed that the response time for the zirconia single-crystal oxygen sensor at high temperatures is 120 milliseconds or even less over its normal operating temperature range [8], whereas the

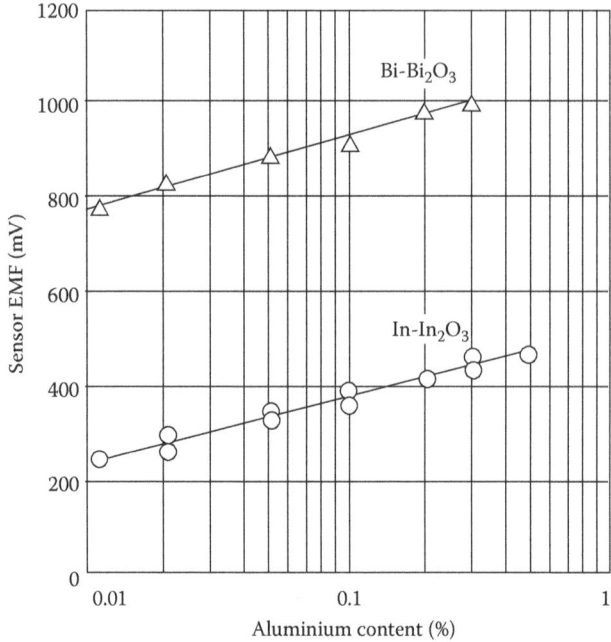

FIGURE 4.11 Measurement of dissolved Al in a zinc plating bath by a zirconia single-crystal sensor with $In+In_2O_3$ and $Bi+Bi_2O_3$ REs. (From Zhuiykov, S., Zirconia single crystal analyser for low-temperature measurements, *Proc. Control and Quality* **11** (1998) 23–37. With permission.)

response time values have exceeded one minute as a typical response of the sensor below 350°C [4, 41].

It is also essential to know the cross-sensitivity of the zirconia single-crystal sensors to other gases. Sensors with porous Pt electrodes are known to be sensitive to gases such as CO at low temperatures [41], and in fact, this cross-sensitivity has been proposed as a principle for carbon monoxide sensors at low temperatures by some researchers [42, 43]. This effect is attributed to the ability of CO to compete successfully with oxygen for adsorption sites on Pt at temperatures from 500°C to 650°C. It was observed that the zirconia single-crystal sensor with thin-film Pt-ZrO_2-Y_2O_3 electrodes is less sensitive to CO than similar polycrystalline sensors with porous Pt electrodes, but small *emf* errors still occur at 300–360°C.

The zirconia single-crystal sensor was also tested for continuous measurement of the oxygen content in different molten metals. For example, Figure 4.11 illustrates the measurement of dissolved Al in zinc, by immersion of the zirconia single-crystal sensors with $In+In_2O_3$ and $Bi+Bi_2O_3$ REs in a zinc plating bath. Consistent measurements have been obtained over a period of 3 months at the temperature range of 450–470°C [8]. It has also been considered that in the case of the polycrystalline zirconia sensors, the accurate *emf* measurement itself is very difficult and, sometimes, not even possible at temperatures of 450–500°C. However, in the case of the zirconia single-crystal sensor with $In+In_2O_3$ and $Bi+Bi_2O_3$ REs, the oxygen potential in low-melting metals can be measured successfully if the melting point of the RE is lower than the melting point of the low-melting metal. Calibration of these oxygen sensors is possible in laboratory tests by using the *emf* measurements in high-purity metal melts with sampling and subsequent oxygen determination by vacuum fusion analysis.

The single-crystal zirconia sensors with $Bi+Bi_2O_3$ REs were also tested in molten metal environments at high temperatures. Sensors for this purpose were immersed in an Al melt contained in the alumina crucible, and the oxygen concentration was measured at temperatures of 700–950°C. Stable output *emfs* were obtained after 8–10 minutes. The zirconia single-crystal sensor with the $Bi+Bi_2O_3$ RE has shown that it can measure traces of oxygen from 0.1 ppm to 100 ppm in the above-mentioned temperature range with a relative measurement error of 10%.

One of the main advantages of zirconia single-crystal oxygen sensors is their lifetime. No phase transition was observed in the single crystal during a 4-month trial at a temperature of 950°C as the oxygen concentration was changed from 250 ppm to 500,000 ppm [8]. The other sensor components have shown no observable chemical or mechanical degradation after such long-term tests at a high temperature. The operating lifetime of zirconia single-crystal sensors at high temperatures is generally up to 3 years, compared with lifetimes from a few weeks to a few months of similar sensors based on polycrystalline zirconia.

Thus, based on the latest research results, the zirconia single-crystal sensor appears to have a high level of reliability and long-term chemical stability and can be used not only for thermodynamic control of oxygen impurity in low-melting metals and alloys, but also for oxygen measurement in high-melting metals and their alloys. Due to its high sensitivity, this sensor can be recommended for applications in nonferrous metallurgy and the semiconductor industry (e.g., when producing the following metals and their alloys: Na, K, Rb, Cs, Ga, Cd, Sn, Pb, Zn, Cu, Bi, and others). It can also be adopted for processing low-melting metals and alloys.

The zirconia single-crystal sensor is used both for periodic variations via short-time dipping into the melts, and for continuous measurement. It is vibration impact proof, needs no precalibration, and can be used for certifying other oxygen impurity control instruments. Small cylinders of the zirconia single crystal, mounted into a ceramic insulating tube, can work at allowable thermal shocks of up to 10°C per second [44]. None of the well-known oxygen sensors based on polycrystalline zirconia can survive when the measuring temperature is high and changes so frequently.

Three important features of these tests are worth reporting, as they have practical implications for the commercialization of zirconia gas sensors. First, it was observed that decreasing the oxidizing potential of the RE by means of the liquid metal–metal oxide electrodes reduces polarization effects at the electrolyte-RE interface considerably. This indicates that there is a chance of a substantial shift of the threshold temperature of operation toward a lower temperature where a meaningful *emf* can be obtained. Second, excellent agreement was observed between the Nernstian and measured *emf*, at any temperature measuring within the range of 360–1000°C. This agreement was found to be consistent over a large number of experiments. Third, the signals obtained from the sensors, working either in gaseous or liquid-metal environments, were constant and were reproducible from cycle to cycle over a large number of tests.

The conclusions concerning the proper design and performance of zirconia single-crystal sensors can be summarized as follows [8]:

1. The area of the electrolyte, which is used to separate compartments with different partial oxygen pressures, should be minimized.
2. The oxygen potential gradient, and thereby the sensor potential, must be minimized by choosing a suitable RE material.
3. The electrolyte size must be decreased.

The latest results, obtained by different research groups around the world in relation to the development of zirconia single-crystal oxygen sensors, have shown that these sensors may be applicable to more complicated applications of oxygen sensors, such as the measurement of oxygen activity in liquid sodium or lithium heat carriers in liquid-metal nuclear facilities [45] as well as in liquid lead-bismuth eutectic alloys, which have been employed as heat carriers in liquid-metal nuclear facilities for submarines [44]. The single-crystal zirconia sensor is only one sensor which can reliably work at the presence of low γ–radiation. Although the analysis of results published in scientific media provided an adequate description of the electrochemical properties of the zirconia single crystal as a superior ionic conductor at low temperatures, more work is necessary to optimize the sensor characteristics which are usually dependent on the sensor design.

A commercial oxygen probe based on zirconia single crystals allows measurement to be taken both in a gaseous environment and in low-melting and high-melting metals and their alloys, some of which were not previously being considered for *in-situ* monitoring. So far, there appear to be no other commercial in-situ oxygen probes with a long lifetime for these demanding applications.

4.1.4 ZIRCONIA SENSORS BASED ON SHAPED EUTECTIC COMPOSITES

Zirconia composites with a naturally assembled structure can be formed by solidification of eutectic melts and the phase interspacing reduced by decreasing the time available for the diffusion of species in the transition from the compositionally homogeneous liquid to the different solid phases [46]. Due to the rapid growth during quenching, the composites have thinner interphase space [47]. The typical lamella sizes also correspond to the size of inhomogeneities and defects not intrinsic to the eutectic pattern, which are inevitably formed during processing and critically control the mechanical properties of the composites. Therefore, there is a strong interest not only in determining the effects of rapid solidification routes in both the eutectic microstructure and the mechanical properties, but also in the ability to use eutectic composites as solid electrolytes for gas sensors. However, the difficulty of obtaining bulk ceramic composites with homogeneous structure increases with solidification velocity. The fabrication of binary eutectics by rapid solidification techniques yielded moderate success in the past, producing partially amorphous or inhomogeneous samples because it is difficult to maintain a homogeneous heat transfer at high cooling rates [48–50]. Therefore, rapid solidification methods were used to fabricate eutectic powders, but not layered eutectic crystals. More recently, Al_2O_3-ZrO_2-Y_2O_3 eutectics with submicron phase spacing have been obtained as fibers (with radiuses of the ZrO_2-Y_2O_3 fibers in the structure from 1 to 2 μm).

The growth of Al_2O_3-$ZrO_2(Y_2O_3)$ eutectic composites by the Stepanov technique form from pure powders in the desired proportions (52.9 wt % Al_2O_3, 42.7 wt % ZrO_2 and 4.4 wt % Y_2O_3) and preliminary data on the mechanical properties and crystal structure have been reported [47]. The chemical resistance of zirconia eutectic composites is three times higher than that of the conventional polycrystalline zirconia electrolytes [47]. In these eutectic composites, the negative influence of impurity segregation on the grain boundaries of zirconia is minimized. This is due to the single-crystal nature of the zirconia fibers in the Al_2O_3-$ZrO_2(Y_2O_3)$ composites. Consequently, the contribution of electronic conductivity to the total conductivity of the solid electrolyte could be decreased. The zirconia single-crystal sensors are still relatively expensive to make, and therefore, alternative oxygen sensors based on Al_2O_3-ZrO_2-Y_2O_3-shaped cylindrical eutectic composites will be more valuable due to less expensive manufacturing technology. There is no doubt that these inexpensive superior ionic conductors for low-temperature oxygen sensors with good ceramic properties will be in great demand. Therefore, oxide-oxide solid electrolyte eutectic composites were considered as potential materials for low-temperature oxygen sensors.

Depending on the die's design, the Al_2O_3-ZrO_2-Y_2O_3 eutectic composites can be grown by the Stepanov technique in various shapes, as shown in Figure 4.12, by using a resistance-heated furnace (~2000°C, vacuum ~10^{-5} torr) equipped with a graphite heater and, after heaters, using an argon atmosphere at pressures of 1.1–1.5

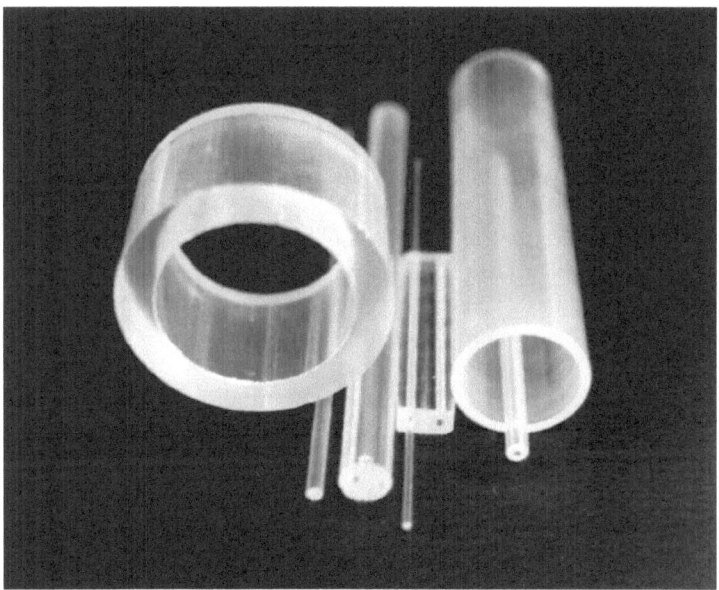

FIGURE 4.12 Different design of Al_2O_3-ZrO_2-Y_2O_3-shaped eutectic composites.

atm. Molybdenum crucibles and sapphire bars as seed crystals should be used. The growth rates are usually within 20–100 mm/h. ROSTOX-N Ltd., Russia, is the manufacturer of these eutectic composites.

After cutting, polishing, and finishing pellets of the eutectic composites with diamond paste, they were assembled to be leak-tight in an alumina insulating tube. A technique developed in the CSIRO Division of Manufacturing and Materials Technology (CMMT), Australia, was employed for this purpose [51]. This technique involves joining a solid electrolyte pellet or cylinder to an alumina tube of the required length and diameter by a high-temperature eutectic welding operation. Thus, the sensors prepared are rugged in construction, have a low leak rate, and are suitable for most industrial and laboratory applications. The cylinder-tube design of the oxygen sensor is a robust form of construction and avoids any possible leaks through the junction of the eutectic composite and insulating tube. Both the SE and RE were made by applying Pt paste on the composite. However, a liquid metal–metal oxide RE is also possible for use in these sensors.

Subsequently, the microstructure and surface topography of the Al_2O_3-ZrO_2-Y_2O_3 eutectic composites investigation was performed by using a field emission scanning electron microscope JEOL JSM-6440F fitted with both a digital-imaging system for electron microscopy and an energy dispersive X-ray detector VOYAGER at Kyushu University, Japan. Specimens of the eutectic crystals were polished using Metadi II diamond-polishing compound (made in the United States by Buehler Ltd.) and were coated with a 30 nm coating of carbon. The chemical composition of the composites was examined using X-ray photoelectron spectroscopy (XPS) on a VG Microlab 310F at RMIT University, Australia. An Al anode unmonochromated X-ray source operated at a power of 300 W and 15 kV excitation voltage was used. The energy of the Al $K\alpha$ line was taken to be 1486.6 eV. The sample was tilted such that the escape electrons were collected by an electron analyzer normal to the sample surface [52]. The circular area of the sample from which escaping electrons were detected was approximately 2 mm in diameter. All spectra were collected in constant analyzer energy (CAE) mode at a pass energy of 20 eV in 0.5 eV steps between data points. The spectrometer was calibrated with a sputtered copper (99.999% pure) sample, and deposited gold on the sample gave Cu $2p_{3/2}$, Cu $_{KLL}$, and Au $4f_{7/2}$ binding energies of 932.60 eV, 334.8 eV (KE), and 84.06 eV, respectively [53]. The base pressure of the analysis chamber was 6×10^{-11} torr before specimens were introduced. The binding energies were calibrated with reference to C 1s at 285.0 eV for hydrocarbon contamination.

The oxygen-sensing properties of sensors based on the Al_2O_3-ZrO_2-Y_2O_3 eutectic composites and polycrystalline zirconia sensors with the same Y_2O_3 concentration in the electrolyte were also investigated [53]. All tests were carried out using nitrogen as a carrier gas. In all measurements, the airflow rate was ~100 cm^3/min. The oxygen-sensing properties of sensors based on the polycrystalline Al_2O_3-ZrO_2-Y_2O_3 solid electrolyte were shown for comparison.

It has been established that the structure of the shaped Al_2O_3-$ZrO_2(Y_2O_3)$ eutectic composite, as well as the structure of composites obtained by other methods of directional solidification in this system, consists of two phases: the matrix, which is alumina, and stabilized zirconia [47]. In this Al_2O_3-ZrO_2 system, alumina is the first phase obtained at growth rates of 20 mm/h and higher. This fact indicates that with increasing pulling rates, when a condition of supercooling is reached, the alumina matrix becomes faceted at the cellular growth front with parts protruding into the melt.

It was reported that Al_2O_3-ZrO_2-Y_2O_3 eutectic composites have a colony microstructure [53]. They showed uniform grain-size distribution of both zirconia and alumina fibers surrounded by some enormous particles of both alumina and zirconia, around 5–10 μm in the submicrometer scale. All specimens have shown zero porosity, which is vital for gas sensor applications. Attainment of a fine, regular structure over the bulk of the crystal and the determination of the growth conditions for which this occurs is one of the main problems in the directional crystallization of eutectic oxide-oxide composites. It is usually assumed that a planar crystallization front is required for the formation of a highly oriented structure in a eutectic system. However, the Al_2O_3-ZrO_2-Y_2O_3 composites displayed the colony structure, and the surface of the crystallization front was cellular. An increase in the growth rate may cause either the faceting of cells or the appearance of dendrites.

Figure 4.13, a, shows the typical XPS survey spectra for the Al_2O_3-ZrO_2-Y_2O_3 composite. It confirms the formation of both the stabilized zirconia with the associated binding energy of the Zr 3d located at 184.16 eV and alumina with the associated binding energy of the Al 2p located at 81.40 eV. Photoelectron peaks for Y 3d at 158.67 eV and O 1s at 531.98 eV were also clearly recorded for all specimens. The peaks for Na 1s at 1073.28 eV and for carbon C 1s at 285 eV were detected due to the contamination of the surface of specimens by organic components at the finishing stage of the manufacturing process. However, after cleaning the surface of the Al_2O_3-$ZrO_2(Y_2O_3)$ eutectic composite with neutral gas, the carbon peak disappeared from the survey spectra (Figure 4.13, b). Table 4.3 summarizes the chemical composition of samples calculated from the experimental data in Figure 4.13. The calculation is based on the integrated area under the assigned element peak and the sensitivity factor for the element [54]. The peak at 531.98 eV is characteristic of metallic oxides and is in agreement with O 1s electron binding energy for ZrO_2 and Al_2O_3. From the calculated atomic percentages of Al and Zr, it can be seen that the implied oxygen content of Al_2O_3 and ZrO_2 is approximately correct for all samples.

The microstructure of the Al_2O_3-ZrO_2-Y_2O_3 eutectic composite is shown in Figure 4.14 outlining the boundary between eutectic colonies [46]. The dark areas are the Al_2O_3 phase, and the white areas are the ZrO_2-Y_2O_3 phase. The radiuses of both Al_2O_3 and ZrO_2-Y_2O_3 grains in this structure vary from 5 to 15 μm. However, in average, the zirconia grains are bigger than the Al_2O_3 grains. This appears to indicate that this structure may be the most preferable for optimization of the conductivity and in terms of the microhardness values and fracture toughness of the composite systems. Further investigation of the phase assemblage of the Al_2O_3-$ZrO_2(Y_2O_3)$ eutectic composite confirmed the presence mainly of two phases, one of which is the phase with the cubic fluorite-type structure of ZrO_2-Y_2O_3, and the other phase is alumina [53].

Sensing properties of the sensors based on Al_2O_3-ZrO_2-Y_2O_3 eutectic composites were investigated in the temperature range of 450–600°C to find the lowest possible temperature at which the output emf of a sensor follows the Nernst equation. Dry

a)

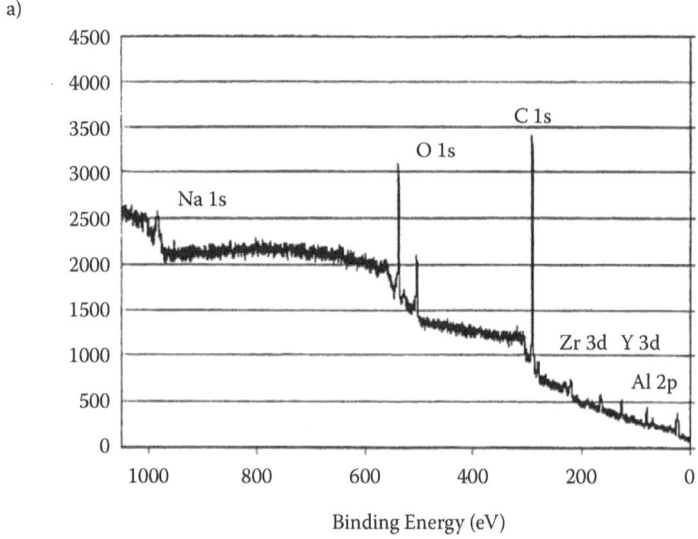

b)

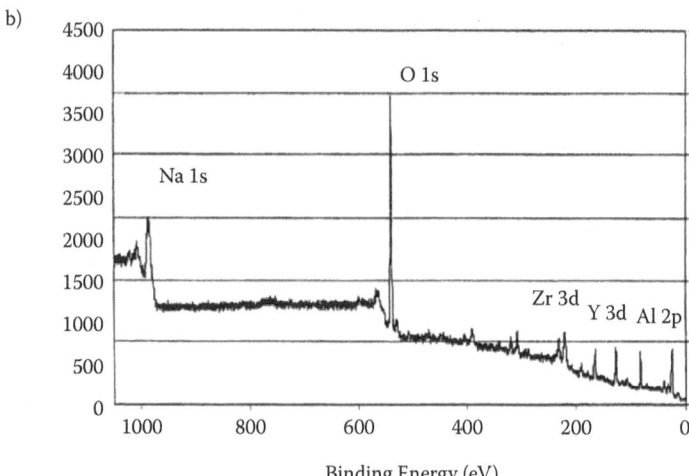

FIGURE 4.13 Typical XPS survey spectra for the Al_2O_3-ZrO_2-Y_2O_3-shaped eutectic composites.

air was used as a reference gas. A N_2 + O_2 mixture with 100 ppm of oxygen was used as a measuring gas. The airflow rate on the SE was equal to the airflow rate on the RE and was ~100 cm^3/min to avoid electrode polarization. The steady-state *emfs* were obtained only after a few minutes. Each *emf* measurement was reversible and was made after a change of temperature. Figure 4.15 illustrates the temperature dependence of the ionic transference number ($\bar{t}_{ion} = emf$ exp. / *emf* theory) calculated for the sensor based on the Al_2O_3-ZrO_2-Y_2O_3 eutectic composite. The $\bar{t}_{ion}$ numbers at the same temperatures for the Al_2O_3-ZrO_2-Y_2O_3 polycrystalline sensor with Pt

TABLE 4.3
Chemical Composition of the Al_2O_3-ZrO_2-Y_2O_3 Eutectic Composite Surface

Peak	Center (eV)	SF	Peak Area	Tx. Function	Norm Area	[AC] %
O 1s	531.98	2.93	65045.837	0.1	2537.88621	63.999
Zr 3d	184.16	7.04	14759.895	0.1	193.28330	15.986
Y 3d	158.67	5.98	7907.681	0.1	144.02052	4.018
Al 2p	81.40	0.54	5155.539	0.1	189.56924	15.997

Source: Data from Zhuiykov, S., 2000.

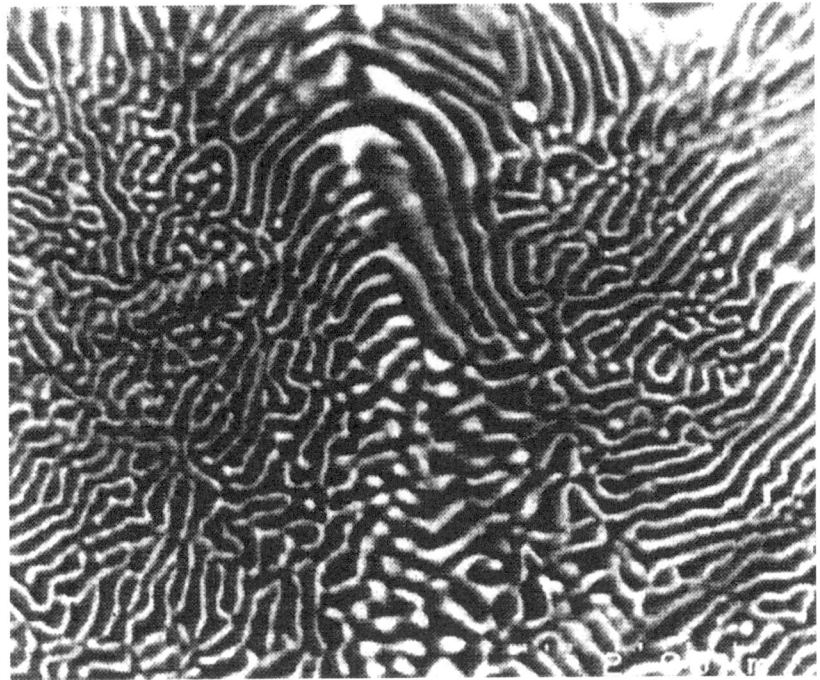

FIGURE 4.14 Typical SEM image of boundary between eutectic colonies. (Reprinted from Calderon-Moreno, J.M. and Yoshimura, M., Stabilization of zirconia lamellae in rapidly solidified alumina-zirconia eutectic composites, *J. Europ. Ceram. Soc.* **25** (2005) 1369–1372, with permission from Elsevier Science.)

electrodes are presented for comparison. It is clear from this figure that Pt, as an electrode material, cannot itself adequately promote the oxygen dissociation on ions at a low temperature range.

As the operating temperature increased, the Nernstian behavior of the sensor based on the Al_2O_3-ZrO_2-Y_2O_3 eutectic became more predominant. The reproducibility of the measured *emf* at temperatures higher than 480°C was reasonably good

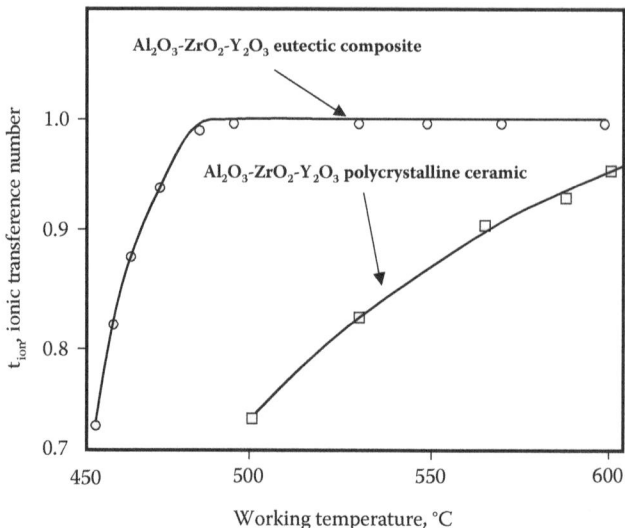

FIGURE 4.15 The temperature dependence of the ionic transference for Al_2O_3-ZrO_2-Y_2O_3 composite eutectic crystal and Al_2O_3-ZrO_2-Y_2O_3 polycrystalline ceramic with Pt electrodes.

($\pm$1mV). The low value of the output *emf* for the polycrystalline zirconia sensors at operating temperatures less than 500°C can be explained by the absence of electrochemical equilibrium on electrodes. The rate of the reaction between electrolyte-electrode and gas phases is generally small at low temperatures. This rate increases with temperature. In comparison with other polycrystalline sensors, the sensor based on the Al_2O_3-ZrO_2-Y_2O_3 eutectic composite with Pt electrodes exhibited good output stability even at a temperature of 480°C. For a similar sensor based on the Al_2O_3-ZrO_2-Y_2O_3 polycrystalline ceramic with Pt electrodes, the stable response at the same temperature requires a much longer time (~10 minutes), which is unsatisfactory for industrial applications. Thus, the sensor based on the Al_2O_3-ZrO_2-Y_2O_3 eutectic composite with Pt electrodes could be used as an alternative oxygen sensor at temperatures as low as 480°C.

The 90% response time was only 55 seconds upon exposure to gas containing 1000 ppm O_2 at 500°C. The recovery time was also rapid. However, the *emf* returned to its original level within ~100 seconds. Although this sensor has much faster response and recovery at high temperatures (less than 1 second at temperatures above 700°C), it almost satisfies the response characteristics demanded for practical sensors working in the temperature range of 450–600°C [53]. Numerous measurements of different O_2 concentrations in this temperature range have shown that the sensor *emf* values are almost linear with the logarithm of oxygen concentration from 0.1 ppm up to 100%. Further testing at high temperatures revealed that the sensors based on the Al_2O_3-ZrO_2-Y_2O_3 eutectic composite with Pt electrodes accurately measure oxygen not only at low but also at high temperatures (550–1000°C). There is a strong correlation between the test results and the calculated *emf* from the Nernst equation in both the high-temperature and low-temperature regimes. This implies that the sensors based on Al_2O_3-ZrO_2-Y_2O_3 eutectic composites, similarly to the zirconia single-crystal sensors, can measure successfully the oxygen concentration from 0.1 ppm not only at low temperatures but also at high temperatures. The investigation of the sensors based on Al_2O_3-ZrO_2-Y_2O_3 eutectic composites with Pt electrodes provides evidence that these composites behave as pure ionic conductors at temperatures as low as 480°C [53]. These facts indirectly confirm the conclusions about the superior ionic conductivity of the zirconia single crystals with Pt electrodes [8], which appeared only at temperatures T ≥ 480°C. Comparison of the results obtained for the sensors based on the Al_2O_3-ZrO_2-Y_2O_3 eutectic composite with the results obtained for the sensors based on the polycrystalline Al_2O_3-ZrO_2-Y_2O_3 electrolyte with Pt electrodes exhibited that the Al_2O_3-ZrO_2-Y_2O_3 eutectic composites appear to be more effective than polycrystalline ceramics in promoting oxygen transfer in the operating temperature range of 480–550°C, and are also less affected by higher temperature exposure.

4.2 IMPEDANCE METHOD FOR THE ANALYSIS OF *IN-SITU* DIAGNOSTICS AND THE CONTROL OF AN ELECTROLYTE/LIQUID-METAL ELECTRODE INTERFACE

Recently, the development of the impedance-based zirconia sensors using the specific spinel-type oxide SE for NO_x measurement at high temperatures allowed detecting the total NO_x concentrations regardless of the NO-NO_2 ratio in a gas mixture [55–57]. Development of the impedancemetric NO_x sensors was described in detail in Chapter 3. Further development of the impedance method could also be applied to the control of the solid electrolyte/liquid-metal electrode interface during measurement of the thermodynamic activity of oxygen in liquid metals measured by zirconia sensors. So far, the zirconia polycrystalline electrolytes or single crystals have been the most widely used electrolytes for these sensors. The critical component of the Nernstian sensor is an oxide ion conducting electrolyte that allows selective and fast diffusion of oxide ions at elevated temperatures. When the high-temperature cubic fluorite phase of zirconia is stabilized by doping with alkali or rare-earth oxides, the dopant cations substitute for the Zr^{+4} sites in the crystal structure, giving rise to the creation of O_2 vacancies in order to maintain the charge neutrality of the crystal. For example, 8 mol % Y_2O_3 in stabilized zirconia contains about 4 mol % O_2 vacancies [58]. Such a high-vacancy concentration facilitates the selective O_2 diffusion via a vacancy diffusion mechanism. Furthermore, the electrical conductivity remains predominantly ionic in nature even at elevated temperatures with practically no concomitant electronic conduction over a wide range of O_2 activities [59]. In other words, the average ionic transference number is near unity (i.e., $\bar{t_i} > 0.999$), indicating that almost all current through the zirconia is carried by oxide ions. The zirconia-based oxygen sensors have three fundamental features: (1) they transform the oxygen chemical potential of the liquid metal to *emf* and, therefore, are sensors of the generating type (they do not require power sources); (2) the magnitude of the electric signal of a sensor does not depend on its size, which forms the fundamental basis for miniaturization; and (3) the measuring *emf* depends on the logarithm of oxygen concentration, which ensures the possibility of measurement of concentrations within several orders of magnitude, for example, from 10^{-1} to 10^{-10} mass % of oxygen in melted lead [44]. According to the Nernst equation, *emf* of the zirconia-based sensors is determined by the thermodynamic properties of the molten metal and the liquid metal–metal oxide RE, and is proportional to the logarithm of the concentration of the mobile species. Therefore, the sensor does not require calibration as long as the total conductivity remains predominantly ionic. The high ionic conductivity ensures that the measurement error of oxygen activity from the electronic contribution to the electrical conductivity of zirconia as well as the chemical reactivity between the electrolyte and electrode systems are within ±5%. Depending on the working temperature, the YSZ-based sensors, therefore, can control extremely low partial oxygen pressures (down to 10^{-12} mass %) with an acceptable level of accuracy [60].

However, the precise measurement of such a low oxygen partial pressure in molten metal applications is impeded by the polarization of electrodes. The practical realization of the zirconia-based sensors to control oxygen impurity in alkaline metals, such as liquid sodium, liquid lithium, and lead-bismuth eutectic, is complicated by the corrosive impact of these melts on the solid electrolyte or by the appearance of products of corrosion on the electrolyte surface which stipulates the blockade of the SE. As a consequence of this, both the polarization of the sensor and its inertness increase [61]. Another problem associated with the use of polycrystalline zirconia with a metal–metal oxide RE is the careful selection of a RE. For example, it has been reported that a Ni-NiO-RE reacts with an inner Pt current conductor and affects the output *emf* of the sensor [16]. Introduction of an alumina isolation layer in the design of oxygen sensors prevented the formation of an alloy between Ni and Pt at high temperatures, leading to enhancement of the performance and stability. However, after repeated tests over a longer period, the sensor performance degrades, possibly owing to grain growth and sintering of the inner RE at a high temperature. Therefore, further combination of the zirconia single-crystal or zirconia eutectic composite with an appropriately selected metal–metal oxide RE specifically for lead-bismuth eutectic alloy applications will allow enhancement of the lifetime of oxygen sensors. This is because a lead-bismuth eutectic alloy has been employed as a heat carrier in liquid-metal nuclear facilities for submarines [44]. The nature of the lead-bismuth heat carrier substantially differs from that of alkali heat carriers. It is a powerful oxidant of structural materials at relatively low O_2 concentration ($< 10^{-3}$ mass %) in liquid metal [61] and is a corrosive dissolvent under lower concentrations.

It is therefore vital to have a method allowing periodic *in-situ* diagnostics of metrological characteristics of oxygen sensors by controlling the electrolyte/liquid-metal interface directly during the lifespan of the sensors. It is especially important when the lifespan of these sensors varies from 10,000 up to 70,000 hours [45]. In addition, to avoid PbO precipitation in coolant for the subcritical transmutation blanket in nuclear installation, it is recommended to control and maintain O_2 concentration from 1.05×10^{-8} to 5.5×10^{-5} wt % in the temperature range of 350–550°C [61]. One of the methods, which could be employed for this purpose, is the developed impedance method [45, 62]. This method allows *in-situ* diagnostics of the zirconia/liquid-metal electrode interface and the level of polarization of the liquid-metal electrode. Some of the zirconia-based oxygen sensors with Pt electrodes, however, work at a relatively low temperature range (450–650°C), where both the electronic and the hole conductivities can make a contribution to the total conductivity of YSZ [63]. The big problem associated with the contribution of electronic and hole conductivities to the total conductivity of YSZ is short-circuiting within the electrolyte. Unfortunately, the impedance methods cannot distinguish the short-circuiting problem from effects from other causes [45]. Furthermore, polycrystalline zirconia with Pt electrodes cannot work precisely at temperatures lower than 450°C, as has been described above. Consequently, to overcome this problem, the combination of a YSZ single crystal with a liquid metal–metal oxide RE has been used in the low-temperature range of 350–550°C [8, 44, 45, 62].

Single-crystal zirconia oxygen probes, described by Equation (4.5), have been used in experiments. Probes were checked for leakage at room temperature before they were sealed. The temperature of the sensors was measured to ±0.5°C by a built-in K-type thermocouple placed adjacent to the RE. The *emf* was measured with a high input impedance (> 10 MΩ) digital multimeter 179 TRMS accurate to ±0.1 mV. The complex impedance was measured by means of a measuring block, consisting of a transformer bridge (device TT-3152), as shown in Figure 4.16. Measurements were made in the frequency range of 1 Hz–100 kHz. The imaginary part (Z") of impedance versus the real part (Z') was plotted in the examined frequency range to obtain the Nyquist plots. As shown in Figure 4.16, the balance adjustment of the bridge was made by selection of the equivalent resistance R. As a result, the impedance module was measured and the phase angle was determined by the magnitude of the differential transformer current, which was measured at the appropriate positions of the switch.

Owing to the difference in oxygen concentration between oxygen-saturated Bi-Bi$_2$O$_3$-RE and liquid metal, a chemical potential difference has been created, resulting in the flow of oxygen ions and subsequently the accumulation of charge. If the sensor electrodes are reversible, when the oxygen exchange between the liquid metal and electrolyte is not impeded, the exchange resistance on the interface electrolyte–liquid metal RE can be ignored. The simplest equivalent circuit for this type of sensor, shown in Figure 4.17, *a*, is the circuit with the parallel connection of the construction capacitance C$_K$ and the solid-electrolyte resistance R$_i$. Both parameters should be measured, as they are unique for each sensor. In the proposed equivalent circuit, the current conductor, the metal hull of the oxygen probe, and liquid-metal electrodes represent the construction capacitance C$_K$. This circuit complicates (see Figure 4.17, *b*) when the blocking reaction layer appears on the surface of the zirconia electrolyte. The following were used on this circuit: R$_f$ is a polarization resistance, and C$_D$ is capacitance of the double electrical layer on the solid electrolyte–liquid metal interface [64].

4.2.1 Galvano-Harmonic Method

The operative impedance Z(p) for oxygen sensors at the absence of polarization can be expressed as follows [45]:

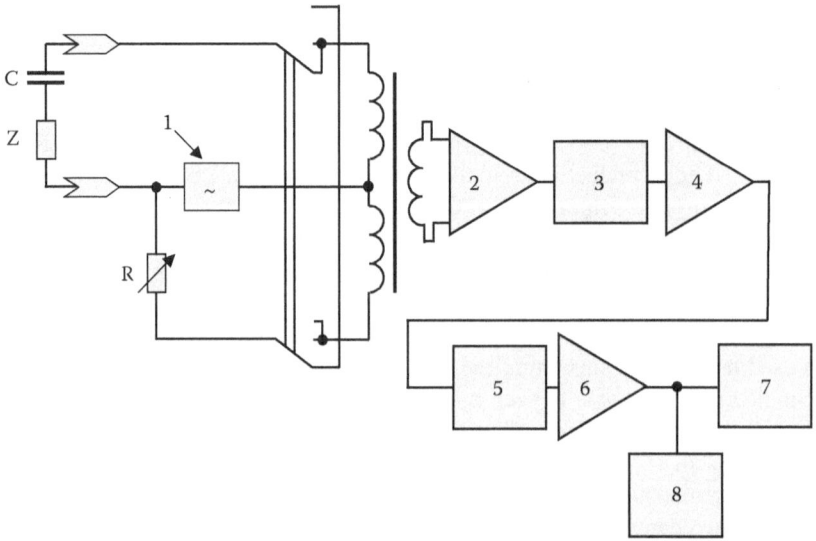

FIGURE 4.16 Instrumentation scheme for impedance measurements: *1*: generator; *2, 4*, and *6*: amplifiers; *3*: attenuator; *5*: filter; *7*: zirconia oxygen sensor; *8*: oscilloscope; *C*: capacitance; and *Z*: electrochemical sensor impedance. (From Zhuiykov, S., *"In-situ"* diagnostics of solid electrolyte sensors measuring oxygen activity in melts by developed impedance method, *Meas. Sci. Technol.* **17** (2006) 1570–1578. With permission.)

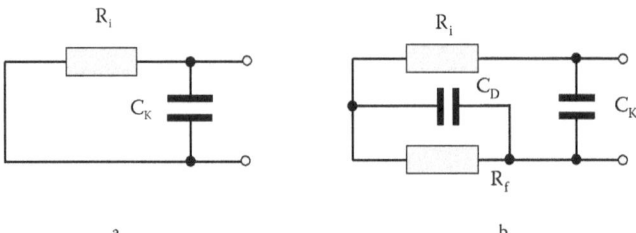

FIGURE 4.17 Equivalent electrical circuit for electrochemical oxygen sensor: *a*: at the absence of polarization; and *b*: polarization of the solid electrolyte–electrode interface. (From Zhuiykov, S., *"In-situ"* diagnostics of solid electrolyte sensors measuring oxygen activity in melts by developed impedance method, *Meas. Sci. Technol.* **17** (2006) 1570–1578. With permission.)

$$Z(p) = \frac{R_i}{R_i C_k p + 1} = \frac{1/c_k}{p + \left(1/R_i C_k\right)} = \frac{l_1}{p + n_1} ,\qquad(4.13)$$

where $l_1 = 1/C_k$; $n_1 = 1/C_k R_i$.

If the step of current I applies on the electrochemical sensor in the galvano-harmonic mode, that is,

$$I(t) = I_0 \sin \omega t, \tag{4.14}$$

or in the operative mode

$$I(p) = I_0 \frac{\omega}{p^2 + \omega^2} \, ,$$

then the operative voltage $E(p)$ of the electrochemical cell, based on the Ohm law, will be

$$E(p) = Z(p)I(p) = \frac{l_1}{p + n_1} I_0 \frac{\omega}{p^2 + \omega^2} \, . \tag{4.15}$$

Based on the rolling functions theorem [65] and in accordance with the linear electrical circuits theory [66], the total voltage in the electrical circuit as the transferral sinusoidal current is applied through it should also be sinusoidal with the same angle frequency ω, that is,

$$E(t) = \frac{l_1 I_0}{n_1^2 + \omega^2} \left(n_1 \sin \omega t - \omega \cos \omega t \right) = E_0 \sin \left(\omega t - \Theta \right), \tag{4.16}$$

where E_0 is the amplitude of sinusoidal voltage and Θ is the angle of the phase transfer between current and voltage, respectively. Equation (4.16) is valid for any moment of time t. Assuming that $\omega t = 0$ and $\omega t = \pi/2$, from Equation (4.16) figure out,

$$\frac{l_1 I_0}{n_1^2 + \omega^2} \omega = E_0 \sin \Theta \, ; \tag{4.17}$$

$$\frac{l_1 I_0}{n_1^2 + \omega^2} n_1 = E_0 \cos \Theta \, . \tag{4.18}$$

From the vector diagram, shown in Figure 4.18, which describes the correspondence between the triangle of voltages and the triangle of resistances, it is seen that

$$E_{react} = E_0 \sin \Theta, \tag{4.19}$$

$$E_{act} = E_0 \cos \Theta. \tag{4.20}$$

Considering the left parts of Equations (4.17), (4.18), (4.19), and (4.20) and after dividing them on the value of current I_0, the equation for impedance components yields the following:

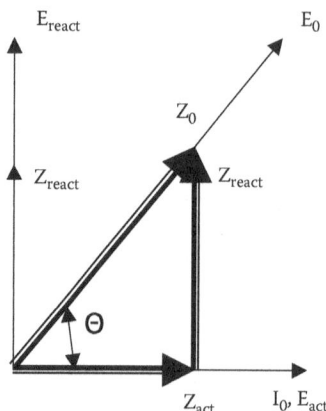

FIGURE 4.18 Vector diagram describing correspondence between the triangle of voltages and the triangle of resistances. (From Zhuiykov, S., "*In-situ*" diagnostics of solid electrolyte sensors measuring oxygen activity in melts by developed impedance method, *Meas. Sci. Technol.* **17** (2006) 1570–1578. With permission.)

$$L_{react} = \frac{l_1 \omega}{n_1^2 + \omega^2} \; ; \tag{4.21}$$

$$L_{act} = \frac{l_1 n_1}{n_1^2 + \omega^2} \; . \tag{4.22}$$

The complex impedance and the total voltage can then be calculated from the following equations:

$$L_0 = \left(L_{react}^2 + L_{act}^2 \right)^{1/2} = \frac{l_1}{\sqrt{n_1^2 + \omega^2}} = \frac{R_i}{\sqrt{1 + \omega^2 R_i^2 c_k^2}} \; ; \tag{4.23}$$

$$E_0 = \left(E_{react}^2 + E_{act}^2 \right)^{1/2} = \frac{R_i I_0}{\sqrt{1 + \omega^2 R_i^2 c_k^2}} \; . \tag{4.24}$$

For evaluation, the character of the semiarc hodograph of impedance described by Equations (4.21) and (4.22) in L_{react} and L_{act} coordinates, and Equation (4.21) divides on correlation (4.22) receiving $L_{react}/L_{act} = \omega/n_1$. Consequently, from this equation follows:

$$\omega = L_{react} \, n_1 / L_{act}. \tag{4.25}$$

Further, taking ω form Equation (4.25) into (4.22), the equation of semiarc in L_{react} and L_{act} coordinates can be expressed as follows:

$$\left(L_{act} - \frac{R_i}{2}\right)^2 + L_{react}^2 = \left(R_i/2\right)^2 . \qquad (4.26)$$

The center of this semiarc lies on the axis L_{act} on the distance of $R_i/2$, and the diameter of the semiarc is equal to R_i, that is, the resistance of zirconia. The construction capacitance C_K value can be found from the condition of maximum of L_{react}, that is,

$$\frac{dL_{react}}{d\omega} = 0 . \qquad (4.27)$$

Working out Equation (4.27), C_K can be obtained as follows:

$$C_K = 1/R_i\,\omega_{max}, \qquad (4.28)$$

which allows drawing the hodograph of the sensor impedance. Figure 4.19 shows calculated (*solid line*) and measured (*dots*) hodographs of impedance for the single-crystal zirconia sensor based on Equations (4.21) and (4.22) at a temperature of 480°C and measuring O_2 concentration in lead ($C_{O2} = 10^{-8}$ mass %), considering different magnitudes of R_i, $C_D = 0.948 \times 10^{-9}$ F and $C_K = 0.474 \times 10^{-10}$ F.

In case of the appearance of the blocking reaction layer on the surface of zirconia with polarization resistance R_f (see Figure 4.17, *b*), the operative impedance equation would read as follows:

$$L(p) = \frac{R_i R_f C_D p + R_i + R_f}{p^2 R_i R_f C_D C_k + p\left(R_i C_k + R_f C_k + R_f C_D\right) + 1} = \frac{kp+1}{p^2 + ap + n} , \qquad (4.29)$$

where $k = 1/C_k$; $l = (R_i + R_f)/R_i R_f C_D C_k$; $a = (R_i C_k + R_f c_k + R_i C_D)/R_i R_f C_D C_k$; and $n = 1/R_i R_f C_D C_k$.

Equation (4.29) expands on a summary of the simplest fractions as follows:

$$\frac{kp+l}{p^2 + ap + n} = \frac{d_1}{p+m_1} + \frac{d_2}{p+m_2} , \qquad (4.30)$$

where $-m_1$ and $-m_2$ are roots of the characteristic equation $p^2 + ap + n = 0$,

$$-m_1 = a/2 + \sqrt{\left(a^2/4\right) - n}, \quad -m_2 = a/2 - \sqrt{\left(a^2/4\right) - n} .$$

Values of the constants d_1 and d_2 in Equation (4.30) can be determined by equalizing coefficients at the equal orders of magnitude p from the left and from the right, that is,

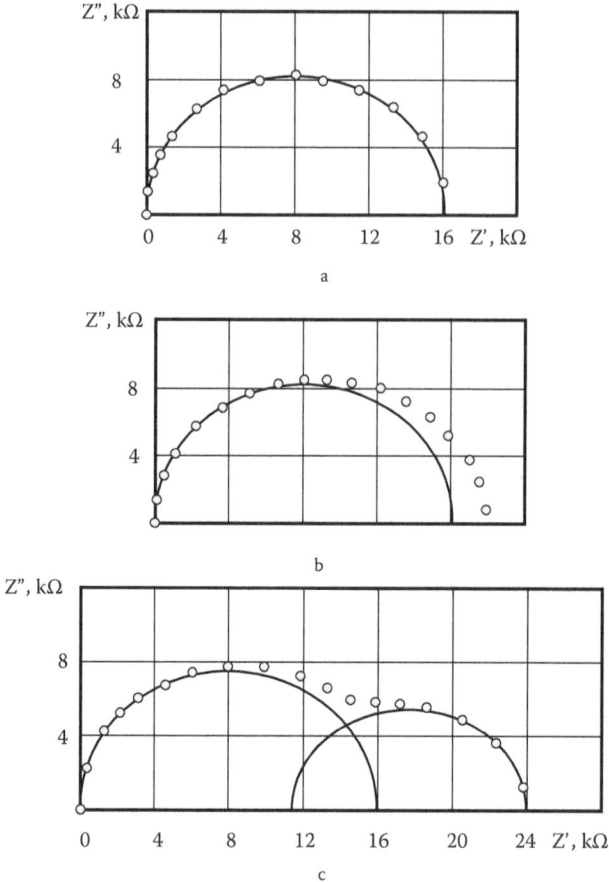

FIGURE 4.19 Calculated (*solid line*) and measured (*dots*) hodographs of impedance for the zirconia-based oxygen sensor at a temperature of 480°C (*a*) at the absence of polarization and (*b*, *c*, and *d*) at the blocking reaction layer appearance on the solid electrolyte/liquid-metal electrode interface at the following magnitudes: R_f; : b – 1.6×10^3; c – 8×10^3; and d – 16×10^3. (From Zhuiykov, S., "*In-situ*" diagnostics of solid electrolyte sensors measuring oxygen activity in melts by developed impedance method, *Meas. Sci. Technol.* **17** (2006) 1570–1578. With permission.)

$$d_1 = (k \, m_1 - l)/(m_1 - m_2); \quad d_2 = (l - m_2 k)/(m_1 - m_2).$$

Considering the behavior of the electrochemical system in the galvano-harmonic mode $[I(p) = I_0 \omega/(p^2 + \omega^2)]$ the operative potential reads,

$$E(p) = I_0 \frac{\omega}{p^2 + \omega^2} \left(\frac{d_1}{p + m_1} + \frac{d_2}{p + m_2} \right). \tag{4.31}$$

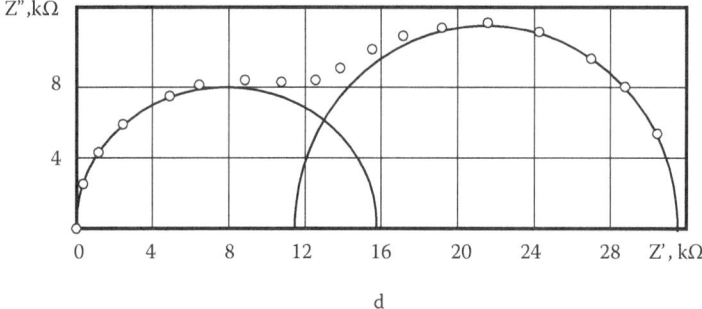

d

FIGURE 4.19 (Continued)

Applying the rolling functions theorem again [65] on Equation (4.31), we can receive the following equations for components of the complex impedance:

$$L_{react} = \frac{d_1\omega}{m_1^2 + \omega^2} + \frac{d_2\omega}{m_2^2 + \omega^2} \; ; \qquad (4.32)$$

$$L_{act} = \frac{d_1 m_1}{m_1^2 + \omega^2} + \frac{d_2 m_2}{m_2^2 + \omega^2} \; . \qquad (4.33)$$

Considering the hodographs of impedance presented in Figure 4.19, it has been noticed that each hodograph represents two semiarcs at $R_i/R_f = 1$, that is, when R_i and R_f have the same magnitude. The radius of the first semiarc is (corresponding to high frequencies) correlating to the radius of the semiarc drawn at the absence of the electrode polarization, that is, at $R_f = 0$ (Figure 4.19, a). Whilst the value of parameter R_f, characterizing the polarization of electrodes, is decreasing, the dimensions of the second semiarc, corresponding to the lower frequencies, are also decreasing. However, the dimensions of the first semiarc remain unchanged. This fact, apparently, shows that the first semiarc of the hodograph of impedance represents parameters R_i and C_K, and the second semiarc represents parameters R_f and C_D, respectively. In fact, the first semiarc provides the value of $R_i = 16 \times 10^3 \; \Omega$ on the axis L_{act}. In the meantime, the second semiarc provides the summarizing value which is equal to $(R_i + R_f)$ on the same axis. The same result is following from Equation (4.32) for the active component of impedance at $\omega \to 0$:

$$L_{act} = \frac{d_1}{m_1} + \frac{d_2}{m_2} = R_i + R_f \; .$$

Then,

$$C_K = 1/\omega_{max\;1} \, R_i; \; C_D = 1/\omega_{max\;2} \, R_f \; .$$

Therefore, the analysis given above shows that the second semiarc of the hodograph of impedance characterized by parameters R_i and C_D always appears when the polarization on the solid electrolyte/liquid-metal electrode interface takes place. If polarization is absent, the hodograph of impedance represents only one semiarc with parameters R_i and C_K.

4.2.2 IMPULSE GALVANIC-STATIC METHOD

This method can also be applied for diagnostics of the solid electrolyte sensors. In this case, $i\,(t) = \text{const}$ or $i\,(p) = I/p$. Operative impedance at the absence of polarization can be determined from the following equation:

$$L(p) = \frac{R_i}{R_i p C_k + 1} = \frac{l_1}{p^2 + n_1},$$

and an operative voltage from the following equation:

$$E(p) = \frac{1}{p}\left(\frac{l_1}{p + n_1}\right) = I l_1\left[\frac{1}{p}\left(\frac{1}{p + n_1}\right)\right]. \tag{4.34}$$

After a few mathematical manipulations, the voltage can be determined from the following correlation:

$$E(t) = IR_i\left[1 - exp\left(-\frac{t}{R_i C_K}\right)\right] \tag{4.35}$$

or, alternatively, from

$$lg\left[IR_i - E(t)\right] = lgIR_i - \frac{t}{2.3\,R_i C_K}. \tag{4.36}$$

It follows from Equation (4.36) that the total voltage $E(t) \to I\,R_i$ at $t \to \infty$. Therefore, if both values for the total voltage $E(\infty)$, corresponding to the saturation of the galvanic-static curve, and values for the current I are known, it is possible to calculate the resistance of the solid electrolyte

$$R_i = E(\infty)/I. \tag{4.37}$$

Function of the total voltage $E(t)$ versus time, calculated in accordance with Equation (4.35) at $R_i = 16 \times 10^3\ \Omega$, $C_K = 0.47 \times 10^{-10}$ F, $I = 10^{-4}$ A, is shown in Figure 4.20.

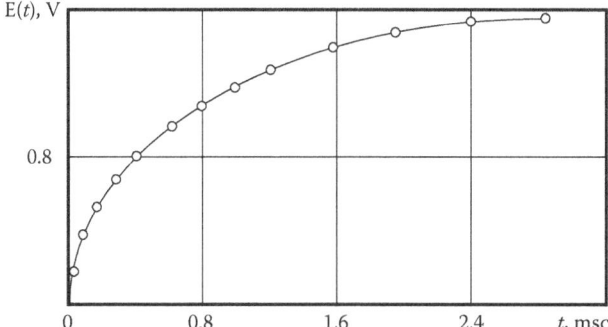

FIGURE 4.20 Calculated and galvano-static curves for an electrochemical sensor at the absence of polarization. (From Zhuiykov, S., "*In-situ*" diagnostics of solid electrolyte sensors measuring oxygen activity in melts by developed impedance method, *Meas. Sci. Technol.* **17** (2006) 1570–1578. With permission.)

In case of the appearance of the blocking reaction layer (Figure 4.17, *b*), the total voltage in the impulse galvanic-static mode at $t \rightarrow \infty$ can be found from the following equation:

$$E\left(\infty\right) = I\left(\frac{d_1}{m_1} + \frac{d_2}{m_2}\right) = I\left(R_i + R_f\right). \tag{4.38}$$

Galvanic-static curves saturate toward the limiting voltage $E\left(\infty\right)$ at $t \rightarrow \infty$, and the limiting voltage itself corresponds to the active resistance, that is, $I\left(R_i + R_f\right)$. Then,

$$\left(R_i + R_f\right) = E\left(\infty\right)/I. \tag{4.39}$$

The step of applied current should be selected such that the limiting voltage in circuit $E(\infty)$ should not exceed the decomposition voltage of zirconia ($\approx 2V$) [62]. Therefore, one of the selection criteria allowing determination of the presence or absence of polarization effects on the YSZ-based electrochemical sensor at the impulse galvanic-static method is the value of the limiting voltage $E(\infty)$. The magnitude of the limiting voltage is equal to IR_i. However, at the presence of the blocking reaction layer, the magnitude of the limit voltage is equal to $I\left(R_i + R_f\right)$.

Considering the possibility of the appearance of the blocking reaction layer on the surface of the solid electrolyte, it is necessary to consider the resistance R_f, which is connected in the parallel to the double electrical layer capacitance C_D on the electrolyte/liquid-metal electrode interface, as shown in Figure 4.17, *b*.

Dependence of the total conductivity $Y\left(\omega\right)$ of the electrical circuit on the round frequency ω can be written as a complex function [62]:

$$Y(\omega) = G(\omega) + j\, B(\omega), \tag{4.40}$$

where $G(\omega)$ is conductivity and $B(\omega)$ is susceptibility of the galvanic circuit. Analyzing the frequency dependence of the admittance Y based on the proposed equivalent electrical scheme (Figure 4.17, b), the following correlations for conductivity G and for susceptibility B can be obtained:

$$GR_i = I - \frac{\alpha}{\alpha^2 + x^2}; \quad BR_i = x\left(\frac{C_K}{C_D} + \frac{I}{\alpha^2 + x^2}\right), \qquad (4.41)$$

where $\alpha \equiv (R_f + R_i)/R_f$, $x \equiv C_D R_i$.

Approbation of the proposed impedance method has been done on the oxygen probes with the structure described above as (4.5) in the temperature range of 380–480°C. Initially, the *emf* of the probe was allowed to attain a stable value. Me-MeO/molten metal *emfs* were within ±1 mV of their theoretical values. The thermodynamic *emf* (Nernst equation) for an oxygen probe with a Bi-Bi$_2$O$_3$ RE is

$$E = (\Delta G_{Bi_2O_3} - RT \ln P_{O_2})/4F, \qquad (4.42)$$

where $\Delta G_{Bi_2O_3}$ is the standard Gibbs energy of formation Bi$_2$O$_3$ [39] and P_{O_2} is the measuring oxygen partial pressure. Previous investigation of properties of the liquid-metal REs has shown that the Bi-Bi$_2$O$_3$ electrodes exhibited neutral behavior toward a YSZ solid electrolyte in the temperature range of 297–497°C [8]. It is therefore possible to expect weak polarization in the above temperature range.

Calculated (*solid line*) and measured (*dots*) hodographs of impedance for the electrochemical oxygen sensor based on a zirconia single crystal with a Bi-Bi$_2$O$_3$ RE measuring of dissolved O$_2$ concentration in lead (C$_{O2}$ = 10^{-8} mass %) at the temperature range of 380–480°C are illustrated in Figure 4.21. The measurements show that the experimental data correspond well to the hodographs calculated in accordance with the parallel connection of construction capacity C$_K$ = 47.4 pF and the volume resistance of the YSZ solid electrolyte R$_i$ (T). The dependence of R$_i$ versus temperature has an exponential character which corresponds to the linear dependence of ln R$_i$ on 1/T. The activation energy of the total ionic conductivity was 0.92 eV during all experiments, which correlates (accurately to 5%) with value for the (ZrO$_2$)$_{0.9}$ – (Y$_2$O$_3$)$_{0.1}$ single crystal. The observed discrepancy between measured and calculated impedances at the lower frequencies region (1–10Hz), shown in Figure 4.21, could be explained by relatively low polarization of the electrochemical cell (4.5). In reality, after analysis of the frequency dependence of the electrochemical cell based on Figure 4.17, b, it was found that for the complete correlation of experimental and calculated impedance values, it is necessary that the polarization resistance of the zirconia electrolyte should be (0.05 – 0.1) R$_i$. This fact is possibly stipulated by the increase of the square of the surface contact between zirconia and liquid metal. Therefore, the square of the surface contact between the zirconia single crystal and liquid metal should be minimized in order to decrease the possibility of the polarization resistance appearance. Several temperature cycles from 380°C to 480°C and back were performed to analyze the reproducibility of the sensor with respect to temperature. The results showed that although hysteresis does exist in the

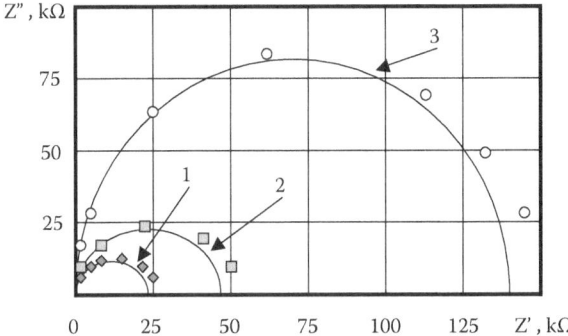

FIGURE 4.21 Calculated (*solid line*) and measured (*dots*) hodographs of impedance for the oxygen sensor based on a $(ZrO_2)_{0.9} - (Y_2O_3)_{0.1}$ single crystal with a $Bi\text{-}Bi_2O_3$ RE: *1*: 480°C; *2*: 430°C; and *3*: 380°C. (From Zhuiykov, S., "*In-situ*" diagnostics of solid electrolyte sensors measuring oxygen activity in melts by developed impedance method, *Meas. Sci. Technol.* **17** (2006) 1570–1578. With permission.)

response time during the reversing temperature change process, it was within 25% for the temperature range of interest. These data indirectly confirmed results of other researchers [58] for impressively fast responses of the zirconia single crystal for these moderate temperatures. The hysteresis might be owing to two major factors: the sensor's response time and the response time of oxygen in liquid lead.

Overall, the analysis of the experimental data allows concluding that the level of the surface polarization of the liquid-metal $Bi\text{-}Bi_2O_3$ electrode is negligible within the temperature range of 380–480°C. The X-ray pattern of the RE did not show any change in the phase assemblage after experiments.

Admittance hodographs for the different R_f/R_i were also calculated for the same working conditions which have been described above for the impedance measurements at $C_K/C_D = 0.05$. The hodographs are shown in Figure 4.22. It is clearly indicated in this figure that the graphical interpretation of the total conductivity changes substantially at the changes of magnitude of R_f/R_i. Furthermore, the susceptibility BR_i, based on the second equation (4.41), at the intersection of the hodographs became sensitive to parameter C_K/C_D. Therefore, magnitudes for both R_f/R_i and C_K/C_D can be determined independently with a high level of accuracy. This fact shows that the admittance measurements can also be successfully applied for diagnostics of the condition of the zirconia-electrode interface directly at the working conditions—in liquid metal. Thus, the simultaneous reduction of R_f/R_i and C_K/C_D at the same working conditions reflects the fact that polarization, and consequently the corrosive impact of the $Bi\text{-}Bi_2O_3$ electrode on the zirconia electrolyte, are absent. In the case of the appearance of the blocking reaction layer on the surface of the solid electrolyte, the ratio R_f/R_i rises and correlation C_K/C_D decreases or remains unchanged. Therefore, the proposed method of impedance measurements allows collecting information about the level of polarization on the zirconia/liquid-metal electrode interface, which is characterized by the second semiarc of the hodograph of impedance. The second semiarc represents parameters R_f as the

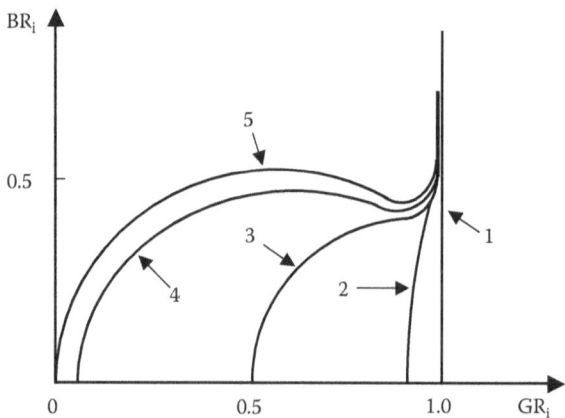

FIGURE 4.22 Admittance hodographs in the electrical circuit at the different magnitudes of R_f / R_i: *1*: 0; *2*: 0.1; *3*: 1.0; *4*: 10; and *5*: 100. (From Zhuiykov, S., "*In-situ*" diagnostics of solid electrolyte sensors measuring oxygen activity in melts by developed impedance method, *Meas. Sci. Technol.* **17** (2006) 1570–1578. With permission.)

polarization resistance and C_D as the capacitance of the double electrical layer on the zirconia–liquid metal interface, respectively.

The proposed impedance method for periodic *in-situ* diagnostics of the solid electrolyte/liquid-metal electrode interface during the life span of the sensors allows periodic inspections of the electrolyte/liquid-metal electrode interface during the lifetime of the sensors and allows making a judgment about the level and the character of electrode polarization. Experiments with liquid-metal Bi-Bi$_2$O$_3$ employed as a RE for zirconia single-crystal oxygen sensors have shown that the level of polarization effects on the electrode-electrolyte interface is negligible at temperatures as low as 380°C. The decrease of the oxidizing potential of the RE by means of careful selection of the appropriate Me-MeO electrode can reduce considerably the polarization effects on the electrolyte/liquid-metal electrode interface. An example of the Bi-Bi$_2$O$_3$ RE has shown that this RE possesses a stable oxidation potential and, consequently, the minimum electric resistance at the zirconia-melt interface. This indicates that the threshold temperature, where a meaningful *emf* can be obtained, can be shifted toward a lower temperature.

The results of the present work may be applicable for diagnostics of oxygen sensors at more complicated applications, such as measurement of oxygen activity in liquid sodium, lithium, or lead-bismuth heat carriers for atomic power plants. Corrosion and mass transfer in nonisothermal lead-bismuth circuits with temperatures of a heat carrier of 300–500°C do usually occur at a concentration of dissolved O$_2$ of 10^{-6} – 10^{-12} mass %. The proposed impedance method is developed for determining the level and the character of polarization at the electrolyte-electrode interface, which ensures a continuous oxide protection of materials against corrosion by means of zirconia sensors in all temperature regimes of exploitation of liquid-metal circuits.

4.3 MEASURING OXYGEN CONCENTRATION IN LEAD-BISMUTH HEAT CARRIERS

The use of liquid-metal heat carriers in nuclear power engineering began in transport facilities in the twentieth century. General Electric Inc. constructed the 50-MW three-loop experimental reactor unit Mark-A with sodium and 22% Na + 78% K alloy, and then the nuclear power facility for the *Seawolf* atomic submarine. In this project, a sodium-graphite reactor on intermediate neutrons was designed and experimentally substantiated. The inlet temperature of its heat carrier was 315°C, and the temperature at the outlet of the active zone was 480°C [67]. This was the result of a 10-year program of the development of sodium and sodium-potassium melts as heat carriers of nuclear power facilities for atomic submarines in the United States. A liquid-metal technology was developed with continuous purification of metallic melts in flow circuits by the method of "cold" traps and with control over the temperature of crystallization of technological admixtures from melts by the method of a "cork" indicator.

The use of Na as a heat carrier at atomic power plants with fast-fission reactors required a substantial updating of the known procedures and the development of essentially new methods and means of liquid-metal technology. Despite the fact that the service life of sodium nuclear power facilities reached 280 reactor-years, some problems of the reactor technology of sodium remain unsolved until now. For example, devices for continuous monitoring of O_2 in the Na in the first and second circuits of atomic power plants have not yet been developed to satisfy demanding industrial needs. A distinctive feature of the exploitation of nuclear power facilities at atomic power plants is the long-term monitoring of the reactor parameters in the nominal regime, which is not characteristic of the nuclear power facilities of submarines. Therefore, the development and optimization of technology of heat carriers for stationary atomic power plants are still topical problems.

The nature of the lead-bismuth heat carrier differs substantially from that of the alkali heat carriers. It is a very powerful oxidant at a relatively low oxygen concentration (less than 10^{-3} mass %) and is also a corrosive dissolvent at low concentrations [44, 58, 67]. The monitoring of the oxygen content in the hear carrier can only be achieved by solid-electrolyte activimetry employing the single-crystal zirconia (hafnia) sensors [8, 44, 58]. The determination of *emf* with the use of these sensors allows measuring the Gibbs energy in a heat carrier and, consequently, the thermodynamic activity of oxygen in the melt from 10^{-1} to 10^{-15} mass %. The unique thermal resistance of the sensor and stability of its metrological characteristics is ensured by the special technology of production of sensitive elements of the sensor [68]. The zirconia single-crystal sensors are highly thermostable and withstood thermal shocks (heating or cooling at the 10°C/sec rate) for many, many cycles in the temperature range of 270–550°C [44]. It was reported that these sensors not only withstood at least 100 cycles under thermal shocks of 50°C/sec at an excessive pressure of liquid metal up to 6 MPa, but also exhibited a record-breaking threshold sensitivity of 10–15 mass % in the lead-bismuth heat carriers at temperatures as low as 300°C [44, 68]. Their life span under the conditions given above is up to 100,000 hours. The manufacturing technology of these sensors will be considered in detail in the next chapter. None of the other zirconia sensors reported in the scientific media to date can measure oxygen activity in molten metals so accurately for so long a time. By using these sensors, it is possible to determine the temperature dependence of the minimum admissible activity of oxygen corresponding to the beginning of the depassivation of steels and the rates of passivation of materials in the lead-bismuth heat carriers.

4.4 REGULATION OF OXYGEN PARTIAL PRESSURE IN MELTS BY ZIRCONIA PUMPS

4.4.1 CHARACTERISTICS OF LAMELLAR OXYGEN PUMPS

In addition to the gas sensors measuring various gases in different environments, zirconia-based electrolytes can also be used as reversible pumps for pumping in and out or for dozing the oxygen concentrations from and into molten metals. The calculating characteristics of such devices will be presented below. For this purpose, the oxygen pump can be represented by the following electrochemical cell:

$$\text{Electrode/Liquid Metal } (P_1) \text{ // } YSZ \text{ // } \text{Gas } (P_2)/\text{Electrode.} \qquad (4.43)$$

The principal scheme of such a lamellar oxygen pump is shown in Figure 4.23. If a YSZ electrolyte possesses only ionic conductivity ($\bar{t_i} > 0.99$), then, based on the Faraday law, the transfer of current through the YSZ electrolyte follows the selective transfer of oxygen through the electrolyte membrane. The *emf* generating in circuit (4.43) can be expressed by the well-known Nernst Equation (3.3), where P_1 and P_2 are the oxygen partial pressure on the liquid metal and the gaseous electrode, respectively. $\bar{t_i}$ is the average ionic transference number.

For the purification of liquid metal from oxygen, the voltage $U > E_0$ has to be applied from the external DC source with the negative polarity on the liquid-metal electrode. The transfer of oxygen ions to the gaseous electrode can result in the following chain reactions:

$$\tfrac{1}{2}O_2 + V + 2e^- \rightarrow O^{2-} // O^{2-} // O^{2-} \rightarrow 2e^- + V + \tfrac{1}{2}O_2 . \qquad (4.44)$$

$$\uparrow \underline{\qquad\qquad \delta \qquad\qquad} \uparrow$$

The electrical current, corresponding to the nonstationary diffusive oxygen flow, decreases during the purification. Consequently, the voltage U is also decreasing. Therefore, the applied voltage U to circuit (4.43) has to be $U = \text{const}$ at the changing current in circuit (4.43). Then the purification voltage has to be chosen from the following inequality:

$$E_0 < U < E^*, \qquad (4.45)$$

where E^* is the maximum *emf* in the circuit without the appearance of the electronic conductivity ($\bar{t_i} \sim 1$).

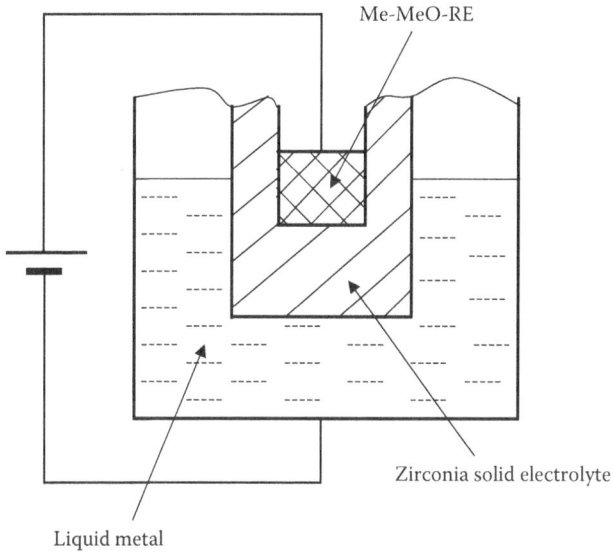

FIGURE 4.23 Principal scheme of the YSZ-based lamellar oxygen pump.

In other cases, the use of the oxygen pumps carries out in galvano-static mode, that is, at the constant current in circuit (4.43). Let us consider the main calculated characteristics of such pumps, which can be represented by the speed and depth of the liquid metal purification from oxygen as well as by the oxygen flow and by the time of the attainment of the set deepness of purification.

4.4.1.1 Potentiometric Mode of the Oxygen Pump

The relative quantity of the removing oxygen from the liquid metal can be expressed by the following equation:

$$\zeta(\tau) = \frac{m(\tau)}{m_0} = \frac{c_0 V - \bar{c}(\tau)V}{c_0 V} = 1 - \frac{\bar{c}(\tau)}{c_0}, \quad (4.46)$$

where V is a volume of the liquid metal, and $\bar{c}(\tau)$ and c_0 are the average and an initial oxygen concentration in metal, respectively.

If the density of the electrical current i corresponds to the diffusive flow on the metal-electrolyte interface j, then the electrochemical balance equation can be presented as follows:

$$m(\tau) = \int_0^\tau j \, d\tau = \frac{1}{2F} \int_0^\tau i \, d\tau. \quad (4.47)$$

The equation for the oxygen flow can be obtained by the differentiation of correlation (4.46) by time τ with consideration of Equation (4.47):

$$j(\tau) = \frac{c_0 V}{S} \times \frac{d}{d\tau}\left[\frac{m(\tau)}{m_0}\right], \tag{4.48}$$

where c is the oxygen concentration, S is the surface of the metal-electrolyte interface, and i is the density of purification current.

The diffusive purification flow from the liquid-metal flat layer with thickness δ will follow the x coordinate toward the metal-electrolyte interface, where the condition $c_1 = const$ should be fulfilled. Coordinate $x = 0$ will be combined with the free surface of the liquid-metal layer. Then, the relative change of the oxygen concentration can be found from the solution of the following nonstationary diffusion equation:

$$\frac{dc}{d\tau} = D\frac{d^2 c}{dx^2}, \tag{4.49}$$

with initial conditions $\tau = 0$, $c(x, 0) = c_0$ $(0 < x < \delta)$, and the following boundary conditions at $\tau > 0$:

$$\frac{\partial c(0,\tau)}{\partial x} = 0; \quad c(\delta,\tau) = c_1. \tag{4.50}$$

The solution of Equation (4.49) for $F_0 > 0.1$ can be obtained as follows [69]:

$$\frac{c(x,\tau) - c_1}{c_0 - c_1} = \sum_{n=1}^{\infty} \frac{2}{\varepsilon_n}(-1)^{n+1} cos\left(\varepsilon_n \frac{x}{\delta}\right) exp\left(-\varepsilon_n^2 F_0\right), \tag{4.51}$$

where $F_0 = D\tau/\delta^2$ is the diffusive Fourier criterion, and $\varepsilon_n = (2n - 1)(\pi/2)$ are roots of the characteristic equation $cos[(2n - 1)(\pi/2)] = 0$.

The average relative oxygen concentration in metal can be expressed as follows:

$$\frac{\bar{c}(\tau)}{c_0} = \frac{1}{\delta}\int_0^\delta \frac{c(x,\tau)}{c_0} dx. \tag{4.52}$$

After substitution of (4.51) into (4.52) and integration, the equation modifies:

$$Q(\tau) = \frac{\overline{c}(\tau) - c_1}{c_0 - c_1} = \sum_{n=1}^{\infty} \frac{2}{\varepsilon_n^2} exp\left(-\varepsilon_n^2 F_0\right). \tag{4.53}$$

For $F_0 > 0.1$ with accuracy 1.5% in Equation (4.53), the first number of series $n = 1$ can be used. Then,

$$Q(\tau) = \frac{8}{\pi^2} exp\left(-\frac{\pi^2}{4} F_0\right). \tag{4.54}$$

In the case of $F_0 < 0.1$ [70],

$$Q(\tau) = 1 - 2\left(\frac{F_0}{\pi}\right)^{0.5}. \tag{4.55}$$

Correlation of Equation (4.54) and Equation (4.55) shows that at $F_0 = 0.1$, the balance corresponds to equation

$$Q_1(\tau)/Q_2(\tau) = 0.985. \tag{4.56}$$

Consequently, the average relative oxygen concentration in the liquid-metal layer can be determined by the Fourier criterion at the boundary conditions considered above.

The speed of the liquid metal purification from oxygen can be expressed by the following equation [71]:

$$v = -m_0 \frac{d}{d\tau}\left[1 - \frac{\overline{c}(\tau)}{c_0}\right] = m_0 \frac{d}{d\tau}\left[\frac{\overline{c}(\tau)}{c_0}\right]. \tag{4.57}$$

Then from (4.54) the following the correlation can be obtained taking into account (4.53):

$$\frac{\overline{c}(\tau)}{c_0} = \frac{1}{\delta}\left[\frac{c_1\delta}{c_0} + \frac{8\delta}{\pi^2}\left(1 - \frac{c_1}{c_0}\right)exp\left(-\frac{\pi^2 F_0}{4}\right)\right]. \tag{4.58}$$

After substitution of (4.58) into (4.57) and the following differentiation,

$$v_1 = -\frac{m_0 D}{\delta^2}\left(1 - \frac{c_1}{c_0}\right)exp\left(-\frac{\pi^2 F_0}{4}\right). \tag{4.59}$$

The correlation between the oxygen concentration in melt with the purification voltage $U = const$ and the equilibrium *emf* in the electrochemical cell (4.43) corresponding to this voltage, should be established. From Equation (3.3) and from condition $U = E_0$, the following appears:

$$U = \bar{t_i} \cdot \frac{RT}{4F} \cdot \ln \frac{P_2}{P_1} \ . \tag{4.60}$$

From Equations (3.3) and (4.60) follows

$$\frac{E_0 - U}{\bar{t_i}} = \frac{RT}{4F} \ln \frac{P_1'}{P_1^0} \ . \tag{4.61}$$

The obtained equation is usually applied for the diluted oxygen solutions in the liquid metals [71]:

$$\frac{c_1}{c_0} = \left(\frac{P_1'}{P_1^0} \right)^{0.5} \ . \tag{4.62}$$

Then Equation (4.61) can be rewritten as

$$\frac{E_0 - U}{\bar{t_i}} = \frac{RT}{4F} \ln \frac{c_1}{c_0} \ , \tag{4.63}$$

and the purification speed can be figured out:

$$v_1 = -\frac{2c_0 SD}{\delta} \alpha_1 exp\left(-\frac{\pi^2 F_0}{4} \right) , \tag{4.64}$$

where

$$\alpha_1 = 1 - 10^{[0.868F(E0-UF)/\bar{t_i}RT]} \ . \tag{4.65}$$

The purification flow can be determined as

$$j = v/S. \tag{4.66}$$

Then

$$j_1 = -\frac{2c_0 D}{\delta}\alpha_1\, exp\left(-\frac{\pi^2 F_0}{4}\right). \qquad (4.67)$$

It can be seen from Equations (4.65) and (4.67) that at $U \gg E_0$, we have $\alpha_1 \approx 1$, that is, the flow and the speed of purification are independent of the voltage of the pump.

The next step is the oxygen flow determination through the electrolyte for $F_0 < 0.1$. The average relative oxygen concentration in melt yields the following:

$$\frac{\overline{c}\,(\tau)}{c_0} = \frac{c_1}{c_0} + \left(1 - \frac{c_1}{c_0}\right)\left[1 - 2\left(\frac{F_0}{\pi}\right)^{0.5}\right]. \qquad (4.68)$$

After substitution of (4.68) into (4.57) and differentiation by τ with consideration of (4.63) and (4.65), the equation for the purification speed is given as

$$v_2 = c_0 S\alpha_1 \left(\frac{D}{\pi\tau}\right)^{0.5}. \qquad (4.69)$$

The purification flow at $F_0 < 0.1$ yields the following:

$$j_2 = c_0\alpha_1 \left(\frac{D}{\pi\tau}\right)^{0.5}, \qquad (4.70)$$

and is also independent on thickness of the liquid-metal layer δ.

The resemblance of the purification speed flows is taking place at $F_0 = 0.2$ and can be determined as

$$\frac{v_1}{v_2} = \frac{j_1}{j_2} = 2\exp\left(-\frac{\pi^2}{4}\right)F_0\left(\pi F_0\right)^{0.5} = 0.97.$$

Consequently, Equation (4.70) can be used for calculation of the oxygen flow in the liquid metal up to $F_0 < 0.2$.

Deepness of the liquid metal purification from the traces of oxygen by the electrochemical cell (4.43) at $\overline{t}_i = 1$ can be determined from Equation (4.46) as the relative quantity of oxygen removed to the time. After substitution of (4.54) or (4.55) into (4.46) with consideration of conditions (4.63), the following equation yields for $F_0 > 0.1$:

$$\zeta_1\left(\tau\right) = \alpha_2\left[1 - \frac{8}{\pi^2}\exp\left(-\frac{\pi^2 F_0}{4}\right)\right]. \qquad (4.71)$$

For $F_0 < 0.1$,

$$\zeta_2(\tau) = 2\alpha_2 \left(\frac{F_0}{\pi} \right)^{0.5}, \qquad (4.72)$$

where $\alpha_2 = 1 - 10^{[0.868F(E0-U)F/RT]}$.

Functions ζ_1 and ζ_2 are growing monotonously up to the limiting value:

$$\lim_{\tau \to \infty} \zeta(\tau) = \alpha_2 = \zeta_\Pi. \qquad (4.73)$$

The achievable deepness of the liquid metal purification ζ_Π from the oxygen by the solid electrolyte can be determined from Equation (4.73). For example, in the case of $F_0 = 0.15$, the following equality $\zeta_1(\tau) = \zeta_2(\tau)$ and consequently,

$$\zeta(0.15) = 0.44 \, \alpha_2. \qquad (4.74)$$

The purification time is calculated at moment $\tau = \tau^*$, to which the set purification deepness can be achieved from the start, that is, then

$$\zeta(\tau^*) = \zeta_0, \qquad (4.75)$$

where ζ_0 is the set purification deepness, for example 90%.

The purification time τ^* at $\zeta > \zeta_0$ can be determined from Equation (4.72):

$$\tau_2^* = \frac{\pi}{4D} \left(\frac{\delta \zeta_0}{\alpha_2} \right)^2. \qquad (4.76)$$

At $\zeta > \zeta_0$ from Equation (4.71) follows,

$$\tau_1^* = -\frac{0.93(\delta^2)}{D} \left[0.09 + \lg \left(1 - \frac{\zeta_0}{\alpha_2} \right) \right]. \qquad (4.77)$$

Equations (4.76) and (4.77) allow the necessary time of achievement for the set purification deepness of the liquid metal from oxygen to be determined.

The irreversible oxygen transfer through the solid electrolyte is possible at $\overline{t}_i < 1$ because in this case the solid electrolyte is also simultaneously represented by the external circuit of the element (3.3). This transfer toward the lower partial pressure takes place in both closed and opened circuits (3.3).

If $P_2 \gg P_1$, then the flow of contamination of the liquid metal by oxygen can be determined as follows:

$$j^- = -\frac{R\sigma T}{F^2 l}\left(1 - \overline{t_i}\right) ln \frac{P_2}{P_1} \cdot \qquad (4.78)$$

Having had an analysis (4.78) similar to the previous analysis of Equations (4.60)–(4.63), the flow of contamination yields

$$j^- = -\frac{4\sigma U}{\overline{t_i} F l}\left(1 - \overline{t_i}\right), \qquad (4.79)$$

where σ is the ionic electroconductivity of the solid electrolyte, and l is the thickness of the solid electrolyte.

It can be seen from Equation (4.79) that the electrolyte parameters σ and l, as well as the mode of the oxygen pump U, can be selected appropriately in order to satisfy $j \gg j^-$. In this case, the flow of contamination of the liquid metal by oxygen can be decreased to the minimum possible level.

Let us consider the case when $j = j^-$ at the low level of flow j^-, that is, at $\overline{t_i} = 1$. Equalizing (4.70) and (4.79), the correlation for the purification time can be given as

$$\left(\tau_2^*\right)^{0.5} = \frac{c_0 \alpha_1 F \overline{t_i}\left(D/\pi\right)^{0.5}}{4\sigma U \left(1 - \overline{t_i}\right)} . \qquad (4.80)$$

In the case of $F_0 > 0.2$,

$$\tau_1^* = -\frac{9,2\,\delta^2}{\pi^2 D} lg \frac{2U\sigma\left(1 - \overline{t_i}\right)\delta}{D l c_0 \alpha_1 F \overline{t_i}} . \qquad (4.81)$$

The equations for calculating the deepness of liquid metal purification from the oxygen are as follows:

For $F_0 > 0.1$:

$$\zeta_1\left(\tau\right) = \alpha_1\left[1 - \frac{8}{\pi^2} exp\left(-\frac{\pi^2}{4} F_0\right)\right] . \qquad (4.82)$$

For $F_0 < 0.1$:

$$\zeta_2\left(\tau\right) = 2\alpha_1\left(\frac{F_0}{\pi}\right)^{0.5} . \qquad (4.83)$$

The constant initial oxygen concentration c_0 can be maintained in the liquid metal on the boundary $x = \delta$ in case of a surplus of oxides in the melt. Then the parameters of purification can be found from the solution of the diffusion Equation (4.49), with initial ($\tau = 0$); $c\,(x, 0) = c_0$, ($0 < x < \delta$) and boundary conditions at $\tau > 0$: $c\,(0, \tau) = c_1$; $c\,(\delta, \tau) = c_0$.

Accordingly, the solution is given as [72]:

$$\frac{c(x,\tau) - c_1}{c_0 - c_1} = \frac{x}{\delta} + \sum_{n=1}^{\infty} (-1)^{n+1} \frac{2}{\mu_n} sin\,\mu_n \left(1 - \frac{x}{\delta} \right) exp\left(-\mu_n^2 F_0 \right), \qquad (4.84)$$

where $\mu_n = n$ are characteristic numbers.

The oxygen flow from the liquid metal on the surface of the electrolyte is given as

$$j = -D \frac{dc(0,\tau)}{dx}. \qquad (4.85)$$

The quantity of oxygen transferring through the surface S:

$$Q = S \int_0^\tau j \, d\tau. \qquad (4.86)$$

After differentiation of (4.84) by x and substitution of the achieved result at $x = 0$ into (4.85), the quantity of oxygen is given as

$$Q = \frac{c_0 \alpha_0 SD}{\delta} \left[\tau - \frac{2\delta}{\pi^2 D} \sum_{n=1}^{\infty} \frac{(-1)^{n+1}}{n^2} cos\,n\pi + \frac{2\delta^2}{\pi^2 D} \sum_{n=1}^{\infty} \frac{(-1)^{n+1}}{n^2} \times \right.$$
$$\left. \times cos\,n\pi\,exp\left(-n^2 \pi^2 F_0 \right) \right] \qquad (4.87)$$

The third item in the square brackets of Equation (4.87) aspires to zero at $\tau \rightarrow \infty$. Therefore, the diffusion process becomes equilibrium from some time τ_e. Consequently, the expenditure of oxygen in the liquid metal at the equilibrated diffusion can be determined as follows:

$$Q_e = \frac{c_0 \alpha_2 SD}{\delta} \left[\tau - \frac{2\delta}{\pi^2 D} \sum_{n=1}^{\infty} \frac{(-1)^{n+1}}{n^2} cos\,(n\pi) \right], \qquad (4.88)$$

where $\tau > \tau_e$.

At $n = 1 \ldots 10$,

$$\frac{2}{\pi} \sum_{n=1}^{\infty} \frac{(-1)^{n+1}}{n^2} \cos(n\pi) \approx -\frac{\pi}{10} .$$
(4.89)

The substitution of (4.89) into (4.88) gives an asymptotic solution for Equation (4.87):

$$Q_e = \frac{c_0 \alpha_2 SD}{\delta} \left(\tau + \frac{\pi \delta^2}{10} \right) .$$
(4.90)

The correlation for the calculation of τ_e is found by equalizing (4.90) to zero:

$$| \tau_e | = (\pi \delta^2 / 10D) .$$
(4.91)

The oxygen flow at the equilibrated diffusion can be determined from Equation (4.90):

$$j = \frac{Q_e}{S\tau} = \frac{Dc_0 \alpha_2}{\delta} \left(1 + \frac{\pi \delta^2}{10D\tau} \right) .$$
(4.92)

Noteworthy is the fact that the obtained equations are also pointed out on the determination method of the coefficients of diffusion and oxygen concentration in liquid metals. For example, after substitution of (4.67) and (4.70), taken by the absolute value, into Equation (4.47) and differentiation by τ for $F_0 > 0.2$, the following appears:

$$\ln i = \ln \frac{Dc_0 \alpha_2 F}{4\delta} - \frac{\pi^2 D\tau}{4\delta^2} .$$
(4.93)

For $F_0 < 0.2$,

$$\ln i = \ln \frac{c_0 \alpha_2 F}{8} \left(\frac{D}{\pi} \right)^{0.5} - \frac{\ln \tau}{2} .$$
(4.94)

The coefficient of oxygen diffusion can be determined from the pitch of the line $i = f(\tau)$ for Equations (4.93) or (4.94). Then, the value of c_0 can be calculated from Equations (4.93) and (4.94).

That's why the solid electrolytes in the electrochemical oxygen pumps not only provide regulation of oxygen concentration in the different technological environments, including molten metals, but are also capable of controlling oxygen concentration constantly. Furthermore, they can measure the changes of the coefficient of oxygen diffusion occurring in the control environments.

4.4.1.2 Galvano-Static Mode of the Oxygen Pump

The DC transfer through the electrochemical cell (3.3) accompanies the continuous decreasing of the oxygen chemical potential on the metal-electrolyte interface and, consequently, oxygen partial pressure. The equilibrium *emf* of the electrochemical cell (3.3) will continuously increase in accordance with the Nernst equation.

The quantity of oxygen removed from the liquid metal through the surface S to the time τ can be expressed as follows:

$$m(\tau) = Si\tau/2F. \qquad (4.95)$$

In this case, the oxygen flow from the metal is given as

$$j = \frac{1}{S}\frac{dm(\tau)}{d\tau} = \frac{i}{2F}. \qquad (4.96)$$

The speed of liquid metal purification from oxygen is as follows:

$$v = \frac{dm(\tau)}{d\tau} = \frac{iS}{2F}. \qquad (4.97)$$

The purification deepness of liquid metal from oxygen at the moment can be found from Equation (4.46) considering (4.95) as

$$\zeta(\tau) = \frac{i\tau}{2Fc_0\delta}. \qquad (4.98)$$

Let us consider the solution of the diffusion Equation (4.49) with initial $\tau = 0$, $c = c_0$ ($0 < x < \delta$), and boundary $c = c_0$ at $\tau \to \infty$; $[dc(0, \tau)/d\tau] = -(i/2DF)$ conditions at $\tau > 0$ for determination of the resource of the oxygen pump in the pure oxygen-ionic conductivity mode. The solution can be found in the following form [72]:

$$c(x,\tau) = c_0 - \frac{i}{F}\left(\frac{\tau}{\pi D}\right)^{0.5} exp\left(-\frac{x^2}{4D\tau}\right) + \frac{ix}{2DF}\left[1 - erf\frac{x}{2(D\tau)^{0.5}}\right], \qquad (4.99)$$

where

$$erf(z) = \pi^{-0.5} \int_{0}^{z} e^{-\Omega^2} d\Omega$$

is the errors integral [71].

The relative oxygen concentration on the metal-electrolyte interface can be figured out from (4.99) at $x = 0$:

$$\frac{c(0,\tau)}{c_0} = 1 - \frac{i}{c_0 F}\left(\frac{\tau}{\pi D}\right)^{0.5}. \tag{4.100}$$

On the other hand, the relative change of oxygen concentration at $x = 0$ can be expressed through an increment of the equilibrium *emf* of the electrochemical cell (3.3):

$$\frac{c(0,\tau)}{c_0} = exp-\left(\frac{2F\Delta E}{RT}\right). \tag{4.101}$$

Then the moment $\tau = \tau^*$, when the Faraday law does not fulfill, can be found from Equations (4.100) and (4.101):

$$\left(\tau^*\right)^{0.5} = \frac{c_0 F\left(\pi D\right)^{0.5}}{i}\left[1 - exp\left(\frac{2F\Delta E}{RT}\right)\right]. \tag{4.102}$$

The deepness of the liquid metal purification from oxygen achievable at the moment τ^* is given as

$$\zeta(\tau^*) = \frac{i\tau^*}{2Fc_0\delta}. \tag{4.103}$$

Therefore, the oxygen concentration on the metal-electrolyte interface is constantly decreasing in time at $i=const$ mode. This fact consequently stipulates the development of an electronic conductivity in the electrochemical cell. Moreover, the prolonged and deep purification of the liquid metal from oxygen is possible only at $\overline{t_i} > 0.99$ in $U = const$ mode of the pump, that is, without influence of an electronic conductivity. As a result, it is more preferable to use the solid electrolyte oxygen pumps in the potentiostatic modes.

The time dependence of the electrochemical cell (4.43) can be achieved during purification process by substitution of (4.100) into (4.101):

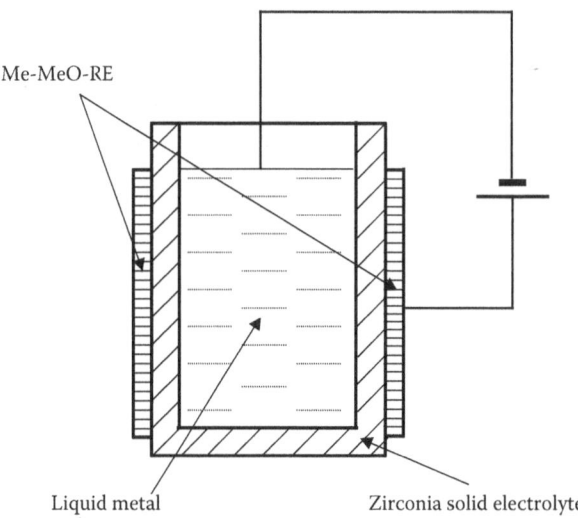

Me-MeO-RE

Liquid metal   Zirconia solid electrolyte

FIGURE 4.24 Principal scheme of the cylindrical oxygen pump.

$$E = E_0 = -\frac{RT}{2F} \ln \frac{\beta^{0.5} - \tau^{0.5}}{\beta^{0.5}} , \qquad (4.104)$$

where $\beta^{0.5} = c_0 F(\pi D)^{0.5}/i$.

The multiplier $c_0(D)^{0.5}$ can be calculated from the experimental dependence $E = f(\tau)$ in accordance with (4.104). Then the c_0 value can be found at the known D. Therefore, the oxygen concentration in the liquid metal can also be determined at the $i = const$ mode of the pump by the method of the equilibrium *emf* measurement.

4.4.2 CHARACTERISTICS OF CYLINDRICAL OXYGEN PUMPS

The cylindrical oxygen pumps based on zirconia electrolytes, in contrast from the lamellar oxygen pumps, can be utilized in such compact designs as test tubes, crucibles, and the like. Therefore, their practical implementation is more accepted by industry. The principal scheme of the cylindrical oxygen pump is shown in Figure 4.24.

4.4.2.1 Potentiometric Mode

The relative changes of the oxygen concentration in the liquid metal are determined from the solution of the diffusion equation in the endless cylinder:

$$\frac{dc}{d\tau} = D\left(\frac{d^2c}{dx^2} + \frac{1}{r} \cdot \frac{dc}{dr}\right). \qquad (4.105)$$

The solution of (4.105) at $\tau > 0$ with initial $\tau = 0$; $c(r, 0) = c_0$ $(0 < r < R)$ and the boundary conditions $[\partial c (0, \tau)/\partial r] = 0$; $c (R , \tau) = c_1$ can be found in the following form:

$$\frac{c(r,\tau)-c_1}{c_0-c_1} = \sum_{n=1}^{\infty}\left[\frac{2I_0\left(\mu_n r/R\right)}{\mu_n I_1\left(\mu_n\right)}\right]exp\left(-\mu_n^2 F_0\right), \qquad (4.106)$$

where $F_0 = D\tau/R^2$ is the diffusive Fourier criterion, I_0 is the Bessel function of the real argument, and μ_n are the roots of the characteristic equation:

$$I_0\left(\mu_n\right) = 0. \qquad (4.107)$$

The value of the average relative oxygen concentration in the liquid metal is given as

$$\frac{\bar{c}(\tau)}{c_0} = \frac{2}{R^2}\int_0^R \frac{c(r,\tau)}{c_0} r\, dr . \qquad (4.108)$$

Substitution of (4.108) into (4.106) gives the following correlation:

$$\frac{\bar{c}(\tau)-c_1}{c_0-c_1} = \sum_{n=1}^{\infty}\frac{4}{\mu_n^2}exp\left(-\mu_n^2 F_0\right). \qquad (4.109)$$

For $F_0 > 0.1$ with accuracy 1.5%, the first member of the row $n = 1$ would be sufficient. Then,

$$\frac{\bar{c}(\tau)-c_1}{c_0-c_1} = \frac{4}{\mu_1^2}exp\left(-\mu_1^2 F_0\right), \qquad (4.110)$$

where $\mu_1 = 2.405$ is the first root of Equation (4.107).

The oxygen flow through the solid electrolyte membrane can be found by combining Equation (4.110) with Equations (4.46) and (4.48):

$$j(\tau) = \frac{2c_0 D}{R}\left(1-\frac{c_1}{c_0}\right)exp\left(-\mu_1^2 F_0\right). \qquad (4.111)$$

Substitution ratio c_1/c_0 by (4.63) yields

$$j(\tau) = \frac{2c_0 D}{R}\alpha\, exp\left(-\mu_1^2 F_0\right), \qquad (4.112)$$

where $\alpha = \alpha_1$ at $\overline{t_i} < 1$, $\alpha_1 = 1 - 10^{[0.868F(E0-U)/tRT]}$, or $\alpha = \alpha_2$ at $\overline{t_i} = 1$,

$$\alpha_2 = 1 - 10^{[0.868F(E0-U)/RT]}.$$

The purification speed is found as $v(\tau) = j(\tau) S$. Consequently, the purification speed at $\overline{t_i} < 1$ after an appropriate substitution is given as

$$v(\tau) = 4\pi h \, c_0 \, \alpha_1 \exp\left(-\mu_1^2 F_0\right), \tag{4.113}$$

where h is the high of the liquid metal and S is the surface of the membrane. For $\overline{t_i} = 1$ condition,

$$v(\tau) = 4\pi h \, c_0 \, \alpha_2 \, exp\left(-\mu_1^2 F_0\right). \tag{4.114}$$

It is seen that at $v \gg E_0$, $\alpha_1 \approx 1$ and $\alpha_2 \approx 1$, that is, both flow and purification speed are independent on the pump voltage. The deepness of the liquid metal purification from oxygen can be found from (4.110) for $F_0 > 0.1$ as

$$\zeta(\tau) = \alpha_2 \left[1 - \frac{4}{\mu_1^2} exp\left(-\mu_1^2 F_0\right)\right]. \tag{4.115}$$

Function $\zeta(\tau)$ is growing monotonously up to the limiting value:

$$\lim_{\tau \to \infty} \zeta(\tau) = \alpha_2 = \zeta_{lim}. \tag{4.116}$$

Therefore, Equation (4.116) allows determining the limiting relevant deepness of the liquid metal purification ζ_{lim} from the oxygen. The purification time τ^* up to the set deepness, for example up to 90% $\zeta(\tau^*) = \zeta_0 = 0.9$ at $F_0 > 0.1$, can be determined from Equation (4.115) as

$$\tau^* = -0.396 \, (R^2/D) \, [0.162 + \lg(1 - (\zeta/\alpha^2))]. \tag{4.117}$$

In the case of $\overline{t_i} < 1$ at $F_0 > 0.1$,

$$\tau^* = -0.396\left(R^2/D\right) \lg \frac{2\sigma UR(1-\overline{t})}{\overline{t} \, F l c_0 \, D \alpha_1}. \tag{4.118}$$

The purification deepness in this case is as follows:

$$\zeta(\tau_1^*) = \alpha_1 \left[1 - \frac{4}{\mu_1^2} exp\left(-\mu_1^2 F_0\right)\right]. \tag{4.119}$$

4.4.2.2 Galvano-Static Mode

The quantity of oxygen m removed from the liquid metal by time τ through the surface $S = 2\pi Rh$, in accordance with the Faraday law at $i = const$, can be determined as

$$m(\tau) = (A\pi Rhi\tau/2F),$$

where $k = A/2F = 8.25 \cdot 10^{-5}$ g/K is the electrochemical equivalent for oxygen [71], and A is an oxygen atom weight.

The oxygen flow from the liquid metal:

$$j = \frac{1}{S}\frac{dm(\tau)}{d\tau} = \frac{Ai}{2F}, \tag{4.120}$$

and the purification speed:

$$v = \frac{dm(\tau)}{d\tau} = \frac{A\pi Rhi}{4F}. \tag{4.121}$$

The deepness of the liquid metal purification from oxygen at the moment τ can be given as

$$\zeta(\tau) = \frac{m(\tau)}{m_0} = \frac{Ai\tau}{2c_0 FR}. \tag{4.122}$$

On the other hand, the diffusion flow, which determines the oxygen transfer from the metal to the boundary $r = R$, is given as

$$j = -D\frac{dc(R,\tau)}{dr}. \tag{4.123}$$

Then equalizing (4.120) and (4.123) together, the oxygen transfer from the metal to the boundary yields,

$$\frac{dc(R,\tau)}{dr} = -\frac{Ai}{2DF}. \tag{4.124}$$

The boundary condition to (4.124) is as follows:

$$\frac{dc(0,\tau)}{dr} = 0. \tag{4.125}$$

The correlation for oxygen diffusion in the cylinder can be determined from the solution of the diffusion Equation (4.105) with initial at $\tau = 0$: $c\,(r, 0) = c_0$ ($0 < r < R$) and boundary conditions (4.124) and (4.125). The solution is given as [71]:

$$c\left(r,\tau\right)-c_0 = -\frac{AiR}{2DF}\left[2F_0 - \frac{1}{4}\left(1-\frac{2r^2}{R^2}\right) - \sum_{n=1}^{\infty}\frac{2I_0\left(\mu_0\,r/R\right)}{\mu_n^2 I_0\left(\mu_n\right)} \times exp\left(-\mu_n^2 F_0\right)\right],$$

(4.126)

where μ_n are the roots of the characteristic equation

$$I_1(\mu_n) = 0. \tag{4.127}$$

Here, I_0 and I_1 are the Bessel function of the real argument.

If $F_0 > 0.1$ is set with accuracy $\pm 1\%$ in Equation (4.126), then the first member ($n = 1$) of the row would be sufficient enough. As a result, on the boundary $r = R$ at the moment $\tau = \tau^*$, the following equation exists:

$$\frac{c^*\left(R,\tau^*\right)}{c_0} = 1 - \frac{AiR_1}{c_0 DF}\left[F_0 + \frac{1}{8} - \frac{1}{\mu_1^2}exp\left(-\mu_1^2 F_0\right)\right], \tag{4.128}$$

where $\mu_1 = 3.832$ is the first root of the characteristic Equation (4.127).

The relative oxygen concentration in Equation (4.128) can be expressed by using the electrophysical parameters of the pump at the initial moment τ_0 and at the moment τ^* when an electronic conductivity appears in the solid electrolyte. The purification time can then be calculated for the obtained correlation:

$$F_0^* - \frac{1}{\mu_1^2}exp\left(-\mu_1^2 F_0^*\right) = \frac{\alpha_3 c_0 DF}{AiR} - \frac{1}{8}, \tag{4.129}$$

where $\alpha_3 = 1 - 10^{[0.868F(E0-E^*)/RT]}$, E^* is the maximum *emf* achievable by the oxygen pump at the moment $\tau = \tau^*$.

Consequently, the maximum deepness of molten metal purification from oxygen can be determined from Equation (4.122) with Equation (4.129).

Therefore, considering the equations given above, it is possible to calculate the reversible zirconia oxygen pumps and to determine their main characteristics during oxygen pumping from the liquid metal into the gaseous environment.

REFERENCES

1. Fergus, J.W., Using chemical sensors to control molten metal processing, *JOM* **52** (2000) 185–190, http://www.tms.org/pubs/journals/JOM/0010/Fergus/Fergus-0010. html.

2. Hong, Y.R. et al., An application of the electrochemical sulfur sensor in steelmaking, *Sens. Actuators B: Chem.* **87** (2002) 13–17.

3. Knevels, J. and Mingnean, F., *Measuring Device for Determining the Oxygen Activity in Molten Metal Melts or Slag Melts*, U.S. Patent No. 6,855,238 (2005).

4. Badwal, S.P.S., Bannister, M.J., and Garret, W.G., Oxygen measurement with SIRO₂ sensors, *J. Phys. E: Sci. Instrum.* **20** (1987) 531–540.

5. Fukatsu, N. et al., Hydrogen sensor for molten metals usable up to 1500K, *Solid State Ionics* **113–115** (1998) 219–227.

6. Vangrunderbeek, J., Lens, P., and Luyten, J., *Sensor for Application in Molten Metals*, U.S. Patent No. 6,514,394 (2003).

7. Fergus, J.W. and Hui, S., Solid electrolyte sensor for measuring magnesium in molten aluminium, *Metall. Mater. Trans. B* **26B** (1995) 1289–1291.

8. Zhuiykov, S., Zirconia single crystal analyser for low-temperature measurements, *Proc. Control and Quality* **11** (1998) 23–37.

9. Fraden, J., *Handbook of Modern Sensors: Physics, Designs and Applications*, 2nd ed., New York, American Institute Physics Press, 1997, 215.

10. Fukaya, T. et. al., *Oxygen Sensor*, U.S. Patent No. 5,393,397 (1995).

11. Janke, D., Fundamental aspects of the electrochemical oxygen measurement in molten iron and iron alloys, in *Applications of Solid Electrolytes*, Eds. T. Takahashi and A. Kozawa, Cleveland, J.E.C. Press, 1980, 154–163.

12. Kurchania, R. and Kale, G.M., Oxygen potential in molten tin and Gibbs energy of formation of SnO₂ employing an oxygen sensor, *J. Mater. Research* **15** (2000) 1576–1582.

13. Kaneko, H., Okamura, T., and Taimatsu, H., Characterization of zirconia oxygen sensors with a molten internal reference for low-temperature operation, *Sens. Actuators B: Chem.* **93** (2003) 205–208.

14. Kaneko, H. et al., Performance of a miniature zirconia oxygen sensor with Pd-PdO internal reference, *Sens. Actuators B: Chem.* **108** (2005) 331–334.

15. Fray, D.J., The use of solid electrolytes as sensors for applications in molten metals, *Solid State Ionics* **86–88** (1996) 1045–1054.

16. Chowdhury, A.K.M.S.S. et al., A rugged oxygen gas sensor with solid reference for high temperature applications, *J. Electrochem. Soc.* **148** (2001) G91–G94.

17. Fray, D.J., Potentiometric gas sensors for use at high temperatures, *Mater. Sci. Tech.* **16** (2000) 237–242.

18. Weyl, A. and Janke, D., High-temperature ionic conduction in multicomponent solid solutions based on HfO₂, *J. Am. Ceram. Soc.* **79** (1996) 2145–2155.

19. Zhuiykov, S., Investigation of conductivity, microstructure and stability of HfO₂-ZrO₂-Y₂O₃-Al₂O₃ electrolyte compositions for high-temperature oxygen measurement, *J. Europ. Ceram. Soc.* **20** (2000) 967–976.

20. Pendit, S.S., Weyl, A., and Junke, D., High-temperature ionic and electronic conduction in zirconate and hafnate compounds, *Solid State Ionics* **69** (1994) 93–99.

21. Kulczycki, A. and Wasiucionek, M., Mechanical and electrical properties of ZrO₂-6.5Y₂O₃ ceramic electrolytes doped with alumina, *Ceram. Inter.* **12** (1986) 181–187.

22. Zhang, T.S. et al., Aging behaviour and ionic conductivity of ceria-based ceramics: A comparative study, *Solid State Ionics* **170** (2004) 209–217.

23. Vlasov, A.N. and Perfiliev, M.V., Ageing of ZrO₂-based solid electrolytes, *Solid Sate Ionics* **25** (1987) 245–253.

24. Iwase, M. et al., Measurement of the parameter *Pe'* for the determination of mixed ionic and *n*-type electronic conduction in commercial zirconia electrolytes, *Trans. Jpn. Inst. Met.* **25** (1984) 43–53.

25. Schmalzrid, H., Ionen- und Electronenleitung in Binaren Oxiden und Ihre Untersuching Mittels EMF-Messungen, *Z. Phys. Chem. Neu Folge* **38** (1963) 87–102.

26. Trubeja, M.F. and Stubican, V.S., Ionic conductivity of the fluorite-type hafnia-R_2O_3 solid solutions, *J. Am. Ceram. Soc.* **74** (1991) 2489–2494.

27. Nowotny, J. and Sorrell, C.C., *Electrical Properties of Oxide Materials*, Zurich, Trans. Tech., 1997.

28. Badwal, S.P.S. and Ciacchi, F.T., Performance of zirconia membrane oxygen sensors at low temperatures with nonstoichiometric oxide electrodes, *J. Appl. Electrochem.* **16** (1986) 28–40.

29. Wepner, W. and Schubert, H., Electrochemical behavior of yttria-stabilized tetragonal zirconia, in *Advances in Ceramics, Science & Technology of Zirconia*, Ed. Hobbs, L.W., American Ceramic Society, 1988, 725–732.

30. Badwal, S.P.S., Bannister, M.J., and Murray, M.J., Non-stoichiometric oxide electrodes for solid state electrochemical devices, *J. Electroanal. Chem.* **168** (1984) 363–382.

31. Ghetta, V. et al., Electrode materials for zirconia sensors working at temperatures lower than 500K, *Sens. Actuators B: Chem.* **13–14** (1993) 27–30.

32. Solier, J.D. et al., Low-temperature ionic conductivity of 9.4-mol%-yttria-stabilized zirconia single crystal, *J. Am. Ceram. Soc.* **72** (1989) 1500–1502.

33. Zhuiykov, S., Talanchuk, P., and Shmatko, B.A., Solid electrolyte sensor for measurement of the oxygen diffusion coefficients in melts, *Herald of the Kiev Polytech. Inst.: Instrumentation* **22** (1992) 49–55.

34. Alisin, V.V. et al., Zirconia-based nanocrystalline material synthesized by directional crystallization from the melt, *Mater. Sci, & Eng.* **25** (2005) 577–583.

35. Cheikh, A. et al., Ionic conductivity of zirconia based ceramics from single crystals to nanostructured polycrystals, *J. Europ. Ceram. Soc.* **21** (2001) 1837–1841.

36. Hartmanova, M. et al., Correlation between microscopic and macroscopic properties of yttria stabilized zirconia: 1. Single crystals, *Solid State Ionics* **136–137** (2000) 107–113.

37. Vladikova, D. et al., Differential impedance analysis of single crystal and polycrystalline yttria stabilized zirconia, *Electroch. Acta* **51** (2006) 1611–1621.

38. Borik, M.A. et al., Partially stabilized zirconia single crystals: Growth from the melt and investigation of the properties, *J. Crystal Growth* **275** (2005) e2173–e2179.

39. Lide, D.R., *Handbook of Chemistry and Physics*, Boston, CRC Press, 1996.

40. Zhuiykov, S., ZrO_2-based partial gas pressures sensors in vacuum: Analysis main components of errors, *Vacuum Tech. & Coating* **2** (2006) 38–44.

41. Burkhard, D.J.M., Hanson, B., and Ulmer, G.C., ZrO_2 oxygen sensors: An evaluation of behavior at temperatures as low as 300°C, *Solid State Ionics* **47** (1991) 169–175.

42. Can, Z.Y. et al., Detection of carbon monoxide by using zirconia oxygen sensor, *Solid State Ionics* **79** (1995) 344–348.

43. Li, N., Tan, T.C., and Zeng, H.C., High-temperature carbon monoxide potentiometric sensor, *J. Electrochem. Soc.* **140** (1993) 1068–1073.

44. Shmatko, B.A. and Rusanov, E.A., Oxide protection of materials in melts of lead and bismuth, *Mater. Sci.* **36** (2000) 689–700.

45. Zhuiykov, S., "*In-situ*" diagnostics of solid electrolyte sensors measuring oxygen activity in melts by developed impedance method, *Meas. Sci. Technol.* **17** (2006) 1570–1578.

46. Calderon-Moreno, J.M. and Yoshimura, M., Stabilization of zirconia lamellae in rapidly solidified alumina-zirconia eutectic composites, *J. Europ. Ceram. Soc.* **25** (2005) 1369–1372.

47. Borodin, V.A., Starostin, M.Y., and Yalovets, T.N., Structure and related mechanical properties of shaped eutectic Al_2O_3–$Zr_2(Y_2O_3)$ composites, *J. Crystal Growth* **104** (1990) 148–153.

48. Calderon-Moreno, J.M. and Yoshimura, M., Rapidly solidified eutectic composites in the system Al_2O_3-Y_2O_3-ZrO_2: Ternary regions in the subsolidus diagram, *Solid State Ionics* **154–155** (2002) 311–317.

49. Lee, J.H. et al., Microstructure of Al_2O_3-ZrO_2 eutectic fibers grown by the micropulling down method, *J. Crystal Growth* **222** (2002) 791–796.

50. Cicka, R., Trnovcova, V., and Starostin, M.Y., Electrical properties of alumina-zirconia eutectic composites, *Solid State Ionics* **148** (2002) 425–429.

51. Bannister, M.J., McKinnon, N.A., and Hughan, R.R., *Oxygen Sensors*, U.S. Patent No. 4,193,857 (1980).

52. Lopato, L.M. et al., Theory, production technology, and properties of powders and fibres: Features of solid solution formation with a fluorite type structure in the system ZrO_2- HfO_2-Y_2O_3 with different synthesis methods, *Powder Metallur. & Metal Ceram.* **45** (2006) 1–7.

53. Zhuiykov, S., Microstructure characterisation and oxygen sensing properties of Al_2O_3-ZrO_2-Y_2O_3 shaped eutectic composites, *Sens. and Mater.* **12** (2000) 117–132.

54. Moulder, J.F. et al., *Handbook of X-ray Photoelectron Spectroscopy*, Eden Prairie, MN, Perkin-Elmer, 1992.

55. Miura, N., Nakatou, M., and Zhuiykov, S., Impedance-based total-NO_x sensor using stabilized zirconia and $ZnCr_2O_4$ sensing electrode operating at high temperature, *Electrochem. Comm.* **4** (2002) 284–287.

56. Miura, N., Nakatou, M., and Zhuiykov, S., Impedancemetric gas sensor based on zirconia solid electrolyte and oxide sensing electrode for detecting total NO_x at high temperature, *Sens. Actuators B: Chem.* **93** (2003) 221–228.

57. Zhuiykov, S. and Miura, N., Solid-state electrochemical gas sensors for emission control, in *Materials for Energy Conversion Devices*, Eds. C.C. Sorrel, J. Nowotny, and S. Sugihara, Cambridge, Woodhead Publishing, 2005, Chapter 12.

58. Van Setten, E. et al., Miniature Nernstian oxygen sensor for deposition and growth environments, *Rev. Scien. Inst.* **73** (2002) 156–161.

59. Nowotny, J. et al., Charge transfer at oxygen/zirconia interface at elevated temperatures. Part 3: Segregation induced interface properties, *Adv. Appl. Ceram.* **104** (2005) 165–173.

60. Konys, J. et al., Development of oxygen meters for use in lead-bismuth, *J. Nuclear Mater.* **296** (2001) 289–294.

61. Li, N., Active control of oxygen in molten lead-bismuth eutectic systems to prevent steel corrosion and coolant contamination, *J. Nuclear Mater.* **300** (2001) 73–81.

62. Zhuiykov, S., Sensors measuring oxygen activity in melts: Development of impedance method for "*in-situ*" diagnostics and control electrolyte/liquid-metal electrode interface, *Ionics* **11** (2005) 352–361.

63. Nowotny, J., Bak, T., and Sorrell, C.C., Charge transfer at oxygen/zirconia interface at elevated temperatures. Part 10: Effect of platinum, *Adv. Appl. Ceram.* **104** (2005) 214–222.

64. Zhou, H. et al., Calculation of the dynamic impedance of the double layer on a planar electrode by the theory of electrokinetics, *J. Colloid and Interface Sci.* **292** (2005) 277–289.

65. Broman, A., *Introduction to Partial Differential Equations: From Fourier Series to Boundary-Value Problems*, New York, Dover, 1989, 192.

66. Hinrichsen, D. and Pritchart, A.J., *Dynamical Systems Theory I: Modelling, State Space Analysis, Stability and Robustness*, New York, Springer-Verlag, 2005, 809.

67. Gromov, B.F., Subbotin, V.I., and Toshinskii, G.I., Application of the melts of the lead-bismuth eutectic and lead as heat carriers at nuclear power facilities, *At. Energy* **73** (1992) 19–24.

68. Krylov, U.V. et al., *Solid Electrolyte Oxygen Sensor and Its Manufacturing Technology*, USSR Patent No. 1,752,069 (1992).

69. Pham, A.Q. and Glass, R.S., Oxygen pumping characteristics of yttria-stabilized-zirconia, *Electrochem. Acta* **43** (1998) 2699–2708.

70. Bebernes, J. and Eberly, D., *Mathematical Problems from Combustion Theory, Applied Mathematical Sciences* **83**, Springer-Verlag, New York, 1990.

71. Shmatko, B.A., Golubkov, S.P., and Talanchuk, P., Theoretical basis of the reversible oxygen pumps on the solid electrolyte membranes, *Electrochem.* **28** (1992) 761–778.

72. Rickert, H., Electrochemische Messung der Sauer – Stoffdiffusion in flussigem Silber und flussigem Kupfer, *Z. Metallkunde* **8** (1968) 635–641.

5 Manufacturing Technologies of Zirconia Gas Sensors

5.1 VACUUM-TIGHT TECHNOLOGIES OF JOINING ZIRCONIA TO CERAMIC INSULATORS

The progress in development of the zirconia-based gas sensors and new and improved traditional materials for SEs and REs have been discovered or invented during the last two decades at a faster rate than ever before. Indeed, both applied technology and the progress with which materials are changing are revolutionary rather than evolutionary. However, specifics of the gas sensor applications together with the dissimilar nature of joining sensor materials restricted the practical applicability of many interesting improvements [1]. In the early twenty-first century, the goal is still to develop sensors with better properties or new and unique combinations of properties, greater reliability combined with lower manufacturing and materials cost, and better environmental compatibility. In other words, the goal is still the same — *to enhance the sensors' performance through greater structures and better processing!* This goal could be and has been achieved by inventing and producing advanced materials for the SE (RE), and by further improvement of processing technologies for interfaces and for sensor construction materials [2, 3]. This is because the lack of suitable joining techniques for dissimilar materials such as metal and ceramic for high-temperature sensor applications is a major hindrance to the utilization of their outstanding properties and lower cost.

Basically, there are only three fundamental ways or methods of joining dissimilar sensor materials: (1) mechanical fastening, (2) welding [4–6], and (3) adhesive bonding [7–10]. These three methods differ in the type of bond created and, consequently, in the nature and strength of the join that is produced. However, mechanical fastening is not suitable for vacuum-tight joining of the sensor materials, and thus only the last two methods will be considered in details.

Similar to welding, the nature of adhesive bonding forces that make a vacuum-tight join is chemical, not mechanical. The atoms (molecules) of the joining materials for both methods bond to each other, lowering the overall free energy of the combination when they do so. In welding, however, the bonds are much stronger than the secondary Van der Waals bonds in adhesive bonding. In welded ceramics, the bonds produced are ionic, covalent, or mixed ionic-covalent, depending on the inherent character of the ceramics being welded [6]. Specifically, one of the first vacuum-tight welding techniques had been developed specifically for the zirconia oxygen sensors in the early 1980s, and it is still employed for manufacturing various

197

zirconia-based sensors [4]. Subsequent to the chemical nature of the bonds involved in welding and, often to a lesser extent, in adhesive bonding, the joints that are produced are permanent and strong. It is therefore impossible to disassemble welded joints, and it is also very difficult to disassemble adhesive bonding joints without damage. Typically, the composition and chemical structure of joining ceramic materials are not seriously disrupted at the adhesive bonding. However, this is not true for welding. Welding usually requires very high temperatures and/or pressure to produce a large number of the primary bonds; thus, the structure and properties of materials being joined are changed.

Apart from these two fundamental methods of joining ceramic materials, there are several secondary methods or options that can be considered as subclasses or variations. Brazing and soldering are subclasses of welding in which, like welding, primary chemical bonds are created. However, in these methods molten filler is caused to flow by capillary forces in a close-fitting joint [11, 12]. Bonds between ceramics are formed without the two ceramic materials' interfaces having to melt. Soldering and brazing differ from each other by the temperature at which their respective required fillers melt. In soldering, fillers melt below 450°C; in brazing, above this temperature [2]. In addition, thermal spraying can be considered as a combination of welding and brazing, in that the insulating materials are caused to bond to a zirconia substrate by being heated and propelled as small particles or vice versa [13–15].

There are numerous obstacles for a successful joining of zirconia to ceramic insulators, the most important of which are the relative inertness of ceramics and the coefficient of thermal expansion (CTE) mismatch. To overcome these obstacles, a very interesting and practical approach has been made for plug-type zirconia sensors [4]. The production of zirconia oxygen sensors for different applications having a pellet of zirconia electrolyte fused into the end of the ceramic insulation tube addressed the problem of extensive microcracks of zirconia. The cracks were of no concern for measurements in molten copper, since they are too tiny to be easily penetrated by the copper. However, gases can readily leak through the cracks, causing errors when sensors are used for gas analysis, particularly the analysis of gases with very low oxygen content (e.g., in the heat-treating industry). It has been experimentally found that among three tubing materials — alumina, aluminous porcelain, and mullite — the least microcracking and, consequently, the most leak-tight assembly is obtained with alumina [16]. Furthermore, the microcracking in the electrolyte makes the pellet susceptible to spalling under thermal cycling or thermal shock conditions, and thus the sensors have restricted their lives under severe *in-situ* gas measurement conditions.

Detailed investigation of the pellet-cracking problem during welding operations has shown that the cracking is, to a considerable degree, a consequence of the rapid heating and cooling cycles involved in the welding. Tensile stresses generated on cooling from the welding temperature, due to the different CTEs of zirconia and alumina, cause cracking of the zirconia pellet. The CTE are 13×10^6 °C^{-1} (average value from ambient to welding temperature) for zirconia and about 10×10^6 °C^{-1} for alumina over the same temperature range [16]. Aluminous porcelain and mullite have an even lower CTE. When the tensile stresses in the zirconia exceed the fracture stress, cracks are formed in the electrolyte. Cracks do not occur in the alumina tube, for ceramics are generally stronger in compression than under tension. Various mechanisms for reducing stresses in the electrolyte were investigated [4]. One approach was to reduce the wall thickness of the tubing to lower the stresses in the zirconia electrolyte and increase those in the tubing. Others were to replace the zirconia-based electrolyte with (1) a hafnia-based electrolyte having a slightly lower CTE, and (2) a partially stabilized zirconia (PSZ) electrolyte having a composition chosen so that it has an average CTE close to that of the tubing. (PSZ electrolytes, it may be noted, are limited in their maximum-use temperature by consequent changes in microstructure and ensuring degradation). All the methods of reducing stresses met with varying quantitative degrees of success but could not be described as totally satisfactory.

The most successful results have been achieved by forming a solid material, which contains an intimate mixture of fine particles of partially stabilized zirconia and a nonelectrolyte ceramic material, such as alumina. If the ratio of the electrolyte and nonelectrolyte phases is chosen appropriately, a strong material is obtained having both satisfactory electrolyte properties and a CTE close to that of the non-electrolyte ceramic material used for the body of the oxygen sensor. The best material of this solid material was found to have a 50/50 wt % ratio between zirconia and alumina [4]. This composite has enabled the construction, by the fusion-sealing technique, of leak-tight "plug-type" sensors, which are suitable for use in measuring the oxygen potential or oxygen content of molten metals and hot gases. Furthermore, it has enabled oxygen sensors having a solid electrolyte tip and a nonelectrolyte body to be fabricated by a technique previously unavailable for this purpose.

The design of the gas sensor made with pellet or disc of composite electrolyte fusion, sealed or otherwise bonded to the end of an alumina cylinder, normally has electrodes mounted to enable *emf* across the pellet or disc to be measured. A protective ceramic or metal sheath may be used around the sensor with apertures to allow the measuring environment to contact the outer surface of the composite electrolyte. The protective metal sheath can be employed for such applications as heat treatment or combustion. In these applications, the metal protective sheath is also used as an outer or forward electrode of the probe. The example of the zirconia-based oxygen probe with a metal protective sheath is given in Figure 5.1. Another advantage of the use of the protective sheath is the ability to protect the composite electrolyte by ceramic fiber filter from the direct impact of soot or any other particles in combustion and heat-treatment applications. In this situation it is essential to have the inner metal current conductor made of the same material as the protective sheath to avoid the high-temperature *emf* generated between the different alloys formed at the two ends of the current conductor wire, in addition to the *emf* generated by the difference of partial oxygen pressures.

However, quite often the use of composite solid electrolyte pellets in the plug-type gas sensors may cause substantial cracking around the welding interface to be formed, as shown in Figure 5.2, *a*. Ordinary ink can be employed for detecting most of the big cracks. Moreover, tiny cracks can also be detected by immersing welded sensors into the water and by applying air pressure (~25 kPa) to the internal part of the plug-type sensors. If no gas bulbs appear within 5 minutes, sensors may be

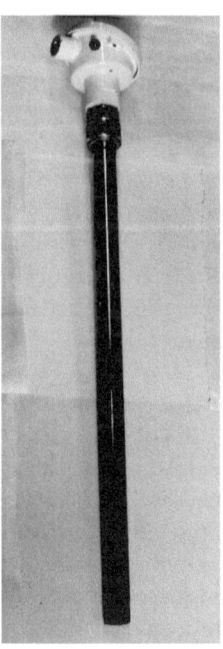

FIGURE 5.1 Zirconia-based oxygen probe with metal protective sheath.

considered crack-free. In addition, for such an industrial application as metal heat treatment, where the crack-free condition of oxygen sensors is an essential prerequisite of their acceptance for use, the performance testing at high temperatures (700–1000°C) in a pure hydrogen atmosphere or a hydrogen-helium gas mixture can subsequently follow the leak testing in water. The hydrogen molecule has the lowest possible size among other gases. In addition, its capability to penetrate through the tiniest cracks in the ceramic interface at elevated temperatures distinguishes hydrogen from other test gases for the final testing of the zirconia-based sensors. In fact, sensors with small leakage detected by the performance testing at high temperatures can still be used in the measuring environments where the presence of small leakage does not affect the measurements. These sensors can still be used in molten metal applications or for controlling the air-fuel ratio in various combustion applications. However, it should also be noted that cracks can — and, in many careless examples of handling, did — develop in the end product even after the final inspection. Mechanical stress or sudden impact during transportation may cause cracks to develop. Thus, extra care should be taken during handling of the dispatched product, and final inspection and testing of sensors should be prepared and organized just before their practical use.

The cracks on the composite solid electrolyte–alumina interface usually appear when the composite solid electrolyte is rapidly quenching after the fusing temperature, which is normally over ~1850°C, down to the temperature around 700–1000°C. Consequently, to avoid cracking, more gradual cooling should be provided around the whole interface area by various techniques. As a result, the

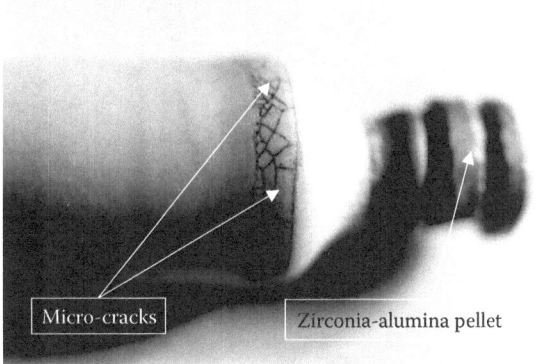

a

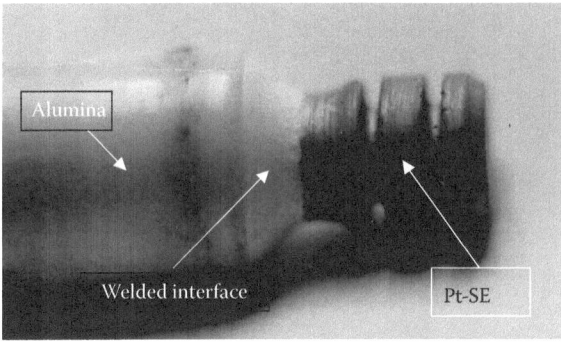

b

FIGURE 5.2 *a*: Microcracks occurred around the welding interface of the zirconia-based sensor; and *b*: crack-free zirconia sensor made by the improved welding technique.

crack-free welding interface between the zirconia pellet and alumina tube can be formed as shown in Figure 5.2, *b*.

The composite solid electrolyte can be produced by an admixture of finely ground stabilized zirconia with alumina powder, followed by consolidating the admixture and firing it to a temperature below 1850°C, which is sufficiently high to produce a high-density, impervious body. Alternatively, slip casting in gypsum molds can be used as the producing technique.

During the last 25 years, the plug-type gas sensors have shown a greater mechanical integrity than other types of zirconia-based sensors of similar construction, and thus have lower leak rates and greater resistance to thermal and/or mechanical shock, factors which produce long sensor lifetimes and enable *in-situ* oxygen probes to be used in applications found to be beyond the capability of the other zirconia-based probes.

Yet another advantageous feature of the plug-type zirconia-based sensors with the composite electrolyte is the capability of these sensors to be formed in the "green"

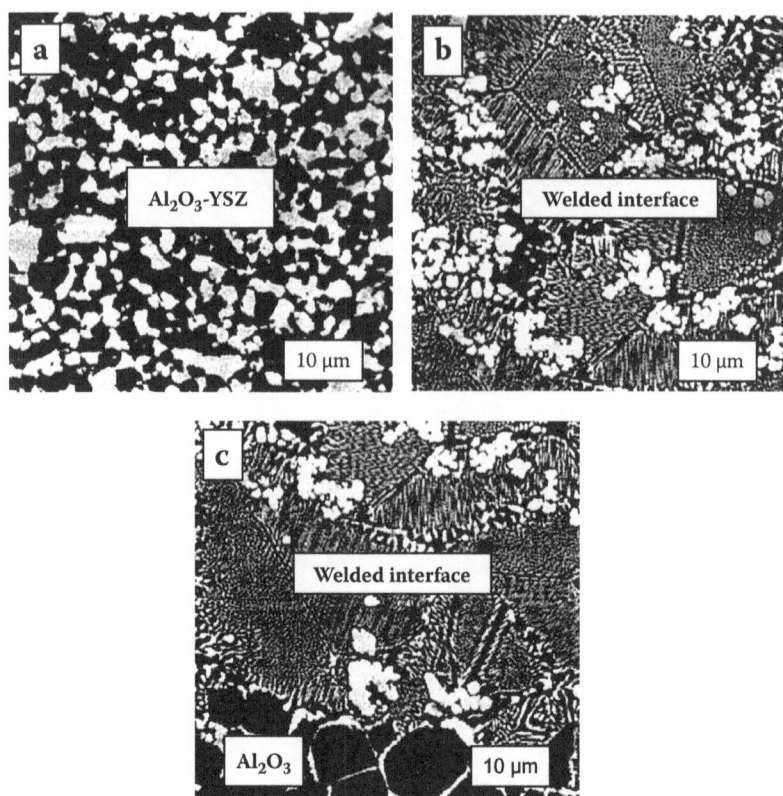

FIGURE 5.3 SEM photographs of (*a*) alumina-zirconia pellet with alumina (dark) and YSZ (white); (*b*) welded region of the sensor; and (*c*) alumina (dark) and welded region interface.

state into various shapes and to be sintered afterwards. For example, some of the automotive exhaust λ-sensors use a generally conical sensor element, constructed entirely of electrolyte material. Sensors of the same shape, but made of a nonelectrolyte ceramic material with a tip of the composite electrolyte at the point of the cone, can readily be fabricated by the "green" sensor construction techniques [4, 6]. Consequently, such sensors perform as efficiently as the λ-sensors in the exhaust gas monitoring systems. Figure 5.3 illustrates SEM micrographs of the different parts of a plug-type oxygen sensor. It is clear from this figure that the pellet was made of a homogeneous polycrystalline mixture of alumina and zirconia with the average grain size from 2 to 5 μm (Figure 5.3*a*). The welded interface, shown in Figure 5.3*b*, is characterized by inhomogeneous eutectic intergrowths which are inevitably formed during quenching. It is interesting to note that the lamella eutectic patterns are always surrounded by alumina inclusions (dark color), which support the view that alumina is the first phase of growth from the melt. In contrast, alumina grains (Figure 5.3*c*) are comparatively large, with the average grain size from 5 to 12 μm, surrounded by zirconia inclusions formed during processing of the sensor.

Therefore, in order to achieve truly optimum properties in a ceramic structure that consists of materials that have been joined together, the materials used in that structure must be designed and engineered for joining. Development of joining techniques must be part of the development of a new composite material. Vacuum-tight joining of the alumina-zirconia ceramics can also be achieved by a super-plastic joining technique. Super-plastic joining has been established as a successful alternative technique for joining ceramic since the mid-1990s [8, 17–22]. It requires minimum surface preparation and generally occurs at lower temperatures than the conventional diffusion bonding [20]. In most super-plastic alumina-zirconia joining, dense pieces are used. It has been reported that nearly fully dense alumina-zirconia joints were made from a sprayed ceramic coating [21]. A joint was made at 1200°C with an interlayer that was 50 vol. % Al_2O_3–50 vol. % ZrO_2, in which the average particle size of both powders was $\leq$ 20 nm. The as-sprayed joint materials were relatively porous following the burnout of organics. SEM revealed that the joint was $\leq$ 50% dense (Figure 5.4). However, following compression at 1300°C or 1350°C, the joints from conventional powders were nearly perfect because the processing temperatures were applied for a relatively short time ($\leq$ 2 hours). Subsequently, the grains within the joints exhibited minimal growth [21]. Even the grain sizes of the nanophase alumina and zirconia remained unchanged by processing at 1200°C, as is clearly shown in Figure 5.5. The resulting fine-grained microstructures corresponded to high strength. Microstructural stability, including grain-size stability, is a primary criterion for super-plastic flow in alumina-zirconia ceramics. Relatively low processing temperatures (1200–1300°C) can minimize grain growth, as can the presence of sufficient concentrations of the second phases. Consequently, alumina-zirconia ceramics exhibit super-plasticity in part because the individual phases pin each other's grain growth [22]. Therefore, further development of the super-plastic technique for joining alumina-zirconia composites is vital for improvement of sensing properties of the planar zirconia sensor structures.

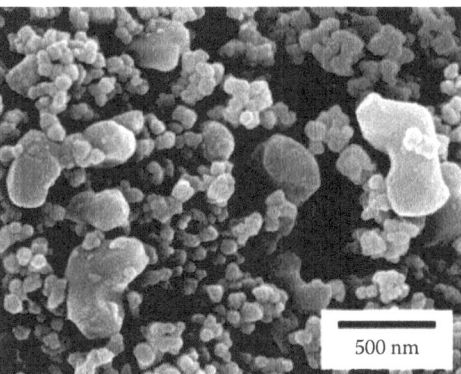

FIGURE 5.4 SEM photograph of 50 vol. % Al_2O_3 – 50 vol. % YSZ joint after removal of the organic binder. (Reprinted from Goretta, K.C. et al., Joining alumina/zirconia ceramics, *Mater. Sci. Eng. A* **341** (2003) 158–162, with permission from Elsevier Science.)

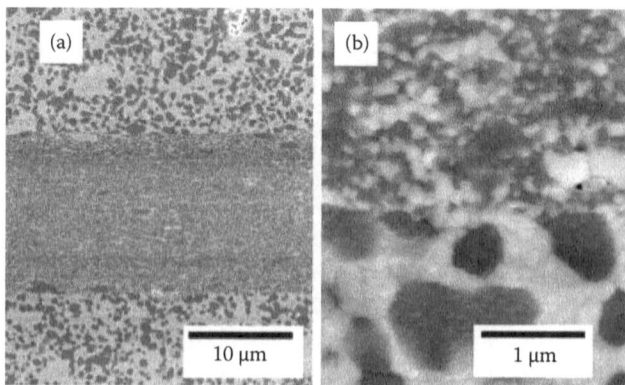

FIGURE 5.5 SEM photomicrographs of (*a*) dense 50 vol. % Al_2O_3 – 50 vol. % YSZ joint made from nanophase powders between 40 vol. % Al_2O_3 – 60 vol. % YSZ compacts (Al_2O_3 is dark phase); and (*b*) interface between joint layer (*top*) and one compact (*bottom*), in which minimal grain growth in joint and high density are apparent. (Reprinted from Goretta, K.C. et al., Joining alumina/zirconia ceramics, *Mater. Sci. Eng. A* **341** (2003) 158–162, with permission from Elsevier Science.)

Another approach to joining ceramics was based on newly developed transient-liquid-phase and liquid-assisted joining methods [11]. Both methods rely on multi-layers, designed to form films at reduced temperatures ($< 450°C$). The liquid films either disappear by interdiffusion or promote ceramic-metal interface formation and concurrent dewetting of the liquid film. If the interdiffusion leads to the disappearance of the liquid film, involving substantial diffusion rather than the much more rapid interstitial diffusion (as with boron in nickel), joining times can be substantial. To reduce the time required for solidification, the bonding temperature and, consequently, the diffusivity should be increased. The Cu/Nb/Cu interlayer system has served as a vehicle for numerous studies of liquid-film-assisted joining. The process can produce ceramic joints with reliably good properties at high temperatures, which is vital for the zirconia-based gas sensors. Assessments of phase diagrams suggest that there are many other candidate interlayer systems, some of which provide much greater solubility of the core metal in the liquid. This could significantly increase the rate of ceramic-core layer growth provided that the interfacial interactions are appropriate. However, both methods mentioned above have not been accepted yet for joining alumina-zirconia ceramics. As has been described in the previous chapter, even small traces of the transient metals in zirconia can lead to a substantial increase of electronic conductivity compromising the integrity and accuracy of measurements.

To overcome this problem, as well as to improve the compatibility of the CTE of the solid electrolyte and insulating ceramic, a $MgAl_2O_4$ spinel in combination with MgO can be used as an insulator in the zirconia-based gas sensors [23]. This insulating material is usually made of ~ 58 wt % of $MgAl_2O_4$ spinel and ~ 42 wt % of MgO by the hot isostatic pressing of row materials. The standard hot-pressing process takes place at $1177–1377°C$ and requires an applied pressure of 40–70 MPa for 45–100 minutes with a subsequent cooling speed of $5–10°C/min$. However, the insufficient density of the insulating tube 3.44–3.46 g/cm³ (theoretical density is 3.58 g/cm³) and the presence of open porosity of 0.3–0.5% are the impediments to the progress of this technology. Furthermore, considering that unfortunately there is no great diversity of ceramic materials with a CTE similar to zirconia available to be reliable insulators for high-temperature zirconia gas sensors, modern designers and materials engineers cannot afford to be parochial in their approach to materials selection. On the other hand, neither does it mean that the existing materials should quickly be abandoned in favor of a "newer" material. Therefore, the basic insulating material ($MgAl_2O_4$–MgO mixture) has been improved by allowing an extra admixture and solid-state diffusion of the ceramic ingredients during hot isostatic pressing. The modified insulating tube represents the mixture of ceramic ingredients in the following wt % [10]:

$MgAl_2O_4$	58.8–69.2
MgO	30–40
CaO-Ga_2O_3 mixture	0.8–1.2

To ensure vacuum-tight joining, the zirconia pellet and, correspondingly, the hole in the elongated insulating tube are made with the central angle 3–4°, as shown in Figure 5.6. The manufacturing technology starts from heating the sensor sample up to 1300°C with the heating rate of 100°C/h. After that is the ceramic sensor elements' exposure to the soaking time of 2 hours with the following temperature increase to 1680°C at the heating rate of 50°C/h. Upon reaching the maximum allowable temperature, the sensor sample has a second soaking time of 5–15 minutes and should be cooled down at the rate of 80°C/h. Figure 5.6 shows the cross-sectional view of the solid electrolyte oxygen sensor. This zirconia sensor consists of the insulating ceramic tube (*1*), which has a conical solid electrolyte pellet (*2*). Me-MeO RE (*3*) is adjacent to the inner surface of the electrolyte pellet. The current conductor (*4*) allows the output signal to be taken from the RE to secondary instrumentation. The insulating tube (*1*) is vacuum-tightly joined to the metal hull of the sensor (*5*), which acts as another current conductor in the measurement of oxygen activity in molten metals.

The present manufacturing technology of the zirconia gas sensors avoids fragmentation in the zirconia connection to the ceramic insulator because the presence of the CaO-Ga_2O_3 mixture in the insulator provides the appearance of the transient-liquid-phase improving solid-state diffusion of ceramic materials during sintering. XRD and thermogravimetric analyses have shown that the appearance of the transient-liquid-phase takes place at 1450°C [10]. The difference in the CTE of the ceramic sensor elements of the oxygen sensor is illustrated in Table 5.1. From the data presented in Table 5.1, it is clear that the discrepancy in the CTE between zirconia and the ceramic insulating material with 1 wt % of CaO-Ga_2O_3 is insignificant for the temperature range of 500–900°C. Moreover, the final density of the isolating tube is 3.49–3.56 g/cm³, which corresponds to 97.5–99.5% of the theoretical density, respectively. The maximum heating rate to the temperature of 1300°C has been selected from conditions of the structural integrity of the ceramic elements. The straight shape of the elongated ceramic tube can be changed if the heating rate

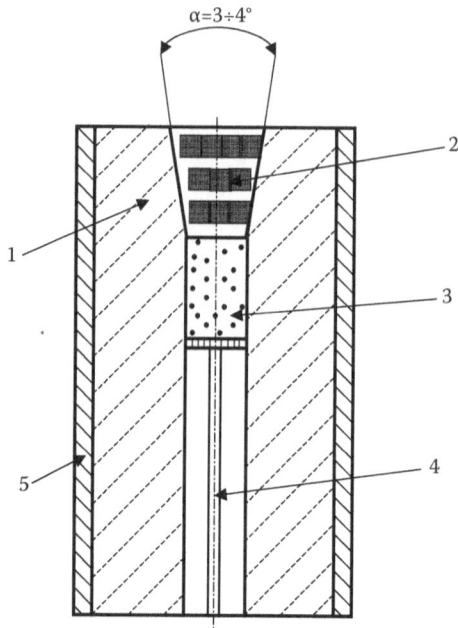

FIGURE 5.6 Cross-sectional view of the oxygen sensor: (*1*) ceramic insulating tube; (*2*) solid electrolyte; (*3*) reference electrode; (*4*) current conductor; and (*5*) metal hull of the sensor.

TABLE 5.1
CTE of the Ceramic Oxygen Sensor Elements

N/N	Chemical Composition	Coefficient of Thermal Expansion ×10⁶, 1/°C at Temperature,° C						
		300	400	500	600	700	800	900
1	ZrO_2-Y_2O_3 (10 mol % Y_2O_3)	8.90	9.18	9.41	9.69	9.91	10.19	10.41
2	$MgAl_2O_4$ + 40 wt % MgO	8.70	9.00	9.28	9.51	9.71	9.95	10.11
3	59 wt % $MgAl_2O_4$ + 40 wt % MgO + 1 wt % (CaO-Ga_2O_3)	8.59	8.88	9.11	9.40	9.61	9.90	10.10

will be more than 100°C/h, which is unacceptable for the following processing. On the contrary, at a lower heating rate the shrinkage rate was found to be inefficient for providing vacuum-tight joining of the solid electrolyte to the insulating ceramic. The duration of soaking time (2 hours) has been experimentally selected to achieve 95% of the theoretical density [10]. After 2 hours, the transient-liquid-phase in the ceramic element disappears and the shrinkage rate decreases substantially. Such behavior completely corresponds to the published data about the joining of ceramics by the transient-liquid-phase [11]. Heating up to 1500°C without the intermediate soaking time at 1300°C was accompanied by the appearance of microcracks on the ceramic interface.

5.2 VACUUM-TIGHT TECHNOLOGIES OF JOINING ZIRCONIA TO SENSOR CONSTRUCTION MATERIALS

Ceramic-metal joining for the construction of zirconia-based gas sensors is the second key component for a wide range of applications of zirconia sensors after joining zirconia to ceramic insulators [24]. So far, several joining techniques have been developed: adhesive joining, friction welding [25], high-energy beam welding [26], microwave joining, ultrasonic welding [27], reaction joining [28], field-assisted bonding [29], brazing [30], diffusion bonding [31, 32], transient-liquid-phase bonding, and partial transient-liquid-phase bonding [12, 33]. Every technology possesses pluses and minuses; however, nowadays brazing and diffusion bonding are the main successful methods for joining zirconia to the sensor construction materials [25]. In addition, partial transient-liquid-phase bonding combining with the advantages of brazing and diffusion bonding has recently been considered as a promising emerging technology for the zirconia-based gas sensors.

The morphology of the metal-ceramic interface depends upon the type of interaction that has occurred. If only physical interaction has occurred, the structure of the metal and ceramic is unchanged. However, if a chemical reaction occurs, the morphology is affected depending upon whether solid-solid or solid-liquid reactions occur and if new interfacial phases are formed. The formation of new interfacial phases alters not only the microstructure but also the physical and mechanical properties. These interfacial phases or reaction product layers are a consequence of the reactions needed to cause wetting of the ceramic. One can consider the reaction product layers as chemical bridges between the metal and ceramic.

It is believed that the advantages or disadvantages of a reaction product layer vary from system to system, and generalization would be difficult. However, a thick reaction product layer tends to weaken the interface due to excessive growth stresses and a brittle nature [12].

As has been stated earlier in this chapter, the CTE mismatch is one of the major obstacles for successful zirconia-insulating ceramics joining. This problem may be even more important for ceramic-metal joining due to the fact that the CTE mismatch during subsequent service of the sensor at high temperatures leads to poor joint strength or failure. It has been proposed that the use of a remaining ductile interlayer, such as active metal brazing, can overcome or reduce the residual stress built up by deformation [12]. Active metal brazing, compared to other joining technologies, has a lower temperature with less influence to joined materials. Therefore, it can be used for precise joining of complex components and heteromaterials. Like all techniques, active metal brazing has its limitations and cannot join all ceramic-metal combinations. For joining ceramics to metals by brazing, the wettability of filler to ceramics is a key factor. In order to improve wettability, on the one hand, it is possible to plate metals directly on ceramics beforehand (Mo-Mn technique) [24]. However, such a conventional metallization process as the Mo-Mn technique is inappropriate for zirconia because it does not have a glassy intergranular phase. Therefore, other techniques such as active metal brazing are required. On the other hand, the active elements are required to be mixed into filler. Such active elements as Ni, Al, Ti, Nb,

Zr, Ta, and Cr are used to react with ceramic to form a reaction layer consequently achieving chemical bonding between ceramics and metals. For example, the addition of Ti to silver-copper alloys promotes low contact angles and hence wettability [30]. The content of the active element in the filler must be reasonable, otherwise the brittleness of the joint will be increased. For instance, the content of Ti in Ag-Cu-Ti fillers should be in the range of 1.5–5.0% [34]. Other examples of fillers are listed in [24].

Although the principle of brazing is the same, the emphasis is different for various metals. With Ag-Cu-Ti, the joining has been realized mainly by reaction with the elimination of porosity in the interlayer. However, with other fillers, the joining is realized mechanically by infiltration of the filler into ceramic. Indeed, Ag-Cu-Ti filler has been successfully used for joining ceramics and metals for a number of years, providing less defects and higher joining strength at low temperatures. The disadvantage of this filler is that this alloy cannot be used at temperatures higher than 500°C. Therefore, the development of intermetallic compounds capable of working at temperatures up to 1000°C at the heating and cooling rate as high as 10°C/min was imperative for further progress in joining ceramic to metal for sensor applications. It was found that Ni_3Al and $NiAl$ intermetallic compounds are considered to be suitable materials for structural applications at elevated temperatures owing to their attractive properties: high melting point, low density, high thermal conductivity, good oxidation and corrosion resistance, and low raw materials cost [7]. Ni_3Al and $NiAl$ can be produced through a termite reaction between nickel and aluminium at a relatively low temperature. Nickel aluminides were produced between the zirconia and the metallic component to achieve the vacuum-tight joining. The morphology, composition, and density of the metal-zirconia joint have been optimized to minimize the CTE mismatch at the metal-ceramic interface, and they are worth considering in detail as they could be used for sensor applications at temperatures up to 1000°C.

The reaction between aluminium and nickel can be controlled by the reaction temperature. Figure 5.7 shows a cross-sectional view of Ni-Al interfaces after being treated in a vacuum at 1000°C, 680°C, and 640°C, respectively. Randomly distributed duplex phases in the reaction product have been produced at 1000°C (Figure 5.7*a*). A nickel aluminide layer with some vertical cracks was formed at 640°C (Figure 5.7*b*). A continuous uniform layer of nickel aluminide was formed between the nickel and the zirconia at 680°C (Figure 5.7*c*). Furthermore, a firm bond between the nickel foil and the zirconia was achieved after the formation of the nickel aluminide layer. Microanalysis indicated that the nickel aluminide layer consists of 43 wt % Ni and 57 wt % Al, and this composition has been maintained constant across the whole nickel aluminide layer [7]. No detectable Al was found in either the Ni foil or the zirconia substrate, and a trace of Ni was found in zirconia in the regions adjacent to the nickel aluminide layer. In accordance with the Ni-Al phase diagram (Figure 5.8), two nickel aluminide phases Ni_2Al_3 and $NiAl$ could be expected in the nickel aluminide layer. However, only the Ni_2Al_3 phase has been detected by the X-ray diffraction spectrum of the reaction layer. The thickness of the formed nickel aluminide layer was found to be around 40–50 μm.

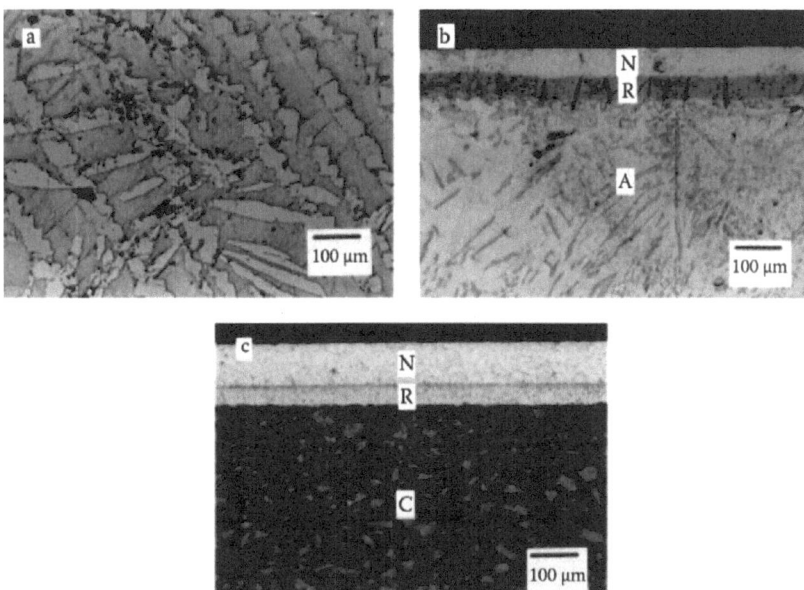

FIGURE 5.7 Optical micrographs of the reaction products between Ni and Al. (*a*) Duplex phases of nickel aluminides formed after a treatment of 1000°C for 1 hour in vacuum; (*b*) a cracked nickel aluminide layer formed after being treated at 640°C for 1 hour in vacuum; and (*c*) a cross-section of a nickel and zirconia joint which was joined together through a nickel aluminide layer at a temperature of 680°C in vacuum. *N*: nickel, *R*: nickel aluminide, *A*: aluminium, and *C*: zirconia. (Reprinted from Mei, J. and Xiao, P., Joining metals to zirconia for high temperature applications, *Scripta Materialia* **40** (1999) 587–594, with permission from Elsevier Science.)

For joining zirconia or other ceramic materials and metals via an intermetallic alloy, the key aspect is to ensure the spreading of liquid Al (the melting point of Al is 660°C) on zirconia, then the reaction between the Al and the Ni form a continuous layer. The eutectic point of the Al-Al$_3$Ni system is 640°C (see the phase diagram in Figure 5.8). Therefore, the working temperature of 680°C has been selected. However, care must be taken to control the reaction temperature since the reaction between Ni and Al could be a combustion reaction [35]. In order to produce a uniform layer of nickel aluminide which bonds to both nickel and zirconia, it is vital to optimize the reaction temperature of Ni and Al. Although Al does not react with zirconia or other ceramics, the high affinity of aluminium to oxygen suggests that aluminium should achieve strong bonding to ceramic. Moreover, the loss of oxygen in the zirconia during the joining process at the vacuum will convert zirconia into a more metallic nature, and Al may well be attracted by the Zr sites in the zirconia. Therefore, chemical bonding between the Al and the zirconia should favor the spreading of liquid Al on the surface of the ceramic. In addition, the application of a gentle force on the zirconia-Al-Ni sandwich during the joining process will also help the spreading of Al on the surface of the electrolyte.

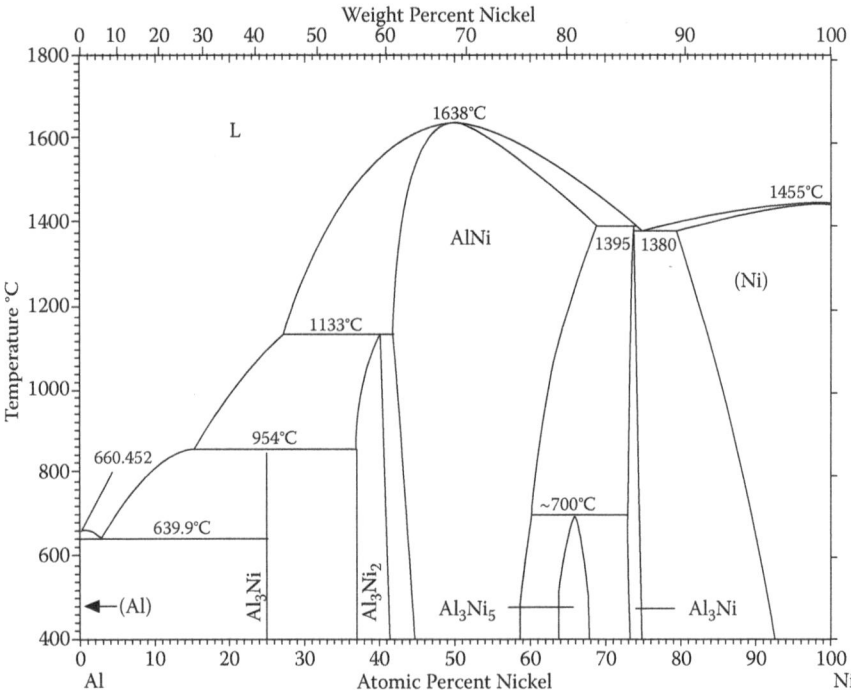

FIGURE 5.8 The binary Ni-Al phase diagram. (Reprinted from Mei, J. and Xiao, P., Joining metals to zirconia for high temperature applications, *Scripta Materialia* **40** (1999) 587–594, with permission from Elsevier Science.)

The advantages of this joining method are not only the low joining temperature, but also the potential high application temperature of the ceramic-metal joint up to 800°C, compared to the upper limit of use temperature of 500°C for Ag-Cu joints, which has been discussed. Nickel aluminides have higher melting points and improved oxidation resistance, compared with that of the Ag-Cu-Ti alloy. As the heat treatment temperature increases, the Ni_2Al_3 formed at 680°C transforms into NiAl, which has an even higher melting point than that of Ni_2Al_3. Further phase transformation from NiAl to Ni_3Al could be predicted, as there is enough Ni source at the nickel aluminide–nickel interface. SEM observation from [7] showed that the thickness of the nickel foil was virtually unchanged, but the thickness of nickel aluminide after the reaction is almost the same as that of aluminium foil before the reaction. Energy dispersive X-ray spectroscopy microanalysis also indicated no traces of diffusion of Al into either the Ni foil or the zirconia substrate, but Ni diffusion into the zirconia was evident. These results suggest that the formation of the nickel aluminide layer might be due to the diffusion of nickel into the aluminium. The enhanced diffusion of Ni as the result of increasing the reaction temperature provides an abundant source of Ni to promote the phase transformation from Ni_2Al_3 to NiAl and then to Ni_3Al. The diffusion of Ni was also confirmed by the fact that a considerable number of voids were found at the Ni side of the joint layer close to the nickel–nickel aluminide interface.

To fabricate reliable metal-ceramic joints for high-temperature zirconia gas sensors, the main challenge is to reduce the thermally induced stresses at the metal-ceramic interface which are due to the difference in the thermal expansion coefficients of the metal and the ceramic to be joined. For example, the CTE at 1000°C ($\times 10^{-6}$ °C^{-1}) of Ni is 18.9, nickel aluminide is 15.6, and zirconia is 11.4. The porous nickel aluminide is treated as a composite, and its CTE was calculated using the rule-of-mixture law $\alpha_c = c_1\alpha_1 + c_2\alpha_2$, where α represents CTE and c represents the content of each component. For the porous Ni_3Al with porosity of 30%, the CTE is very similar to zirconia and is 10.9 [7]. The stress at the interface is generated during both the joining process and the thermal cycling. The stress level is not only determined by the difference in the CTEs of the ceramic and the metal being joined, but also controlled by the fabrication temperature and the temperature range of thermal cycling. The use of the porous nickel aluminide layer considerably reduces the difference in the CTEs of the nickel aluminide layer and the zirconia, and so reduces the thermal stress at the nickel aluminide–zirconia interface. Moreover, there was a large difference of processing temperature in the fabrication of the different types of joints, that is, 680°C for fabricating the Ni-NiAl-zirconia joint and 1250°C for fabricating the Fecralloy-porous Ni_3Al-zirconia joint. However, creep and phase transformation occurring in the nickel aluminide layer, especially in the porous layer, could have reduced the strains caused by the temperature change during the fabrication process and thermal cycling. The porous layer of nickel aluminide also increased the crack resistance of the metal-ceramic joints, since the pores in the nickel aluminide can trap the crack propagation. Therefore, the metal-ceramic joints with the porous layer showed significant improvement of stability during thermal cycling at elevated temperatures.

Diffusion bonding is a suitable technology to achieve a compact joint by diffusion of atoms as well as chemical interactions between materials or interlayer and joining materials. The diffusion of atoms at interface is carried out by several mechanisms, including movement of clearance atoms, movement of vacancies, and so on. Therefore, the surface of materials to be joined must be clean and flat (roughness less than 0.4 μm). The primary variables in the joining process are pressure, temperature, and time. These variables are not independent from each other, and the effect of each variable can be optimized, outlining the important considerations associated with each variable. Joining time can be from a few hours at mild temperature (0.6 T_m, T_m is a melting point of metal to be joined to ceramic) to several minutes at high temperature (0.8 T_m). Diffusion bonding can be achieved with an inserted interlayer or without an interlayer. The interlayer can consequently reduce cracking, relax the thermal residual stress, and improve the joining strength. The interlayer is usually made of element active ceramics [25]. The advantage of diffusion bonding without an interlayer is a very thin reactive layer with a high joining strength. However, high pressure and joining time have to be optimized based on the materials joining. Figure 5.9 illustrates the microdeformation process at the ceramic-metal interface of diffusion bonding without an interlayer. In order to achieve high joining strength without an interlayer, high temperature and applied

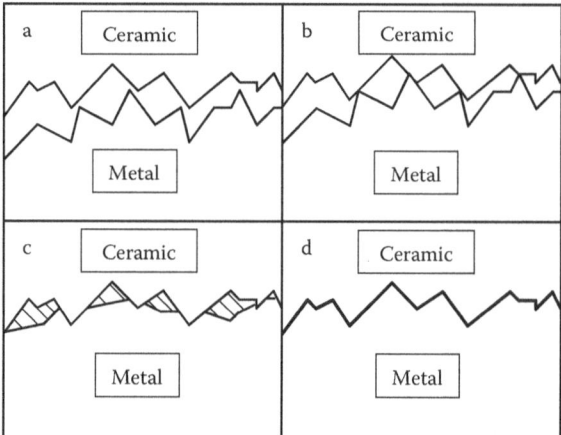

FIGURE 5.9 Different stages of the diffusion bonding without an interlayer at the ceramic-metal interface.

pressure are required to deform metal and eliminate the cavities at the interface. It is clear that this joining method is inappropriate for joining ceramics with high elastic modules to metals. Temperature is a very important variable, and the dependence of strength on temperature can be explained by considering the behavior of the materials in the vicinity of the metal-ceramic interface. Temperature increases interaction across the metal-ceramic interface by increasing the mobility of the atoms and also the mobility of dislocations in the metal during joining. Diffusion and infiltration are thermally assisted processes, and so an increase in temperature is followed by an increase in their respective rates. Increasing temperature can cause a nonwetting-to-wetting transition for liquid metals in contact with ceramics, usually at 1100°C [12]. Unfortunately, the diffusion joining is unsuitable for joining thin metal parts and ceramic components. Moreover, when the joining temperature is too high, brittle compounds can be formed at the joint interface, and their structure, distribution, and thickness will influence joining strength.

Moreover, although much research has been done toward the development of practical technologies for vacuum-tight joining of zirconia to other construction materials of the sensors, there is still a long way to go. Unfortunately, there are neither national nor international testing standards for the joining of ceramics to metals. There also are no testing requirements for how to assess each individual join. Consequently, data from the different sources cannot be compared properly. Therefore, the development of appropriate national and international testing standards as well as further improvement of joining technologies will be priorities for research work for many years to come.

Finally, vacuum-tight joining of sensor construction materials will continue to advance. Designers will continue to place new and greater demands on sensor structures and the materials of SEs (REs). Better performance, higher quality, greater reliability, higher productivity, longer life, and lower cost will be inseparable goals in future sensor designs. Both zirconia and SE materials will continue to be more highly engineered. Design optimization will demand greater and more exotic combinations of materials. The boundaries between materials and structure will become blurred as intelligent enhancement of zirconia and SE (RE) materials become a reality rather than a catchphrase. Consequently, an understanding of joining will have to evolve at least at a comparable pace.

5.3 NANOTECHNOLOGIES FOR ZIRCONIA GAS SENSORS

In reduced dimensions to nanosize, SE materials in gas sensors display characteristics quite different from their bulk behavior. The reduced dimensionality of a system has a profound influence on its physical and electrochemical behavior, more specifically for the nanostructured materials, where the size is comparable to the size of the fundamental physical quantities. Recent technological advances have provided fabrication routes and strategies to reproducibly develop and study reduced-dimensional systems. A recent study of nanostructured materials has projected tremendous potential toward the development of new gas-sensing devices and sensor designs with unique capabilities [36–58]. The SE materials for solid-state gas sensors are electrochemically compatible, chemically inert, but capable of altering electronic properties in the presence of selectively targeted gaseous molecules, and dimensionally compatible with SE molecules, and they have interesting electronic characteristics, thus rendering them as potential electrochemical gas sensors. These new classes of nanostructured SEs (nanowires, nanorods, and nanotubes) have an anisotropy that provides unique properties, which are expected to be critical for the function and integration of nanoscaled devices. The structure of these SEs, characterized by a dense network of closely connected particles, makes the problem of understanding electrical transport properties very complicated. In fact, the electronic properties depend strongly on their conditions of synthesis and also on their morphology. Consequently, the integration of electrochemical sensors and nanostructured materials for a SE requires information to be induced across the interface in a consistent and reproducible format. Recent advances in the field of nanotechnology and processing have resulted in solid-state gas sensors offering unprecedented compatibility of inorganic materials with the chemical gaseous agents, thus enabling stable, direct, and reproducible screening and detection [52–61].

The nanotechnology realm has traditionally been defined as lying in dimensions between 0.1 and 100 *nm*. Figure 5.10 shows the dimensional compatibility of chemical agents to the nanostructured materials of SEs for solid-state gas sensors. So far, nanotechnology has been applied to various industries such as textile, medicine and health, computing, transportation, aeronautics and space exploration, the environment, and so on. In the last decade, however, the specific demand for gas detection and monitoring has emerged, especially as the awareness of the need to protect the environment has grown. Furthermore, the demand for gas sensors has also been driven by the environmental legislation in most of the world's developed countries. Gas sensors using SEs in the form of nanostructured and nanoporous materials are applied in numerous fields of application [40, 51, 56, 58, 61].

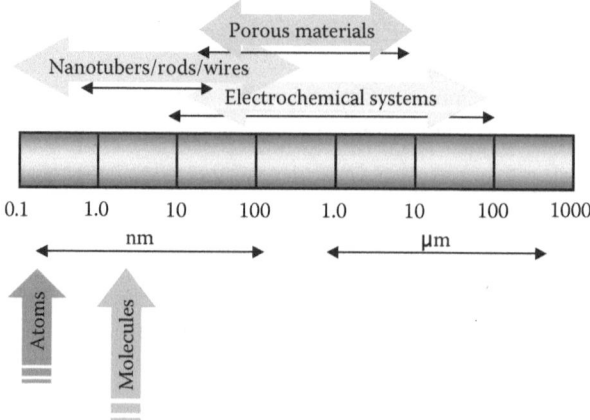

FIGURE 5.10 Dimensional compatibility of chemical agents to the nanostructured materials.

The answer to why nanotechnology has been used for the development of the solid-state gas sensor is that increasing surface-to-volume ratio increases grain size dependence and ultimately provides higher sensitivity. This is evident for the semiconductor gas sensors detecting air pollutants, since the reactions at grain boundaries of the SE and complete depletion of carriers in the grains can strongly modify the material transport properties. These sensors are able to operate with a high level of stability under deleterious conditions, including chemical and/or thermal attack. Basically, the semiconductor gas sensors used for detecting air pollutants are usually produced simply by coating a nanostructured sensing (metal-oxide) layer on a substrate with two electrodes. Typical materials are SnO_2, ZnO, TiO_2, MoO_3, V_2O_5, and WO_3 with typical operating temperatures of 200–400°C [57]. The general sensing mechanism for the semiconductor gas sensor is a change in the resistance (or conductance) of the sensor when it is exposed to pollutant gas, relative to the sensor resistance in background air. The sensor resistance is the best-known sensor output signal and is in most cases determined at constant operation temperature and by DC measurement [40]. Therefore, most of the progress for the last 10–15 years in the development of the nanostructured SE has been reported for semiconductor gas sensors.

On the other hand, the influence of a high surface-to-volume ratio on the sensitivity of the zirconia-based electrochemical gas sensors is complex and can differ from one type of sensor to another. Usually, a high surface-to-volume ratio of the SE provides better sensitivity toward the measuring gas [62, 63]. However, the reproducibility of the measurement would not be stable enough [64]. Specifically, in the ordinary zirconia sensors measuring O_2 partial pressure in the gaseous environment, a high surface-to-volume ratio of both Pt electrodes enhances the sensor sensitivity owing to the increases in the TPB and the number of active Pt sites for dissociative reaction of O_2 at the TPB: $O_2 + 4e^- \rightarrow 2O^{2-}$. However, for the zirconia-based mixed-potential NO_2 sensors with an oxide-SE, a high surface-to-volume ratio decreases the NO_2 sensitivity substantially [65]. During diffusion toward the zirconia-SE interface, NO_2 can be further dissociated to NO and oxygen ions O^{2-} before reaching the TPB. It depends on two reasons: (1) the exhaust gas may reach equilibrium (at temperatures higher than 600°C, the content of NO in the equilibrated total NO_x mixture is more than 90%), and (2) the dense matrix of nanosized SE can accelerate NO_2 decomposition. The larger grains of the SE, produced on zirconia by traditional thin- and thick-film technology, may have fewer reaction sites at the TPB during NO_2 measurement. Thus, in the case of using the traditional technologies for SE sintering, the catalytic activity to the cathodic reaction of NO_2 would be lower compared with the high surface-to-volume ratio, which is usually observed in the nanosized oxide-SEs [65]. In contrast, a low surface-to-volume ratio provides high bulk porosity and, consequently, much less contacting points for NO_2. Thus, NO_2 can reach the zirconia-electrode interface without serious decomposition to NO. In addition, such a geometrical factor of sensor design as the thickness of the SE also influences the sensitivity of the mixed-potential NO_2 sensor. It has been experimentally proven that the sensor with a thickness of the oxide-SE in nanometers provides much higher NO_2 sensitivity than the sensor with a thickness of the same material of the SE in micrometers [66].

Typical SEM images of the zirconia-based NO_2 sensor attached with a nanostructured NiO-SE sintered at 1000°C are shown in Figure 5.11. Thickness of such a nanostructured SE can vary from 25 to 150 nm. In the SE presented in this figure, the thickness was about 100 nm and consisted of colonies of NiO grains uniformly distributed throughout the surface and bulk of the SE.

Despite much attention and intensive investigation of the various physical and chemical properties in nanosized systems, there is still an obvious lack of information available on their reactivity and transport properties, specifically if the sintered SE represents two nanosized oxides. Concerning the reactivity of such systems, one can see that the majority of research work has been dealing with surface reactions in supported catalytic systems, ion exchange, and intercalation processes but not with solid-state synthesis. It was found that the manifestation of the "nanofactor" depends on the nature of the oxide partner, in particular on its solid-state dispersal ability, surface mobility, temperature, and type of experiment [39]. Some of the results can occur due to a "trivial" size factor, that is, an increase in interaction contact area in powder mixtures $NiO^n + MoO_3$ and $NiO^{m,n} + Al_2O_3^{m,n}$ (if at least one of the initial oxides is nanoscaled), and low conductivities of nanostructured NiO and Al_2O_3. However, some data cannot be explained on a conventional basis. This assertion relates to the stabilization of the α-Bi_2O_3 low-temperature polymorph in contact with $Al_2O_3^n$ due to the action of a strong surface force of the latter. It has also been reported [39] that the internal mechanism of stabilization is not understood at the moment. The apparent dispersion of NiO on the $Al_2O_3^n$ surface can be regarded as another example of "true" size effect, resulting in the significant change of the conductivity level in NiO. Moreover, the observed diffusion permeability of the $NiMoO_4$ layer grown at the surface of nanosized NiO^n could also be considered as a manifestation of "true" size effect. Finally, the data on reactivity demonstrate the complex, sometimes unpredictable behavior of oxide systems containing nanostructured components.

Both literature as well as research experience further suggest that additional impurities need to be added to conventional binary metal oxides to stabilize the

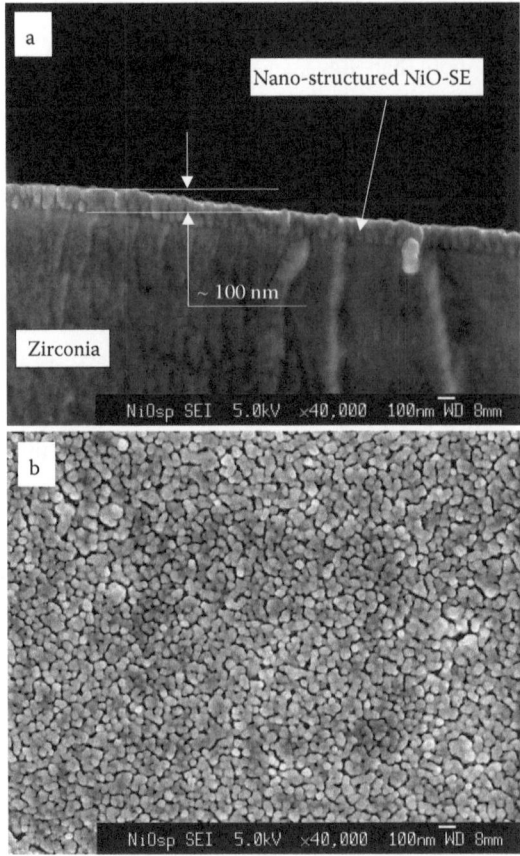

FIGURE 5.11 SEM images of a (*a*) cross-sectional view and (*b*) surface view of the zirconia-based NO$_2$ sensor attached with a nanostructured NiO-SE.

porous nanocrystal morphology of the SEs. The literature also suggests that the fundamental properties of sensors are strongly dependent on the technologies (thin- and thick-film, etc.) and techniques used for the realization. Then new deposition techniques and optimized deposition processes, which allow controlling of the material characteristics and consequently their performances, need to be investigated. Among the many techniques that had been used for the deposition of metal-oxide-SEs for the zirconia-based gas sensors are sputtering, chemical vapor deposition (CVD), thermal evaporation, spray pyrolysis, and the like [57]. Nanoparticles of Al$_2$O$_3$ and ZrO$_2$ in the size of ~20 *nm* can also be achieved by screen printing and micropowder injection molding [46, 55]. Sol-gel, as an alternative technique, seems to offer some specific advantages for obtaining nanostructured SEs for various gas sensors [49, 53, 54]. These are simplicity, flexibility, low cost, ease of use on large substrates, and the capability to modify the composition by the addition of dopants and modifiers. A typical structure of the nanostructured NiO-SE with additives of 10 wt % CuO deposited on zirconia substrate is shown in Figure 5.12*c* [66]. This

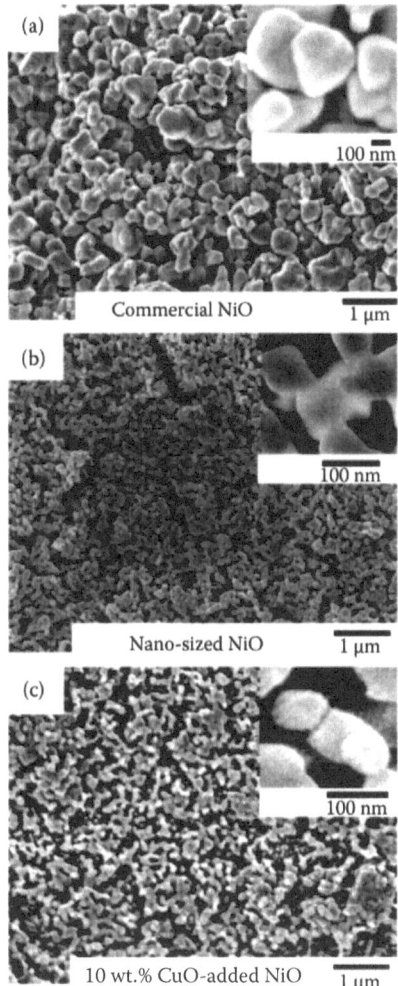

FIGURE 5.12 Typical structure of the nanostructured NiO-SE: (*a*) commercial pure NiO; (*b*) synthesized pure NiO; and (*c*) 10 wt % CuO-added synthesized NiO. (From Plashnitsa, V.V., Ueda, T., and Miura, N., Improvement of NO_2 sensing performances by an additional second component to the nano-structured NiO sensing electrode of YSZ-based mixed-potential-type sensor, *Int. J. Appl. Ceram. Tech.* **3** (2006) 127–133. With permission.)

SE material yielded a more porous structure with NiO grain size maintained between 80 to 120 *nm*. It has also been reported that a fine CuO additive was homogeneously distributed throughout the nanostructured NiO matrix with an average size of ~100 *nm*, whereas the average size of the commercially available NiO (Figure 5.12*a*) was ~700 *nm* [66]. The positive influence of the two-oxide SE was in the enhancement of NO_2 sensitivity of the zirconia-based sensor at high temperatures.

Figure 5.13 illustrates the dependence of Δemf in the NO_2 concentration of the sensor attached with NiO-CuO-SEs at 800°C in the presence of 5 vol. % H_2O. Wet

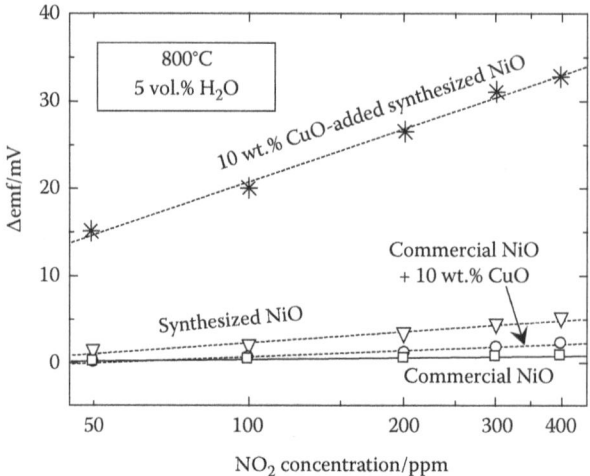

FIGURE 5.13 Dependence of Δemf on the NO_2 concentration for the zirconia-based sensors attached with different NiO-based SEs at 800°C in the presence of 5 vol. % H_2O. (From Plashnitsa, V.V., Ueda, T., and Miura, N., Improvement of NO_2 sensing performances by an additional second component to the nano-structured NiO sensing electrode of YSZ-based mixed-potential-type sensor, *Int. J. Appl. Ceram. Tech.* **3** (2006) 127–133. With permission.)

conditions always exist in car exhausts, and therefore it is essential to establish both high sensitivity and rapid response-recovery time at high temperatures for zirconia gas sensors. It is clearly shown in this picture that the NO_2 sensitivity for the sensor attached with an NiO-CuO-SE has been improved substantially in comparison with the same sensor attached with a commercial NiO-SE. The response-recovery time was estimated to be less than 3 minutes at 800°C. The results obtained strongly suggested that the addition of 10 wt % CuO to nanostructured NiO degrades the catalytic activity for anodic reaction of oxygen, which has been supported by the measured polarization curves. This degradation was also confirmed by the results of complex-impedance measurements [66] and ultimately was responsible for such a big improvement in NO_2 sensitivity. Interestingly, the optimization of the two-oxide nanostructure of the SE leads not only to challenges in the sensor sensitivity at high temperatures, but also to the overall improvement of the sensor's selectivity. Figure 5.14 shows the real-time response transients to various gases, such as CO, C_3H_8, NO, and NO_2 (400 ppm each) at 800°C in the humid conditions explained above. The zirconia sensor attached with an NiO-CuO-SE has shown an excellent selectivity to NO_2 and has practically no cross-sensitivity to any other gases.

5.4 LIMITATIONS OF EXISTING TECHNOLOGIES AND FUTURE TRENDS

Nowadays, a lot of expectations are placed on nanotechnology. Others are afraid that it is just another hype or buzzword. However, unlike, for example, dotcoms and telecoms, nanotechnology is to a greater extent enabling technology with a solid,

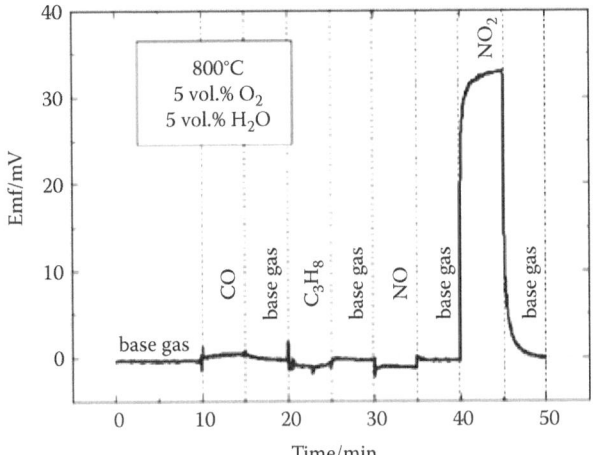

FIGURE 5.14 Cross-sensitivities to various gases, such as CO, C_3H_8, NO, and NO_2 (400 ppm each), at 800°C in humid conditions for the sensor attached with a 10 wt % CuO-added synthesized NiO-SE. (From Plashnitsa, V.V., Ueda, T., and Miura, N., Improvement of NO_2 sensing performances by an additional second component to the nano-structured NiO sensing electrode of YSZ-based mixed-potential-type sensor, *Int. J. Appl. Ceram. Tech.* **3** (2006) 127–133. With permission.)

broad-based scientific background, which will have various implementation possibilities for gas sensors. Nanotechnology now perhaps receives the greatest interest in the scientific world, but obviously such revolutions are waiting for their time to break through.

Bulk nanostructured materials are solids with nanosized microstructure. Their basic units are usually nanoparticles. Several properties of nanoparticles are useful for applications in electrochemical sensors [67]. However, their catalytic behavior is one of the most important. The high ratio of surface atoms with free valences to the total atoms has led to the catalytic activity of nanostructured SEs being used in electrochemical reactions. The catalytic properties of nanoparticles could decrease the overpotential of electrochemical reactions and even provide reversibility of redox reactions, which are irreversible at the bulk metal SE [68]. Multilayers of conductive nanoparticles assembled on electrode surfaces produce a high porous surface with a controlled microenvironment. These structures could be thought of as assemblies of nanoelectrodes with controllable areas.

Basically, there are two fundamental approaches to fabricating nanomaterials [69]. The "bottom-up" approach represents the concept of constructing a nanomaterial from basic building elements, that is, atoms or molecules. This approach illustrates the possibility of creating materials of SEs with exactly the properties desired. The second approach, the "top-down" method, involves restructuring a bulk material in order to create a nanostructure. Inert gas condensation, considered a bottom-up technique, was the first method used to intentionally construct a nanostructured material, and has become a widespread means of producing nanostructured metals, alloys, intermetallics, ceramic oxides, and composites

[70]. Mechanical alloying, another commonly used method to produce nanostructured materials, provides an example of a top-down method.

However, despite a variety of existing technologies producing nanostructured elements of the zirconia gas sensors, there is a huge undeniable problem associated with the high-temperature applications of these sensors. The vast majority of zirconia gas sensors work at temperatures over 600°C for a long time, which inevitably affects the properties of SEs and SE–solid electrolyte interfaces. Longtime exposure to high temperatures causes growth of the grains' size and grain boundaries for most of the oxide materials and/or composites employed as SEs today. Following the longtime exposure to high operating temperatures, the TPB at the SE-zirconia interface inevitably changes and ultimately reduces gas sensitivity of the sensor. A typical scenario for any new zirconia-based electrochemical gas sensor attached with a nanostructured SE consists of the following main stages:

1. *Stage of stabilization.* After the first 12–48 hours of working at a temperature higher than 600°C, the output *emf* of the zirconia sensor stabilizes. The signal is usually stable, and the sensor sensitivity is unchangeable for all concentrations of the measuring gas. The accuracy of measurement is also stable for the whole measuring temperature range. This stage is characterized by the stabilization of the SE-zirconia interface to the certain level of sensitivity.

2. *Stage of stability.* It usually lasts from 3 weeks to 4–6 months depending on the properties of SE material and the manufacturing technology used for attaching the SE to the zirconia substrate. This stage is characterized by precise, stable measurements of the gas concentrations at different measuring, repeatable temperatures with a high accuracy level.

3. *Stage of decay.* This stage is characterized by gradual, steady, and irreversible reduction of the sensor output *emf*, which leads to decreases in the sensor's sensitivity. Sometimes it is even difficult to recognize the beginning of this stage, but the longer the sensor works at high temperatures, the clearer the degradation in the sensor's *emf* is, which is caused by the changes at the TPB. The duration of this stage may vary from a few months up to several years.

Therefore, the most interesting results achieved with the nanostructured oxide and/or composite SEs have been reported to be the stability of high temperature measurements from 3 weeks to a couple of months. There are a few exceptions where the sensor-*emf* growths from month to month more or less stabilize after the first 3 months of operation at 700°C. The example of such an oxide-SE is the $ZnFe_2O_4$-SE for a NO_2 high-temperature sensor [71, 72]. However, overall, hundreds of publications report the incremental improvements in the SE's properties, electrode's modifications, and sensor's selectivity, and/or the technological improvements achieved for the relatively short period of time. Unfortunately, in spite of a great amount of works in the area, the approach for the SE materials selection is still highly empirical and proprietary. Furthermore, there has been a misunderstanding regarding the concept of "long-term stability" in many publications to date.

Usually, long-term stability has been reported only for several months of the sensor's operation at high temperatures with lack of supporting evidences relative to the real long-term stability. In contrast, as far as industry is concerned, the long-term stability starts only after one year of operation (about 10,000 hours). Thus, the claim of long-term stability in the majority of works published to date is irrelevant, and, consequently, only a few zirconia-based gas sensors with nanostructured SEs have been commercialized so far. Moreover, it is rare to see the comprehensive study which would lead to the serious conclusions supported by extensive experimental data. Researchers are trying to "save face" and not report the degradation in the sensor characteristics after initial bright publications. On the other hand, industry is still reluctant to accept the preliminary results of nanotechnology, which have not yet passed the significant impact of the long-term stability tests, which would change the foundations of nanoengineering and nanotechnology in contrast to the conventional manufacturing methods. Contrary to laboratory tests, industrial long-term stability tests possess "screw-the-product" attitude toward developed sensors in order to verify their durability.

It is foreseen that the design of methods of grain-size stabilization during the long-term exploitation of nanoscale zirconia gas sensors will gain priority over the design of methods producing nanoscaled materials with minimal grain size. Some studies [61, 72–74] have considered the possibility of introducing microadditives of various impurities in order to limit the mobility of adatoms and stabilize grain size. For example, impregnating Pt into the matrix of the $ZnFe_2O_4$-SE has enhanced response-recovery time substantially for the zirconia-based NO_x sensor [72]. Both experimental research and theoretical simulations have established that in case of the presence of a second phase, this effect takes place when the average domain size is comparable to the average interparticle distance. Therefore, the control of impurities in metal oxide is one way to achieve the particular grain size, morphology, and structural stability necessary in practical applications. However, it is well-known that even small quantities of the material additives used for grain-size stabilization can affect catalytic and adsorption surface properties of SEs. Consequently, although the grain size may be stabilized, there may be a strong change in both the electrophysical and the gas-sensing properties of oxide-SEs and the TPB. However, this research is still in the early stages; therefore, the detailed studies of applied metal oxides are needed to develop real technology for the stabilization of nanoscaled materials.

Furthermore, the results given above testify that the further development of technologies for nanostructured SE (RE) materials really has great possibilities to improve selectivity, to increase sensitivity, to decrease the response and recovery times, to shift the maximum of sensitivity in the range of lower operation temperatures, and to design nanostructured gas sensors with new consumer properties. However, because of a lack of understanding of the nature and mechanism of gas sensitivity, the structural engineering method of developing new materials for SEs (REs) dominates and remains an empiric one, especially in the field of catalytic additives selection. As it follows from the analysis of contemporary works devoted to solid electrolyte gas sensors, the understanding of the nature and mechanism of many processes responsible for the effectiveness of gas detection reactions has to be reconsidered in order to obtain a comprehensive picture of all processes that occur at the TPB [75]. Therefore, though the addition of a second phase seems to be a feasible approach for achieving high sensitivity and selectivity, the process of choosing the additive becomes a challenge due to the lack of basic understanding.

Consequently, nanotechnology will of necessity change, and the different nanotechnologies will change to different degrees, not just in different ways. It is clear that increased efforts in basic studies for better understanding of sensing mechanisms become an important condition of progress achievement in the elaboration of solid electrolyte gas sensors acceptable for practical use. However, it is necessary to note that the science and technology of metal-oxide gas sensors have advanced considerably in the last decade. The theoretical understanding of the surface processes, involved in the reactions of gas detection, is developing, and therefore a detailed understanding of the elementary processes underlying sensor behavior now seems achievable. Finally, entirely new approaches will be required to deal with the increasingly blurred boundary between materials for SEs (REs) and solid-electrolyte structures epitomized by nanocrystalline and "smart" materials.

REFERENCES

1. Gobina, E., Gas sensors and gas metering: Applications and markets, *BCC Report* (2005) 75.
2. Messler, R.W., The challenges for joining to keep pace with advancing materials and designs, *Mater. & Design* **16** (1995) 261–269.
3. Akbar, S., Dutta P., and Lee, C., High-temperature ceramic gas sensors: A review, *Int. J. Appl. Ceram. Techn.* **3** (2006) 302–311.
4. Bannister, M.J. et al., *Oxygen Sensors*, U.S. Patent No. 4,193,857 (1980).
5. Oh, S. et. al., Multilayer ionic devices fabricated by the plasma-spray method, *Solid State Ionics* **53–56** (1992) 90–94.
6. Nicholas, M.G., *Joining of Ceramics*, London, Chapman & Hall, 1990, 207.
7. Mei, J. and Xiao, P., Joining metals to zirconia for high temperature applications, *Scripta Materialia* **40** (1999) 587–594.
8. Gutierrez-Mora, F. et. al., Influence of internal stresses on superplastic joining of zirconia-toughened alumina, *Acta Mater.* **50** (2002) 3475–3486.
9. Aravindan, S. and Krishnamurthy, R., Joining of ceramic composites by microwave heating, *Mater. Letters* **38** (1999) 245–249.
10. Krylov, U.V. et al., *Solid Electrolyte Oxygen Sensor and Its Fabrication Technology*, USSR Patent No. 1,752,069 (1991).
11. Sugar, J.D. et al., Transient-liquid-phase and liquid-film-assisted joining of ceramics, *J. Europ. Ceram. Soc.* **26** (2006) 363–372.
12. Jadoon, A.K., Ralph, B., and Hornsby, P.R., Metal to ceramic joining via a metallic interlayer bonding technique, *J. Mater. Proc. Techn.* **152** (2004) 257–265.
13. Chang, J.T. et al., Deposition of yttria-stabilized zirconia films using arc ion plating, *Surf. & Coat. Techn.* **200** (2005) 1401–1406.
14. Hanson, W.B., Ironside, K.I., and Fernie, J.A., Active metal brazing of zirconia, *Acta Materialia* **48** (2000) 4673–4676.
15. Brinkiene, K. and Kezelis, R., Effect of alumina addition on the microstructure of plasma sprayed YSZ, *J. Europ. Ceram. Soc.* **25** (2005) 2181–2184.

16. Gitzen, W.H., *Alumina as a Ceramic Material, The American Ceramic Society* 1970 241.

17. Lesuer, D.R., Wadsworth, J., and Nieh, T.G., Forming of superplastic ceramics, *Ceram. Int.* **22** (1996) 381–388.

18. Mortimer, A.G. and Reed, G.P., Development of a robust electrochemical oxygen sensor, *Sens. Actuators B: Chem.* **24** (1995) 328–335.

19. Mohamed, F.A. and Li, Y., Creep and super-plasticity in nanocrystalline materials: Current understanding and future prospects, *Mater. Sci. Eng. A* **298** (2001) 1–15.

20. Gutierrez-Mora, F. et al., Joining advanced ceramics by plastic flow, *Ceram. Int.* **30** (2004) 1945–1948.

21. Goretta, K.C. et al., Joining alumina/zirconia ceramics, *Mater. Sci. Eng. A* **341** (2003) 158–162.

22. Boniecki, M. et al., Superplastic joining of alumina and zirconia ceramics, *J. Europ. Ceram. Soc.* **27** (2007) 1351–1355.

23. Zhuiykov, S., Zirconia single crystal analyser for low-temperature measurements, *Proc. Contr. Qual.* **11** (1998) 23–37.

24. Zhanh, Y. et al., Progress in joining ceramics to metals, *Int. J. Iron & Steel Res.* **13** (2006) 1–5.

25. Nicholas, M.G., *Joining Structure Ceramics in Designing Interfaces for Technological Applications*, London, Elsevier, 1989, 287.

26. Fernie, J.A. et al., Progress in joining of advanced materials, *Weld. & Mater. Fabr.* **59** (1991) 179–184.

27. Matsuoka, S., Ultrasonic welding of ceramic/metals using inserts, *J. Mater. Proc. Techn.* **75** (1998) 259–265.

28. Miyamoto, T., Ceramic-to-metal welding by a pressured combustion reaction, *J. Mater. Res.* **1** (1986) 7–9.

29. Ming, Q.S. and Sexual, J., Joining mechanism of field-assisted bonding of electrolytes to metals, *J. Mechan. Eng.* **18** (2002) 1–5.

30. Hanson, W.B., Ironside, K.I., and Fernie, J.A., Active metal brazing of zirconia, *Acta Mater.* **48** (2000) 4673–4676.

31. Novikov, V.G., Diffusion bonding dissimilar materials in aerospace technology, *Weld. Int.* **65** (1995) 477–478.

32. Muolo, M.L. et al., Wetting, spreading and joining in the alumina–zirconia–Inconel 738 system, *Scripta Mater.* **50** (2004) 325–330.

33. MacDonald, W.D. and Eager, T.W., Transient liquid phase bonding, *Ann. Rev. Mater. Sci.* **22** (1992) 23–46.

34. Vianco, P.T. et al., A barrier layer approach to limit Ti scavenging in FeNiCo/Ag-Cu-Ti/Al_2O_3 active braze joints, *Weld. Journal* **82** (2003) 252–262.

35. Locatelli, M.R., Tomsia, A.P., and Nakashma, K., New strategies for joining ceramics for high-temperature applications, *Key Eng. Mater.* **111** (1995) 157–190.

36. Hotovy, I. et al., NiO-based nanostructured thin films with Pt surface modification for gas detection, *Thin Solid Films* **515** (2006) 658–661.

37. Lee, K.D., Preparation and electrochromic properties of WO_3 coating deposited by the sol-gel method, *Sol. Ener. Mat. & Sol. Cells* **57** (1999) 21–30.

38. Rothschild, A. and Komem, Y., The effect of grain size on the sensitivity of nanocrystalline metal-oxide gas sensors, *J. Appl. Phys.* **95** (2004) 6374–6380.

39. Neiman, A. et. al., Solid state interactions in nano-sized oxides, *Solid State Ionics* **177** (2006) 403–410.

40. Zhuiykov, S., Wlodarski, W., and Li, Y., Nanocrystalline V_2O_5-TiO_2 thin-films for oxygen sensing prepared by sol-gel process, *Sens. Actuators B, Chem.* **77** (2001) 484–490.

41. Menil, F., Debeda, H., and Lucat, C., Screen-printed thick-films: From materials to functional devices, *J. Europ. Ceram. Soc.* **25** (2005) 2105–2113.

42. Tan, O.K. et. al., Nanostructured oxides by high-energy ball milling technique: Application as gas sensing materials, *Solid State Ionics* **172** (2004) 309–316.

43. Piticescu, R., Monty, C., and Millers, D., Hydrothermal synthesis of nanostructured zirconia materials: Present state and future prospects, *Sens. Actuators B: Chem.* **109** (2005) 102–106.

44. Yang, H. et al., Synthesis of $ZnFe_2O_4$ nanocrystallites by mechanochemical reaction, *J. Physics & Chem. Solids*, **65** (2004) 1329–1332.

45. Čyvienë, J. and Dudonis, J., Preparation of $Zr_{1-x}Al_xO_2$ by annealing of ZrO_x/Al thin films in the air atmosphere, *Mater. Sci. Eng.: C* **26** (2006) 1102–1105.

46. Zhang, Y. et. al., A study of the process parameters for yttria-stabilized zirconia electrolyte films prepared by screen-printing, *J. Power Sources* **160** (2006) 1065–1073.

47. Čyvienë, J., Laurikaitis, M., and Dudonis, J., Deposition of nanocomposite Zr–ZrO_2 films by reactive cathodic vacuum arc evaporation, *Mater. Sci. Eng. B* **118** (2005) 238–241.

48. Kaya, C. and Butler, E.G., Near net-shape manufacturing of alumina/zirconia high temperature ceramics with fine scale aligned multiphase microstructures using co-extrusion, *J. Mater. Proc. Tech.* **135** (2003) 137–143.

49. Klinger, R.E., Thin film deposition technologies, and structure/property relationships applied to solid state ionic conductors, *Solid State Ionics* **52** (1992) 249.

50. Huisman, W., Graule, T., and Gauckler, L.J., Centrifugal slip casting of zirconia (TZP), *J. Europ. Ceram. Soc.* **13** (1994) 33–39.

51. Badwal, S.P.S. and Rajendran, S., Effect of micro- and nano-structures on the properties of ionic conductors, *Solid State Ionics* **70–71** (1994) 83–95.

52. Anderson, M.A. and Xu, Q., *Metal Oxide Porous Ceramic Membranes with Small Pore Sizes*, U.S. Patent No. 5,104,539 (1992).

53. Chiba, R. et. al., Ionic conductivity and morphology in Sc_2O_3 and Al_2O_3 doped ZrO_2 films prepared by the sol-gel method, *Solid State Ionics* **104** (1997) 259–266.

54. Livage, J. et al., Sol-gel synthesis of oxide materials, *Acta Materialia* **46** (1998) 743–750.

55. Zauner, R., Micro powder injection moulding, *Microelectr. Eng.* **83** (2006) 1442–1444.

56. Kimura, T. and Goto, T., Ir–YSZ nano-composite electrodes for oxygen sensors, *Surf. Coat. Tech.* **198** (2005) 36–39.

57. Vaseashta, A. and Dimitrova-Malinovska, D., Nanostructured and nanoscale devices, sensors and detectors, *Sci. Techn. Adv. Mater.* **6** (2005) 312–318.

58. Carotta, M.C., et al., Nanostructured thick-film gas sensors for atmospheric pollutant monitoring: Quantitative analysis on field tests, *Sens. Actuators B: Chem.* **76** (2001) 336–342.

59. Menil, F., Debeda, H., and Lukat, C., Screen-printed thick films: From materials to functional devices, *J. Eur. Ceram. Soc.* **25** (2005) 2105–2113.

60. Wu, P.Y. et al., Low-temperature synthesis of zinc oxide nanoparticles, *Int. J. Appl. Ceram. Techn.* **3** (2006) 272–278.

61. Korotchensev, G., Gas response control through structural and chemical modification of metal oxide films: State of the art and approaches, *Sens. Actuators B: Chem.* **107** (2005) 209–232.

62. Comini, E. et al., CO sensing properties of W–Mo and tin oxide RGTO multiple layers structures, *Sens. Actuators B: Chem.* **95** (2003) 157–161.

63. Nowonty, J., Bak, T., and Sorrell, C.C., Charge transfer at oxygen/zirconia interface at elevated temperatures. Part 7: Effect of surface processing, *Adv. Appl. Ceram.* **104** (2005) 195–199.

64. Zhuiykov, S., Mathematical modeling of YSZ-based potentiometric gas sensors with oxide sensing electrodes. Part II: Complete and numerical models for analysis of sensor characteristics, *Sens. Actuators B: Chem.* **120** (2007) 645–656.

65. Elumalai, P. et al., Sensing characteristics of YSZ-based mixed-potential-type planar NO_x sensor using NiO sensing electrodes sintered at different temperatures, *J. Electrochem. Soc.* **152** (2005) H95–H101.

66. Plashnitsa, V.V., Ueda, T., and Miura, N., Improvement of NO_2 sensing performances by an additional second component to the nano-structured NiO sensing electrode of YSZ-based mixed-potential-type sensor, *Int. J. Appl. Ceram. Tech.* **3** (2006) 127–133.

67. Chevallier, L. et al., Non-Nernstian planar sensors based on YSZ with Ta (10 at. %)-doped nanosized titania as a sensing electrode for high-temperature applications, *Int. J. Appl. Ceram. Tech.* **3** (2006) 393–400.

68. Riu, J., Maroto, A., and Xavier Rius, F., Nanosensors in environmental analysis, *Talanta*, **69** (2006) 288–301.

69. Puurunen, K. and Vasara, P., Opportunities for utilising nanotechnology in reaching near-zero emissions in the paper industry, *J. Cleaner Prod.* (in press).

70. Lane, R., Craig, B., and Babcock, W., Material ease: Materials engineering with nature's building blocks, *Amptiac* **6** (2002) 31–36.

71. Zhuiykov, S. et al., Stabilized zirconia-based NO_x sensor using $ZnFe_2O_4$ sensing electrode, *Electrochem Solid-State Lett.* **4** (2001) H19–H21.

72. Zhuiykov, S. et al., High-temperature NO_x sensors using zirconia solid electrolyte and zinc-family oxide sensing electrode, *Solid State Ionics* **152–153** (2002) 801–807.

73. Holody, P.R.J., Soltis, R.E., and Hangas, J., Limiting particle growth in platinum/tin oxide nanocomposites, *Scripta Mater.* **22** (2001) 1821–1824.

74. Commini, E., Metal oxide nano-crystals for gas sensing, *Anal. Chimica Acta* **568** (2006) 28–40.

75. Zhuiykov, S., Mathematic model of electrochemical sensors with distributed temporal and spatial parameters and its transformation to models of the real YSZ-based sensors, *Ionics* **12** (2006) 135–148.

6 Errors of Measurement of Zirconia Gas Sensors

6.1 BASES OF ERRORS THEORY IN RELATION TO ELECTROCHEMICAL GAS SENSORS

In accordance with BS EN 61207-1:1994, the error is the difference between the device reading and the real value of the measuring parameter [1]. Indeed, if the real value is equal to x and the measured value is equal to x_m, then the absolute error of the measuring device is $\Delta x = x_m - x$. In this case the error manifests itself in the units of the physical or chemical measuring parameter. However, it is not always convenient to manifest error in the absolute values because, in reality, the absolute value of the error indicates neither the quality of the measuring instrument nor the quality of the measuring process. In fact, the accuracy of zirconia gas sensors is determined by a combination of internal errors (due to random and systematic effects); fluctuation of external sensitivities such as temperature, presence of moisture, and combustible gases in the measuring environment; and uncertainty of the gas mixtures used to calibrate it. The term *uncertainty* should be used to characterize the inaccuracy of an obtained result, while the term *error* is used to characterize the components of the uncertainty.

Theoretically, in relation to the zirconia gas sensors, the logarithmic relationship should permit signal resolution at the extremes of the partial pressure differential. In reality, interference from reducing gases that are present even in high-purity inert gases will cause false low readings. The external influences can be sufficiently minimized by conducting the measurement under well-controlled conditions. Unfortunately, even under the controlled conditions, any combustible gases in the process (CO, H_xC_y, H_2, etc.) will burn with oxygen at the sensor electrodes, consuming oxygen. Due to the fact that Pt is an excellent catalyst material, Pt electrodes of the zirconia sensor act in the high normal operating temperature to catalyze reactions between oxygen and other gas constituents in the gas sample, such as *ppm* or *ppb* levels of residual hydrocarbon vapors. In atmospheres such as incinerators or heavy fuel-fired units, high-molecular-weight combustibles result in more serious error. One molecule of propane, for example, consumes 5 molecules of oxygen ($C_3H_8 + 5O_2 \leftrightarrow 2CO_2 + 4H_2O$). A similar situation has been observed for oxygen in the presence of carbon monoxide, for example 4% O_2 in the presence of 2% CO: $O_2 + 2CO \leftrightarrow 2CO_2$. Each molecule of O_2 reacts with 2 molecules of CO on the surface of the Pt-SE, so the zirconia sensor will read 3% rather than 4%. Reactions between the reducing gas and oxygen molecules will take place in near perfect stoichiometric balance because of the superb catalytic properties of the Pt-SE. However, if even minuscule amounts of residual oil vapors are present, such as the vapors emitted from a human fingerprint, a more dramatic error will result from the stoichiometric proportioning. Two $C_{12}H_{26}$ heavy hydrocarbon molecules will consume 37 oxygen molecules. Therefore, if the background gas contains 10 ppm of the hydrocarbon, a nearly 200 ppm measurement error by the zirconia sensor will result.

The uncertainty of commercial calibration gas mixtures is typically ±1% of the amount fraction, and the contribution of this to the overall measurement uncertainty is much greater than the intrinsic error of the oxygen sensor for amount fractions of oxygen that are 0.1 ppm and above [2]. Consequently, test procedures that require calibrated gas mixtures will be limited by the uncertainty of those gas mixtures, and therefore cannot be used to test the accuracy of oxygen and other zirconia-based gas sensors, assuming the uncertainty of the sensor calibration is no worse than that of the test mixtures. At present, the uncertainty of commercial calibration gas mixtures is a limitation to verifying the accuracy of these gas sensors.

Based on the above facts, conception of the *relative error* of the sensor can be introduced, which can be determined as $\gamma_x = \Delta x/x$. This appraisal, expressing as a rule in percentage (i.e., $\gamma_x = (\Delta x/x) \times 100\%$), is a more comprehensive characteristic of the accuracy of the sensor. The relative error γ_x is the function of the measuring parameter x. It is therefore impossible to point out the single exact characteristic of the sensor, which would somehow characterize the error. Consequently, the concept of the *reduction error* γ is introduced, which represents the ratio between the maximum value of the absolute error Δx_{max} and the maximum value of the measuring parameter x_{max}, that is, $\gamma = \Delta x_{max}/x_{max}$.

If the zirconia gas sensor can be represented by the structural scheme, shown in Figure 6.1, and realizing the function of the sensor $y = f(x)$, then the presence of linear dependence between the argument (measuring parameter) x and the output signal y can be written as follows:

$$\Delta x = \Delta y/k; \quad \Delta x k = \Delta y; \quad \Delta x/x = \Delta y/y = \Delta x/x_{max}, \tag{6.1}$$

where k is the transformation coefficient (amplification, weakening, etc.).

Unfortunately, there is no ideal linear dependence between argument x and function y. Nevertheless, the equalities (6.1) are quite acceptable in the overwhelming

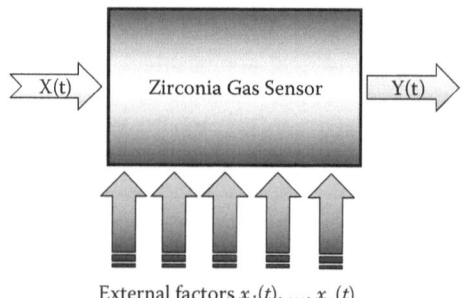

External factors $x_1(t), ..., x_n(t)$

FIGURE 6.1 Structural scheme of the zirconia gas sensor with influence of external factors.

majority of cases. Therefore, by the manner of expression, they can be divided on absolute, relative, and reduction errors.

All errors can also be divided into two independent classes by the influence on the resulting accuracy of measurement: *error of zero* and *error of sensitivity*. Let's consider these two classes in details. In the general view, the zirconia sensor function can be represented as follows:

$$y = k\,(x, x_1, x_2, \ldots, x_n)\, x + b\,(x_1, x_2, \ldots, x_n), \qquad (6.2)$$

where $k\,(x, x_1, x_2, \ldots, x_n)$ is the transformation coefficient of the sensor, which is the function of the measuring parameter x and whole complex of the influencing parameters $x_1, x_2, \ldots, x_n$; $b\,(x_1, x_2, \ldots, x_n)$ is an initial quantity of the output signal (at $x = 0$). In the general case, it is also a function of values $x_1, x_2, \ldots, x_n$.

Equation (6.2) can be rewritten in more compact form as a function of time:

$$y\,(t) = kx(t) + b, \qquad (6.3)$$

considering that both k and b are complex functions of the various arguments. Figure 6.2 illustrates the graphical interpretation of the zirconia gas sensor function with a impact of the external influencing factors such as instability of the reference pressure, inaccuracy of setting and measurement of the sensor temperature, *emf* variations, and so on. Segment b by the x axis corresponds to the initial level of the output signal of the sensor at the absence of the input signal (i.e., $x = 0$). The pitch of line *1* corresponds to the sensitivity of the zirconia sensor to the measuring parameter x (equal at the linear case to coefficient k). Line *1* represents the case when both coefficients k and b are the constant values, independent of time and influencing factors. However, owing to the inconsistency of the whole complex of influential factors (parameters) in the real measuring environments, coefficients k and b are changing.

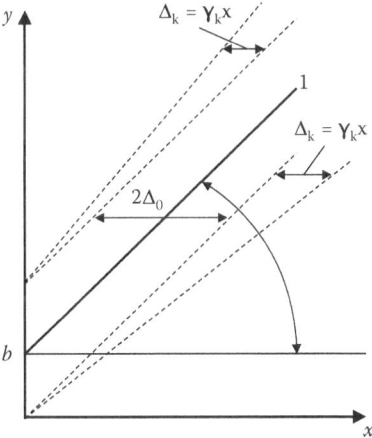

FIGURE 6.2 Graphical interpretation of the zirconia gas sensor function.

The real zirconia gas sensor function is located between two dotted lines due to the inconsistency of both sensitivity and the initial level of output signal of the sensor. The uncertainty of measurements is restricted by the value $2\Delta_0 + 2\gamma_x$, where Δ_0 is the absolute error of the zero level, or additive absolute error; $\gamma_x x$ is the absolute sensitivity error, or multiplicative absolute error; and γ_x is the relative sensitivity error, or relative multiplicative error.

If some of the errors given above can be dominated for the real zirconia-based sensors, then the other constituents can be ignored. If only the additive constituent part takes place, then the zirconia gas sensor function assumes $y = k(x \pm \Delta_0)$, where the current value of the absolute sensor error, equal to $\Delta = \Delta_0$, is independent of the value of the measuring parameter. The value of the relative error $\gamma = \gamma_0 = \Delta_0 / x$ is inversely proportional to the value of the measuring parameter, the sensor error growths up to 100% at $x = \Delta_0$, that is, it is impossible to make a measurement in this case. The value of the measuring parameter x, equal to the value of the relative error of zero, is acceptedly called the sensitivity threshold of the sensor.

If only the multiplicative constituent part is present, then the function of the zirconia gas sensor assumes $y = k(1 \pm \gamma_k) x$, where $\gamma = \gamma_k$ is the current value of the relative error. In this case, the function of the measuring parameter of the sensor is the absolute value of the error.

The situation where both additive and multiplicative constituents of the error are present is the most common case and has been accepted in practice. For all that, the function of the zirconia gas sensor can be written as $y = k(1 \pm \gamma_k) \times (x \pm \Delta_0)$, and the absolute and the relative errors of the sensor are determined by $\Delta = \Delta_0 + \gamma_k x$ and $\gamma = \gamma_k + (\Delta_0/x)$, respectively.

All correlations considered above and their graphical interpretations are combined into Table 6.1 for the convenience of use. The standardization of the sensor errors takes place in accordance with those correlations given in Table 6.1. The first two rows in this table illustrate the method of standardization with the aid of the monomial equations. The standardization method, with the help of the binomial equation, is presented in the third row of the table. Generally speaking, there are more complex equations for standardization of the errors of various gas sensors. However, in the vast majority of cases, the estimates presented above are sufficient enough for the zirconia-based gas sensors.

Speaking of errors, the instantaneous value of errors has been considered so far. In contrast, due to inertness, most of the zirconia gas sensors possess a difference between the instantaneous values of the measuring parameter and the results of measurement. This difference can vary depending on how fast the input signal can be changed and how fast the characteristics of the sensors can be changed, especially at low temperatures. Consequently, the errors can further be divided into *dynamic*, *static*, and *progressing* [3]. To some extent, such classification is conditional and only necessary for those advantages which stipulate the development of a simplified model of each type.

At the same time, the analysis of the cause-effect relationship at the consideration of the zirconia sensor errors allows both designer and consumer not only to compile a more correct model of the resulting error, but also to decrease it by appropriate

TABLE 6.1
Standardization of the Sensor Errors

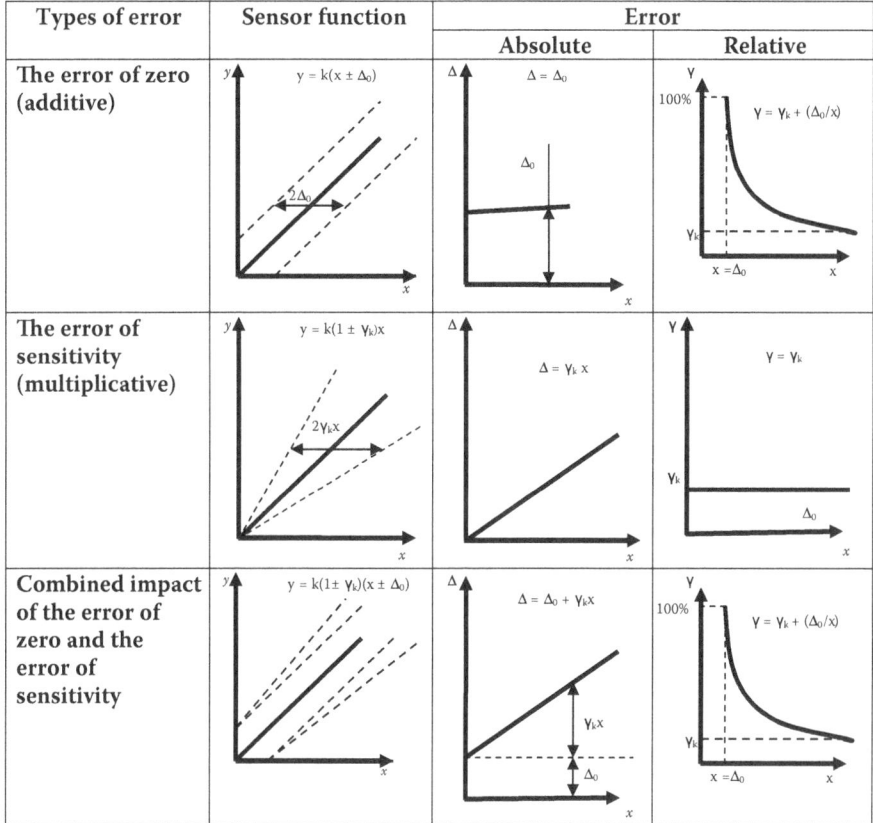

Types of error	Sensor function	Error	
		Absolute	Relative
The error of zero (additive)	$y = k(x \pm \Delta_0)$	$\Delta = \Delta_0$	$\gamma = \gamma_k + (\Delta_0/x)$
The error of sensitivity (multiplicative)	$y = k(1 \pm \gamma_k)x$	$\Delta = \gamma_k x$	$\gamma = \gamma_k$
Combined impact of the error of zero and the error of sensitivity	$y = k(1 \pm \gamma_k)(x \pm \Delta_0)$	$\Delta = \Delta_0 + \gamma_k x$	$\gamma = \gamma_k + (\Delta_0/x)$

means. Considering the reasons of appearance of the measurement errors, two main circumstances should be highlighted:

1. On the one hand, reasons causing the errors' appearance are inseparably connected to the nonideality of the physical and electrochemical characteristics of the elements of zirconia gas sensors and to the relative inconstancy of their working conditions. This follows the fact that the measurement process distorts the sideline physical and chemical effects subordinate to the strict enough appropriateness. The clear example of this can be represented by the temperature dependence of physical and electrochemical properties of the SE materials.
2. On the other hand, the reason causing the errors' appearance can be the imperfection of knowledge about the physical and chemical processes, and their kinetics, accompanying the gas sensor work.

Samples of various batches of the zirconia gas sensors can have different characteristics owing to instability of the technological processes involved in the manufacturing of these sensors. Errors arising from these factors usually have an accidental character. Therefore, the process of their correction is highly difficult.

In compliance with the reasons of appearance, all errors can be divided into systematic and accidental. *Systematic errors* are the errors possessing the determined function connection with the source of their cause, and the error function itself and its arguments are known. *Accidental errors* are the errors caused by the combined actions of the influencing (destabilizing) factors, and owing to their uncertainty their functional connection with the source of errors cannot be determined.

It should be noted that the accurate boundaries between the sorts of errors mentioned above are very difficult to establish since the possibility of redistribution of errors between considered groups is likely to occur in the various measurement situations. For example, if the zirconia gas sensor works in a wide, frequently changing temperature range and its precise temperature characteristic as well as the zero offset between SE and RE are known, then the error of measurement is a systematic one. Indeed, if the zero offset is known and can be kept close to zero within the whole measuring temperature range, then the influence of deviation of the zero offset on measurement results would be corrected. However, if the changes of zero offset are unknown within the whole measuring temperature range and they change accidentally, then the establishment of the zero offset correction is impossible. Consequently, in this case the systematic error caused by the zero offset between SE and RE of the sensor becomes an accidental error due to uncertainty of the influencing factor.

6.2 ANALYSIS OF SYSTEMATIC ERRORS OF ZIRCONIA GAS SENSORS

Development of ionic conductors based on stabilized zirconia has reached a level of maturity, where most of the research on such materials concentrates mainly on obtaining incremental empirical improvements in conductivity by better processing control and refinement of the microstructure of the solid electrolyte and SE. Further increases in the conductivity are important in terms of enhancing the efficiency of systems such as O_2 sensors, zirconia-based mixed-potential gas sensors, electrochemical oxygen pumps, heating elements, and fuel cells [4–7]. The systematic errors, as have been considered before, are errors with a known determined functional connection with the source of their cause, and the conformity of their appearance can be definitely described.

Considering that the influential quantities are independent, the absolute systematic error can be found as a full differential of the complex linear function (6.2). Then,

$$dy = x \sum_{i=1}^{n} \frac{\partial k}{\partial x_i} dx_i + \sum_{i=1}^{n} \frac{\partial b}{\partial x_i} dx_i . \qquad (6.4)$$

Having divided (6.4) on y, the value of the relative systematic error can be calculated as follows:

$$\gamma_y = \frac{dy}{y} = \frac{x}{y} \sum_{i=1}^{n} \frac{\partial k}{\partial x_i} dx_i + \sum_{i=1}^{n} \frac{1}{y} \frac{\partial b}{\partial x_i} dx_i , \qquad (6.5)$$

where k is the transformation coefficient.

In Equation (6.5) correlation,

$$\frac{x}{y} \frac{\partial k}{\partial x_i} = \frac{\partial k/k}{\partial x_i}$$

defines the relative changing of the transformation coefficient, arising from the changes of influential factor x_i. This correlation is called the relative multiplicative sensitivity of the sensor to the influential factor x_i, and it is designated as S_{ki}. Let's introduce correlation

$$\frac{1}{y_{lim}} \frac{\partial b}{\partial x_i} ,$$

which characterizes the relative changes of the initial level of the sensor output signal (*emf*) arising from the changes of influential factor x_i, where y_{lim} is the value of the output signal of the sensor corresponding to the limit of sensor measurements. This correlation is defined as an additive sensitivity of the sensor to the influential factor x_i, and it is designated as S_{bi}. Using the designated parameters introduced above and transferring them to the final increments in Equation (6.5), the following correlation yields

$$\gamma_y = \sum_{i=1}^{n} S_{ki} \Delta x_i + \left(y_{lim}/y \right) \sum_{i=1}^{n} S_{bi} \Delta x_i . \qquad (6.6)$$

Quantities Δx_i define the deviation of influential factors from the values at which the calibration of the gas sensor was taking place (determination of sensitivity S and the initial level of the output signal b). Having known the relative multiplicative and additive sensitivities of the sensor to the influential factors and the deviation of influential factors from the calibration values, the systematic error of the sensor can be calculated and appropriate corrections to the measurement results can be introduced.

For analysis of the systematic errors of the sensor, it is necessary to know the following data:

- Correlation for the zirconia gas sensor function
- List of the influential factors and the parameters of their work functions
- Dependence of arguments on the influential factors involving into the zirconia sensor function

6.3 ANALYSIS OF THE MAIN COMPONENTS OF ERRORS OF ZIRCONIA GAS SENSORS

At present, considerable research has been devoted to understanding the influence of various stabilizing oxides on the ionic conductivity and the defect structure of zirconia. However, much less attention has been allocated to investigation and analysis of the main components of errors for the zirconia-based sensors measuring partial gas pressures in vacuum, in spite of the fact that the numerous technological processes require precise measurement of gaseous components in the diluted atmosphere. Therefore, the analysis of the main components of errors of the zirconia gas sensor as well as the relative error of measurement for the zirconia electrolyte will be presented for sensors measuring oxygen in a vacuum from 1×10^{-8} to 1×10^5 Pa. Specifically, the contribution of each component to the total relative error of measurement should be investigated, and the limit relative error of measurement should be analyzed and calculated for the zirconia sensor measuring partial oxygen pressure in the temperature range of 500–1000°C. Similar analysis can be applied for other zirconia-based sensors available to date.

The planar ZrO_2-based oxygen sensor configuration is presented in Figure 6.3. This planar sensor was fabricated at KASTEC, Kyushu University, Japan, by using a YSZ plate (8 wt % Y_2O_3-doped, 10×10 mm; 0.2 mm thickness). *Pt* paste (Tanaka Kikinzoku, Japan) was printed on both sides of the ZrO_2 plate and was fired at 1000°C for 2 hours in air. On one side of the zirconia plate, two rectangular *Pt* stripes were formed as the SE of the sensor; and on the other side, six narrow *Pt* stripes were formed as a base (current collector) for the thin-film metal–metal oxide RE. Both SE and RE were exposed simultaneously to the measuring environment. Figure 6.4 shows the interactions scheme of the measuring gas with the planar thin-film ZrO_2-based sensor, which has been developed on the basis of analysis of the electrode processes in the electrochemical systems [8]: Stage 1 is the multicomponent diffusion of gaseous species through the porous SE and RE, allowing for the effect of nonuniform gas pressure and sensor temperature, and Knudsen diffusion; and Stage 2 is adsorption-desorption and partial dissociation of oxygen molecules $(O_2 \rightarrow 2O^{2-} + 4e^-)$ on the gas-SE interface [9]. Furthermore, oxygen ions O^{2-} can subdissociate to subions O^-, which can diffuse on the metal-oxide-electrolyte and gas-electrolyte boundaries toward the SE-ZrO_2-gas interface. Stage 3 is the electrochemical reaction of oxygen, which initiates changing the capacity of the double electric layer on the solid electrolyte-electrode interface and, consequently, altering the potential of the SE (thin-film *Pt* SE). It should be noted that oxygen may be oxidized in different forms — O_2^-, O_2^{2-}, O^{2-}, O^-, and O — depending on the working temperature, the oxygen partial pressure, and whether the oxide ME is a reversible

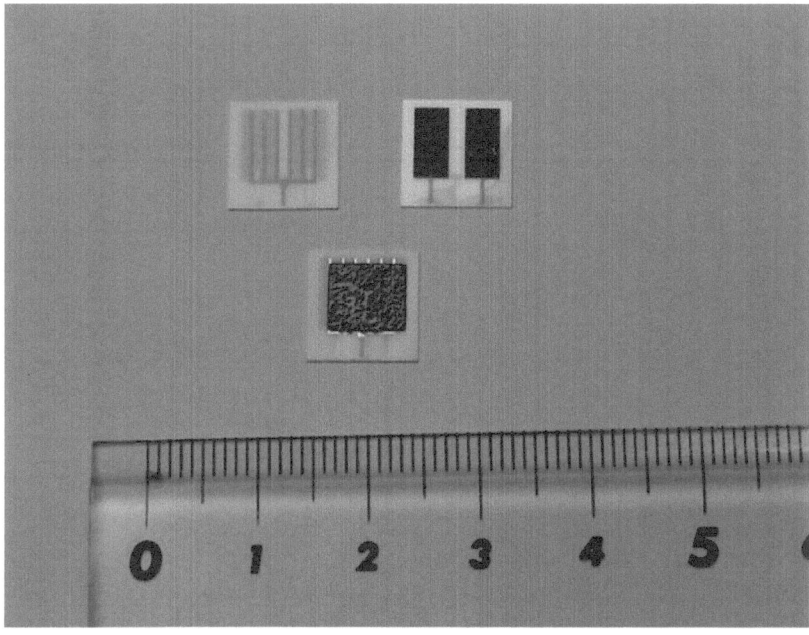

FIGURE 6.3 Front and back views of the planar zirconia-based oxygen sensor with the Pt-SE and Fe-FeO-RE.

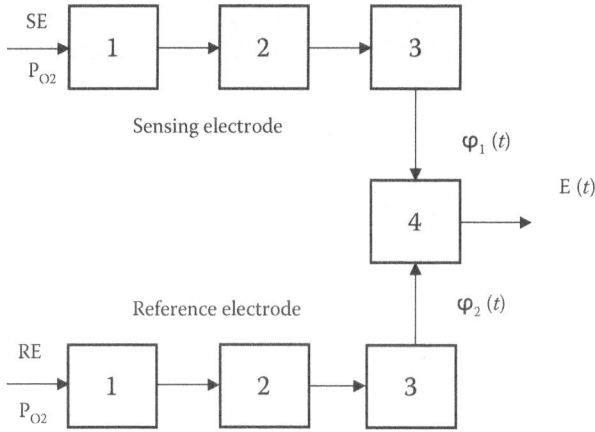

FIGURE 6.4 Schematic presentation of interaction of the zirconia-based solid electrolyte sensor with a gaseous environment: Stages 1–4 of interaction.

(irreversible) oxygen electrode. The output informative parameter of the sensor in Stage 4 is the difference of potentials of the SE and RE: $E(t) = \varphi_1(t) - \varphi_2(t)$.

One of the main assumptions for the thin-film ZrO_2-based sensors is the negligible interinfluence of the thin-film layers. This means that the mechanical, physical,

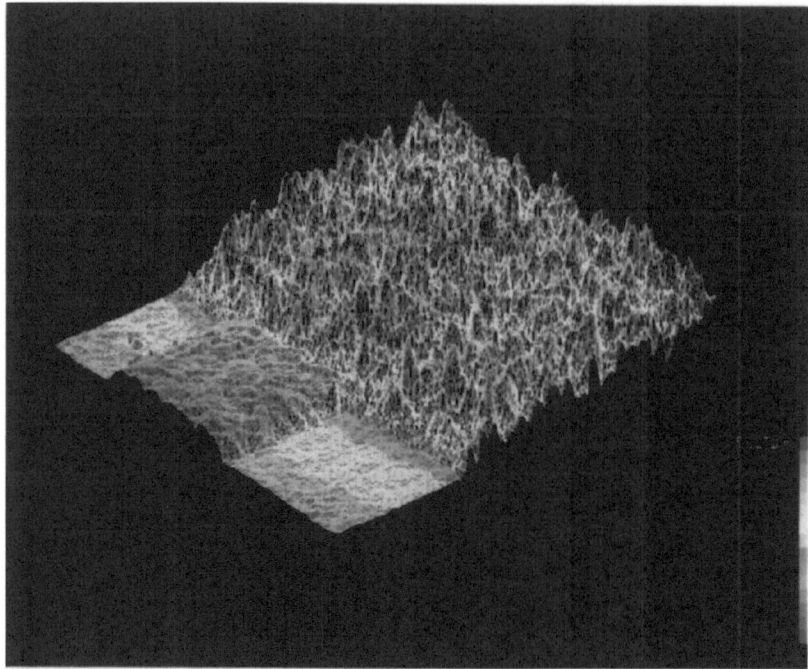

FIGURE 6.5 Three-dimensional SEM image of the parts of the SE and current conductor of the zirconia-based sensor.

chemical, and electrostatic components of their interactions are close to zero [10]. In order to achieve that, the following requirements must be implemented: the purity of row materials for thin films should be ultra-high (99.999%), and the row materials must have compatible coefficients of the thermal expansion and parameters of the crystalline structure. They should also be characterized by the minimum value of the contact difference of potentials. Figure 6.5 shows a three-dimensional image of the parts of the SE and current conductor of the zirconia-based sensor.

Let's consider the work of the solid electrolyte cell (SEC) using either a metal–metal oxide mixture (Fe-FeO) or an air RE at the temperature range of 500–1000°C. The highest limit of measurement will be the atmospheric pressure ($\sim 1.3 \times 10^5$ Pa). The lowest limit depends on conditions of operation of the sensor at which the output signal (*emf*), gathered by the proceeding of electrochemical reactions, corresponds to the thermodynamic value. In order to reach these conditions it is necessary that, at first, the load circuit should not interfere with the work of the SEC, and, at second, the kinetic of the electrode reactions should avoid the appearance of polarization, which is possible as a result of the development of two types of oxygen permeability in the SEC: (1) oxygen anions transfer, which establishes as a compensative anticurrent to the electrons transfer at the mixed conductivity of the SEC; and (2) molecular diffusion of the gaseous oxygen on the open porosity of the SE. Based on the fact that the electrochemical reaction has many stages (adsorption of oxygen molecules on the SE, their dissociation of atoms and ionization, diffusion, etc., as was described in chapter 2), we will consider the oxygen diffusion on the open electrode pores toward the TPB *electrolyte-electrode-gas* as a limiting reaction at lowered pressure [8]. For the minimum pressure at which work of the SEC is possible, the calculation of number of oxygen ions capable of providing working current of the SEC exceeding the parasitic current on the secondary measuring devices (for electrometer B7-30, this current is no more than 2×10^{-14} A) was done. The electric charge, arising from oxygen ions on the boundary of the solid electrolyte-electrode to the unit of time (q), is equal to 2×10^{-14} coulomb. Using this value, the quantity of necessary oxygen ions was found by equation $n = q / 2e$, where e is the charge of the electron. It was 6.25×10^3 pieces. When the calculated quantity of the oxygen ions in accordance with dependence [11],

$$P = n \sqrt{2\pi m k_B T} \Big/ \chi \left(1 - n_{ad}/n_0\right)^2 \qquad (6.7)$$

corresponds to the pressure of 1.5×10^{-14} Pa. In Equation (6.7), m is the molecular weight, k_B is the Boltzmann constant, T is the absolute temperature, χ is the probability of adsorption, n_{ad} is the number of adsorbed gaseous molecules, and n_0 is the number of adsorption centers.

It is vital to provide an equilibrium condition of the SEC for calculation of its static characteristic. For an equilibrium condition, the current picking out into the measuring device should be on several orders of magnitude less than the exchange current on the SE. It follows from the assumption above that the minimum measuring pressure should be increased in comparison with the calculated value on 2–3 orders of magnitude. This pressure therefore should presumably be $\sim 10^{-11}$ Pa. Based on the facts that the properties of the real SEC differ slightly from the properties of the ideal SEC and that both adhesion and ionization coefficients are not equal to 1, consequently, the lowest limit of measurement should not exceed 10^{-10} Pa. Fulfilment of the above-mentioned restrictions secures the equilibrium condition of the SEC at which the following Wagner equation can be used for calculation of the static characteristic:

$$E = \frac{1}{4F} \int_{\mu o2(\mathrm{I})}^{\mu o2(\mathrm{II})} t_{ion} \, d\mu o_2 \, , \qquad (6.8)$$

where μo_2 (II) and μo_2 (I) are the oxygen chemical potential of the SE and RE, respectively; F is the Faraday constant; and t_{ion} is the true ionic transference number, defining part of the ionic conductivity in the total conductivity of the SEC: $t_{ion} = \sigma_i/(\sigma_i + \sigma_e + \sigma_p)$, where σ_i, σ_e, and σ_p are ionic, electronic, and hole conductivities, respectively.

It is well-known that

$$E = \overline{t}_{ion}\big/4F\left(\mu o_2\left(\mathrm{I}\right) - \mu o_2\left(\mathrm{II}\right)\right) = RT \, \overline{t}_{ion}\big/4F \, ln\left(Po_2\left(\mathrm{I}\right) - Po_2\left(\mathrm{II}\right)\right), \qquad (6.9)$$

because $\mu o_2 = \mu o_2{}^\circ + RT \ln Po_2$, where $\mu o_2{}^\circ$ is the standard value of the chemical oxygen potential, T is temperature, R is universal gas constant, and Po_2 (II) and Po_2 (I) are oxygen partial pressures on the SE and RE, respectively.

Let's consider $\bar{t}_{ion}$ within the above-mentioned ranges of temperatures and pressures. The true ionic transference number based on [10] can be expressed according to the following equation:

$$t_{ion} = [1 + (P/P_p)^{1/4} + (P_e/P)^{1/4}]^{-1}, \tag{6.10}$$

where P_e and P_p are the electron and the hole transference parameters, respectively, that is, the oxygen partial pressure at which $\sigma_i = \sigma_e$ and $\sigma_i = \sigma_p$. At relatively low oxygen partial pressures when the hole conductivity can be ignored, the dependence of the average ionic transference number $\bar{t}_{ion}$ on the oxygen partial pressure P_{O2} is as follows:

$$t_{ion} = 4\left\{ \ln\left[\left(Po_2\left(\text{II}\right)^{1/4} + \left(P_e\right)^{1/4}\right] - \ln\left[\left(Po_2\left(\text{I}\right)^{1/4} + \left(P_e\right)^{1/4}\right]\right\}\right/\left(\ln Po_2\left(\text{II}\right) - Po_2\left(\text{I}\right)\right),$$

$$\tag{6.11}$$

where P_e in the SEC, consisting of $0.85\ ZrO_2 + 0.15\ CaO$, is calculated by equation $\lg P_e = -60.5 \times 10^3/T + 19.5$. Parameter P_e changes from 10^{-79} to 10^{-21} Pa for the temperature range of 500–1000°C when the oxygen partial pressure Po_2 on the lowest measurement limit changes from 10^{-36} to 10^{-10} Pa at the same temperatures (for the ideal gas). Therefore, the value of P_e can be ignored, that is, t_{ion} is always equal to 1 at the low partial pressure for the entire temperature range. The dependence of the average ionic transference number $\bar{t}_{ion}$ on the oxygen partial pressure P_{O2} at higher pressures, when the hole conductivity can appear in the SEC, and is calculated by the following equation:

$$\bar{t}_{ion} = \left[\int_{Po_2(\text{I})}^{Po_2(\text{II})} t_{ion} dP/P\right]\Big/\left(\ln Po_2\left(\text{II}\right) - \ln Po_2\left(\text{I}\right)\right), \tag{6.12}$$

can be expressed as

$$\bar{t}_{ion} = 4\left[\ln\left(Po_2\left(\text{II}\right)\right)^{1/4}\left[\left(Po_2\left(\text{I}\right)\right)^{1/4} + \left(P_p\right)^{1/4}\right] - \ln\left(Po_2\left(\text{I}\right)\right)^{1/4}\right.$$

$$\left.\left[\left(Po_2\left(\text{II}\right)\right)^{1/4} + \left(P_p\right)^{1/4}\right]\right]\Big/\left(\ln Po_2\left(\text{II}\right) - \ln Po_2\left(\text{I}\right)\right) \tag{6.13}$$

where P_p calculates by equation $\lg P_p = 28 \times 10^3/T - 16$ for the SEC consisting of $0.85\ ZrO_2 + 0.15\ CaO$. Parameter P_p changes from 10^{37} to 10^{11} Pa at the above-mentioned temperature range. $P_p \gg Po_2$ (II) and $P_p \gg Po_2$ (I) at temperatures of

500–700°C and $\bar{t}_{ion}$ = 1. At the temperature of 1000°C, P_p = 9.7 × 10^{10} Pa, Po_2 (I) = 8.9 × 10^4 Pa (oxidizing potential of air), and $\bar{t}_{ion}$ = 0.996.

Therefore, the ZrO$_2$-based SEC can be considered as a pure ionic conductor within the above-mentioned ranges of temperatures and pressures, and its static characteristic can be expressed from (6.9):

$$Po_2 \text{ (II)} = Po_2 \text{ (I) } exp \text{ (}- RT/4F\text{)}. \qquad (6.14)$$

For evaluation of the influence of each parameter from the static characteristic of the SEC on the accuracy of measurement, Equation (6.14) can be differentiated, and the functions and their differentials will be substituted by their increments:

$$\Delta Po_2 \text{ (II)} = Po_2 \text{ (II) } [\Delta Po_2 \text{ (I)}/Po_2 \text{ (I)} + (-4F/RT)_0 \, \Delta E + (4EF/RT^2)_0 \, \Delta T].$$
$$(6.15)$$

Index 0 means the nominal parameter value, that is, the value at which the measurement mode has no error. The influence coefficients are as follows:

$$A_{Po2(I)} = 1/Po_2 \text{ (I) } [Pa^{-1}]$$

and

$$A_E = - 4F/RT \; [V^{-1}] \text{ and } A_T = 4EF/RT^{\,2} \; [°C^{-1}].$$

For evaluation of the limit values of errors of measuring pressures, the total error of measurement and its components will be determined at $Po_2\text{(II)}/Po_2\text{(I)} \rightarrow 1_+$, 10, and 100 in the temperature range of 500–1000°C.

6.3.1 ERROR STIPULATED BY THE REFERENCE PRESSURE INSTABILITY

The contribution of the reference oxygen partial pressure Po_2(I) into the results of measurement can vary and usually depends on the RE type. If an air RE is used (ambient air), the oxygen concentration C, in relation to the normal conditions ($C_{n.c.}$ = 20.8%), can change from 20.3% to 20.9%. Consequently, the error of measurement (δPo_2(II))$_{Po2 \, (I)}$ stipulated by this component can be approximately ~3%. However, the exchange chambers can be used for reducing the component of error of measurement stipulated by the reference air. Then, (δPo_2(II))$_{Po2 \, (I)}$ can be reduced down to 0.6–1.0%.

If the metal–metal oxide mixture is used as a RE for the ZrO$_2$-based oxygen sensor, the oxygen pressure can be calculated by the following equation: ln P_{O2} = 2$\Delta G/RT$, where ΔG is the standard Gibbs potential of formation of a MeO by the reaction Me + O$_2$ = MeO. The error contributed by this component depends on the inaccuracy of the temperature setting, and the error of determination of the standard Gibbs potential. Variation of ΔG is determined by the thermodynamic methods by means of measurement of the thermal effect of the reaction Me + O$_2$ = MeO and thermocapacity of its reagents and products. Divergence in determination of the standard Gibbs potential for the reaction $2Fe + O_2 = 2FeO$ is $\pm$ 420 J/mol [12]. In this case, the error can be expressed as

$$(\delta Po_2(II))_{Po2(I)} = 2/RT(\Delta G) - 2\Delta G \Delta T/RT^2, \qquad (6.16)$$

with the influence coefficients $B_{\Delta G} = 2/RT$ [mol/J] and $B_T = - 2/RT^2$ [°C^{-1}]. Then the value of $(\delta Po_2(II))_{Po2(I)}$ calculates by the following equation:

$$\left(\delta Po_2\left(II\right)\right)_{Po2(I)} = \sqrt{\left[B_{\Delta G}\Delta\left(\Delta G\right)\right]^2 + \left[B_T\Delta T\right]^2}. \qquad (6.17)$$

Both components decrease with the temperature growth. As a consequence of this, the summary error contributed by this component changes from 7.0% at 500°C to 4.2% at 1000°C at the deviation of the temperature of the RE on ±0.5°C.

6.3.2 ERROR STIPULATED BY THE VARIATIONS OF *EMF*

This error's component (the second item of (6.15)) consists of several items: error by the current in the measuring circuit, error by the nonionic component of conductivity, error by the nonisothermic character of electrodes in flatness, error stipulated by the thermoelectric effect, and error by the secondary measuring device.

Error by the current in the measuring circuit is equal to the voltage drop on the SEC calculated by equation $(\delta Po_2(I))_{E1} = A_E I R_{SEC}/(R_{inp} + R_{SEC})$, where A_E is the influence coefficient by *emf*, I is the strength of current in the measuring circuit, R_{imp} is the input impedance of the secondary measuring device, and R_{SEC} is the impedance of the SEC. $(\delta Po_2(I))_{E1}$ represents only $10^{-8} - 10^{-12}\%$ at $R_{inp} = 10^{14}$ Ω and R_{SEC} equal to dozens of Ω, and this error can be ignored in further calculation.

As has been shown above, the average ionic transference number $\bar{t}$ at the above-mentioned ranges of oxygen pressures and temperatures is equal to 1, and consequently, this error, stipulated by the nonionic component of conductivity $(\delta Po_2(I))_{E2}$, can also be ignored.

The component of error by the nonisothermic character of electrodes in flatness calculates by equation $(\delta Po_2(I))_{E3} = A_E R(T_{eq} - T_0)/4F \cdot \ln (Po_2(II)/Po_2(I))$, where

$$T_{eq} = \int_s T\left(x, y\right)\sigma\left(x, y\right)ds \bigg/ \int_s \sigma\left(x, y\right)ds$$

is the equivalent temperature; $T (x, y)$ and $\sigma (x, y)$ are functions of distribution T and σ by flatness S; and T_0 is the temperature at which the thermodynamic *emf* has been calculated. At the temperature uniformly distributed by the flatness of electrodes, this component of error $(\delta Po_2(I))_{E3}$ can be ignored.

The component of error, stipulated by the thermoelectric effect, appears at the heterogeneity of the temperature field of the SEC:

$(\delta Po_2(\text{I}))_{E4} = A_E \, a\Delta T + b \, ln \, (T_1/T_2) + 5R(T_1 \, ln \, T_1 - T_2 \, ln \, T_2)/8F + R\Delta T \, ln \, Po_2(\text{I})/4F,$

where $a = (-0.9 \pm 0.1)$ mV/°C; $b = (50 \pm 10)$ mV; and T_1 and T_2 are the temperature of the SE and RE, respectively. Numerical calculations of this component of error for the most unfavorable conditions ($\Delta T = T_1 - T_2 = 10$°C and $T_1 = 500$°C) have shown that at measurement of the sensor *emf*, the heterogeneity of the temperature field of the SEC, stipulated by the thermoelectric effect, is equal to 25 mV, which corresponds to $(\delta Po_2(\text{I}))_{E4} = 202\%$. However, at $\Delta T = 0.5$°C and $T_1 = 500$°C, this component of error decreases to $(\delta Po_2(\text{I}))_{E4} = 10.5\%$. Furthermore, if the SEC temperature T_1 can reach 1000°C, the component of error by the thermoelectric effect can be decreased to $(\delta Po_2(\text{I}))_{E4} = 2.2\%$.

The limit allowable main error of the electrometer B7-30 is calculated by equation $\Delta E_5 = \pm (A + 0.01E)$, where A is the electrometer discredit error; E is the real value of the measuring parameter. Based on the fact that the electrometer B7-30 possesses a multiple error of 1% and an additive error equal to the two lowest ranks within the whole measurement range, it is preferable to use the SEC generating small *emf* ($Po_2(\text{II})/Po_2(\text{I}) \rightarrow 1_+$), with the condition that the error of measurement stipulated by the additive component will not be higher than the set value. The error by the secondary measuring device aspires to 1% at the deviation of ratio $Po_2(\text{II})/Po_2(\text{I})$ from 1 on parts of percent within the whole temperature range. The absolute value of the measuring *emf*, at this condition, does not exceed 20 mV. Consequently, the error component by the secondary measuring device $(\delta Po_2(\text{I})_{E4}$ remains around 1% within the whole working measurements range.

6.3.3 ERROR STIPULATED BY INACCURACY OF SETTING AND MEASUREMENT OF THE SEC TEMPERATURE

This component of error can be determined by equation $(\delta Po_2(\text{I}))_T = A_T\Delta T$. Numerical calculations of this component have shown that at the ratio $Po_2(\text{II})/Po_2(\text{I}) \rightarrow 1_+$, inaccuracy of the temperature setting $\Delta T = 0.5$°C and the measuring temperature changes from 500°C to 1000°C, and the $(\delta Po_2(\text{I}))_T$ changes from 8.7×10^{-5} to $3.9 \times 10^{-5}\%$.

The evaluation of the total error of measurement considering mutual independence of the error's components can be calculated by the following equation:

$$\left(\delta Po_2(\text{I})\right)_\Sigma = \sqrt{\sum_{i=1}^{n}\left(\delta Po_2(\text{I})\right)_i^2}, \qquad (6.18)$$

where $(\delta Po_2(\text{I}))_i$ are the error's components.

Numerical calculations of the total error of measurement for the ZrO_2-based oxygen partial pressures sensor have shown that at the ratio $Po_2(\text{II})/Po_2(\text{I}) \rightarrow 1_+$, inaccuracy of the temperature setting $\Delta T = 0.5$°C, the total error of measurement $(\delta Po_2(\text{I}))_\Sigma$ changes from 12.65% to 4.84% as the measuring temperature changes from 500°C to 1000°C. Figure 6.6 shows the contribution of different components

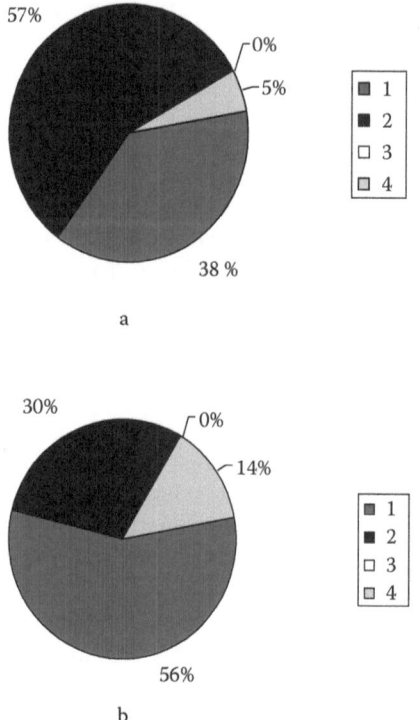

FIGURE 6.6 Contribution of different components into the total error of measurement for the zirconia-based oxygen sensor at 500°C (*a*) and 1000°C (*b*) at the temperature deviation of the RE (± 0.5°C). *1*: component of error by instability of setting reference oxygen pressure by Me-MeO RE; *2*: component of error by the thermoelectric effect; *3*: component of error by inaccuracy of setting and measuring of the SEC temperature; and *4*: component of error by the secondary measuring device.

into the total error of measurement for the ZrO_2-based oxygen sensor at 500°C (*a*) and 1000°C (*b*) at the temperature deviation of the RE (± 0.5°C). As has been clearly illustrated in this figure, the biggest contribution into the total error of measurement at the working temperature of 500°C comes from the error's component based on the thermoelectrical effect, which appears at the heterogeneity of the temperature field of the SEC. However, as the working temperature grows, the influence of the thermoelectrical effect on the total error of measurement decreases, and, consequently, contribution of this component of error goes down to 30% at 1000°C (Figure 6.6*b*). On the other hand, the component of error by instability of setting oxygen reference pressure by Me-MeO RE increases with temperature, and in our case (Fe-FeO-RE) at 1000°C it is almost 56% of the total error of measurement. Therefore, the careful selection of suitable material for the Me-MeO mixture can further decrease the influence of this component on the total error of measurement.

Consequently, considering specific features of the vacuum measurements, the produced analysis of the total error of measurement and the main components of errors for the zirconia-based solid electrolyte sensors has shown that, at the ratio $Po_2(II)/Po_2(I) \rightarrow 1_+$ and inaccuracy of the temperature setting $\Delta T = 0.5°C$, the total error of measurement $(\delta Po_2(I))_\Sigma$ changes from 12.65% to 4.84% within the measuring temperature range of 500–1000°C. The results obtained raised the expectations of creating the SEC partial gas pressures with high enough metrological characteristics in relation to the vacuum measurements. The present analysis is also admissible for the SEC with conductivity by other ions. Hence, it can be concluded that the opportunity in principle exists for zirconia sensors to be used as a working measurement means for partial pressures of the gaseous components in the vacuum range of 1×10^{-8} to 1×10^5 Pa.

6.4 CALCULATION OF ERRORS ON THE BASIS OF EXPERIMENTAL DATA

Errors of any sensors, including zirconia gas sensors, can be determined only by specifically planned tests. Appropriately designed experiments combined with the correct processing of data obtained during these tests guarantee the receipt of the objective data characterizing the accuracy of the sensor.

The questions related to the secondary equipment selection, calculation of the number of points (cross-sections), and number of the cycles of tests and their sequence are worked out at the planning stage of an experiment. Let's consider the sequence of the zirconia gas sensor test and calculate its errors based on the results of the experimental data without the detailed consideration of the questions related to the planning of these tests (their interpretation will be done in Chapter 7).

Figure 6.7 illustrates the zirconia gas sensor function at the constant temperature with slight hysteresis of the measurement results. Basically, this function represents the dependence of output sensor *emf* on the measuring gas concentration at the logarithmic scale at axis x, corresponding to the well-known Nernst equation. Based on the function given in Figure 6.7, it is assumed that the gas sensor possesses a nonlinearity error of the sensor function and hysteresis. It is also assumed that the present zirconia gas sensor works at unfavorable conditions when the influence of the temperature fluctuations, the presence of humidity in the measuring gas, and so on are changing both initial level b of the output *emf* and the pitch of the sensor function. The sensor's error has to be calculated for such working conditions.

The first step in this investigation is based on determination of the zirconia sensor function at the normal working conditions. The normal working conditions represent the laboratory conditions, that is, normal unipolarly distributed working temperature throughout the SEC, absence of vibration, temperature fluctuations and humidity in the laboratory-calibrated and -certified gas mixtures, and so on.

The real values of all parameters characterizing the test conditions must be recorded in the test protocol and should be unchangeable during the whole duration of the experiment. Calibrated gas mixtures with fixed concentrations of the measuring gas are used for setting the standard values of the measuring gas $X_1, \ldots, X_n$. Fixed measured values of the output signal $Y_1, \ldots, Y_n$ are recorded. Further, parameter X has to be gradually decreased toward the value X = 0. This experiment has

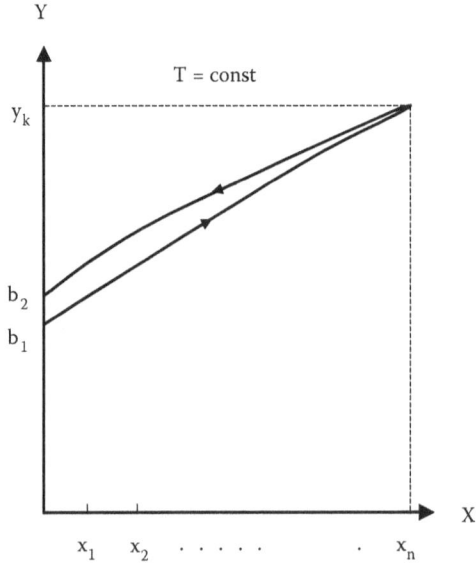

FIGURE 6.7 Nonlinear zirconia gas sensor function with hysteresis.

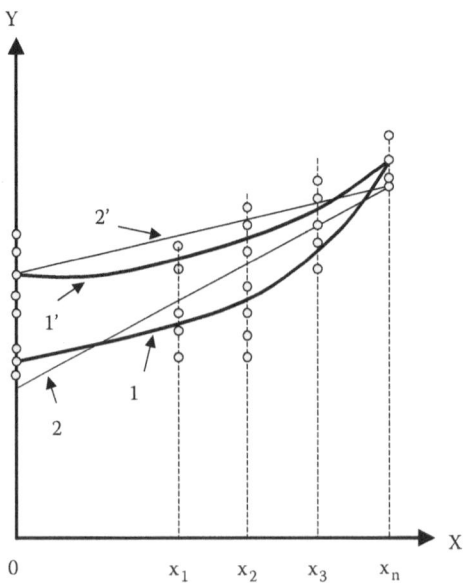

FIGURE 6.8 Real and approximating sensor functions.

to be repeated at least three times. Then, the average values of the output signal $\bar{Y}$ for the direct and reverse changing of the sensor function can be calculated based on the experimental data received (Figure 6.8). These calculations can be done by the following equations:

$$\bar{Y}_i = \sum_1^n Y_i/n; \quad \bar{Y}_i' = \sum_1^n Y_i'/n \ ,$$

where $\bar{Y}_i$ and $\bar{Y}_i'$ are the average values of the output sensor signal at each point for direct and reverse movements, respectively, and n is the number of points in each cross-section of the sensor function.

Assume that the approximation function is a linear one and can be expressed as

$$Y_a = kX_i + b_a, \tag{6.19}$$

where X_i is some value of the measuring parameter in the present selection. Owing to the fact that the real sensor function is a nonlinear one, parameter X_i corresponds not to the output signal Y_a as follows from Equation (6.19), but rather to some value Y_j. Consequently, the measurement error appears with the absolute value $\Delta = |Y_a - Y_i| = |kX_i + b_a - Y_i|$. Considering that the distribution law of the measuring parameter X_i within the work range is the uniform one, the dispersion can be written as follows:

$$D_y = 1/n \sum_0^n \left(kX_i + b_a - \bar{Y}_i\right)^2 , \tag{6.20}$$

where n is the number of cross-sections taken on the sensor function.

Equation (6.20) is a function of k and b_a; therefore, the dispersion D_y should be minimized depending on k and b_a parameters. In other words, it is necessary to calculate parameters of the approximating regression line.

The following system of two equations is obtained after differentiation of Equation (6.20) by k and b_a parameters:

$$\sum_0^n \left(kX_i + b_a - \bar{Y}_i\right)X_i = 0;$$

$$\sum_0^n \left(kX_i + b_a - \bar{Y}_i\right) = 0,$$

which can be worked out in relation to k and b_a as follows:

$$k = \frac{(n+1)\sum_{0}^{n}\overline{Y_i}X_i - \sum_{0}^{n}\overline{Y_i}\sum_{0}^{n}X_i}{(n+1)\sum_{0}^{n}X_i^2 - \left(\sum_{0}^{n}X_i\right)^2} \; ;$$

and

$$b_a = \frac{\sum_{0}^{n}\overline{Y_i}\sum_{0}^{n}X_i^2 - \sum_{0}^{n}\overline{Y_i}X_i\sum_{0}^{n}X_i}{(n+1)\sum_{0}^{n}X_i^2 - \left(\sum_{0}^{n}X_i\right)^2} \; .$$

By substituting values k and b_a into Equation (6.20), the dispersion value minimized from nonlinearity can be obtained and can be expressed in the relative units as follows:

$$\gamma^2 = (1/n)\sum_{0}^{n}\left(1-\left(\left(X_i - b_a\right)/kX_i\right)\right)^2 \; .$$

After determination of pitches k, k' and the initial values b_1, b' of the approximating function for the direct and reverse movements, the dispersion of the output signal aroused by nonlinearity can be worked out from the following equations:

$$D_y = (1/n)\sum_{1}^{n}\left(kX_i + b - \overline{Y_i}\right)^2 \; ;$$

$$D_y' = (1/n)\sum_{1}^{n}\left(kX_i + b' - \overline{Y_i}\right)^2 \; .$$

Dispersion of the output signal aroused by the hysteresis of the sensor function D_h can be determined as

$$b_a = \frac{\sum_{0}^{n}\left[\left(k-k'\right)X_i + \left(b-b'\right)\right]^2}{12n} \; .$$

Then the total dispersion in the absolute and relative units yields

$$D_\Sigma = D_h + (D_y + D'_y)/^2 \; ; \; \gamma^2_\Sigma = D_\Sigma/y^2_n \; ,$$

where y_n is the normal value of the sensor output signal.

Thus, based on the presented calculation, the dispersions resulted by pitch of the lines 2 and 2' (Figure 6.8) with the pitch of the real sensor functions 1 and 1' as well as inconsistency of the lines 2 and 2' can be found.

Analysis of the statistical data (presented as dots in Figure 6.8) shows that the separate selections by each section have some discrepancy. This discrepancy of calibration points is explained by the presence of the so-called error of the measurement variations, which is more often known as the *laboratory error*.

Movement of the real sensor function (curves 1 and 1') along the present statistic should be established for the laboratory error determination. Then the sum of the square deviations of each selection from the average value should be calculated. This calculation can be done for direct and reverse movements of the sensor function as follows:

$$b_i = \frac{\sum_0^n Y_i \sum_0^n X_i^2 - \sum_0^n (Y_i X_i) \sum_0^n X_i}{(n+1)\sum_0^n X_i^2 - \left(\sum_0^n X_i\right)^2} \; ;$$

$$D_{0\,var} = \sum_1^K \left[\left(b_i - M_b\right)^2\right]\Big/(K-1),$$

where n is a number of cross-sections, b_i is the initial value of the output signal for i^{th} cycle, $M_b = \sum_1^K b_i/K$ is the average number of b obtained during K cycles of measurement, and $D_{0\,var}$ is the additive dispersion arising from the variation of the sensor output signal. In the relative units, it is expressed as

$$\gamma^2_{0\,var} = D_{0\,var}/Y^2_n \; .$$

The relative multiplicative dispersion arising from the sensor output signal is as follows:

$$\gamma^2_{k\,var} = \sum_1^K \left[\left(k_i - M_k\right)^2\right]\Big/(K-1)M_k^2 \; ,$$

where

$$k_i = \frac{(n+1)\sum_{0}^{n} Y_i X_i - \sum_{0}^{n} Y_i \sum_{0}^{n} X_i}{(n+1)\sum_{0}^{n} X_i^2 - \left(\sum_{0}^{n} X_i\right)^2} \; ;$$

$$M_b = \sum_{1}^{K} b_i / K \, .$$

Thus, all components of error peculiar to the sensor work at the laboratory conditions have been determined.

The next stage of investigation is the determination of both additive and multiplicative dispersions affected by the various influencing factors. The test setup as well as the order of experiment remain the same. The only difference is based on the fact that the zirconia gas sensor has to be affected by one of the influencing factors. For example, considering the error of setting working temperature, the sensor has to be placed into the laboratory furnace, which sets up temperature $T = T_0 + \Delta T_1 + \Delta T_2$, where T_0 is the initial temperature value, ΔT_1 is the zero offset between SE and RE ($\Delta T_1 \neq 0$), and ΔT_2 is the error of setting the required temperature by the laboratory furnace ($\Delta T_2 \neq 0$), respectively. ΔT_2 value for the working temperature range is usually available from the furnace manual. Then, the sensor has to be kept at temperature T for a certain time until stabilization of the sensor temperature T_s = const takes place.

After stabilization of the set temperature, the zirconia gas sensor has to be graduated to exactly the same volume and sequence as at determination of those sensor errors in laboratory conditions. The relative additive and multiplicative sensitivities to the temperature factor are calculated by experimental data:

$$S_{0\,xr} = \overline{S}_{0\,xr} / Y_n \; ; \quad S_{k\,xr} = \overline{S}_{k\,xr} / K \, ,$$

where

$$\overline{S}_{0\,xr} = \sum_{1}^{m} S_{0\,rj} / m \; ; \quad \overline{S}_{k\,xr} = (1/m) \sum_{1}^{m} \left[(k_{xrj} - k) / (X_{rj} - X_0) \right] \; ;$$

$$S_{0\,rj} = [(b_{rj} - b_0)/(X_{rj} - X_0)];$$

$$b_{rj} = \frac{\sum_{0}^{n} Y_{xi} \sum_{0}^{n} X_i^2 - \sum_{0}^{n}(X_i Y_{xi}) \sum_{0}^{n} X_i}{(n+1)\sum_{0}^{n} X_i^2 - \left(\sum_{0}^{n} X_i\right)^2} \; ;$$

and

$$k_{xrj} = \frac{(n+1)\sum_{0}^{n}(Y_{xi} X_i) - \sum_{0}^{n} Y_{ki} \sum_{0}^{n} X_i}{(n+1)\sum_{0}^{n} X_i^2 - \left(\sum_{0}^{n} X_i\right)^2} \; ;$$

where m is the number of fixed point steps for changes of the influencing factor X_r from X_{rj} to X_{r0}, X_i is the value of the input measuring signal (concentration of measuring gas), and Y_{xi} is the value of the output sensor signal (*emf*) at the fixed value of the input signal at the presence of the influencing factor X_r.

The additive and multiplicative dispersions of the output signal can be determined as follows:

$$\gamma^2_{0\,xr} = S^2_{0\,xr} D\,(X_r), \tag{6.21}$$

$$\gamma^2_{k\,xr} = S^2_{k\,xr} D\,(X_r). \tag{6.22}$$

Total reduction additive and multiplicative dispersions on the influencing factor's activity are evaluated by the following equations:

$$\gamma^2_0 = \sum_{1}^{L} \gamma^2_{0\,xr} ; \quad \gamma^2_K = \sum_{1}^{L} \gamma^2_{kxr} \; .$$

Dispersion of the influencing factor $D(X_r)$ is present as a factor in (6.21) and (6.22). For its calculation, it is essential to know the distribution law of the accidental value X_r. However, if data about the distribution law are unavailable (quite a common situation for the development of new zirconia gas sensors), the designer must decide for him or herself which distribution law belongs to each influencing factor, depending on the sensor's purpose and potential applications. Furthermore, if the information regarding the value of influencing factor $D(X_r)$ is set only by boundaries of the scope of changes and no other information is available, then the distribution law of the probability density must be assumed as the uniform one. In this case, the dispersion of the influencing factor $D(X_r)$ is calculated as follows:

$$D_{Xi} = (X_{ih} - X_{il})^2/12,$$

where X_{ih} and X_{il} are the highest and lowest boundaries of the scope of changes of the influencing factor, respectively.

Contrarily, if the distribution law of the influencing factor is closed to the normal distribution law, then the following equation should be applied for the dispersion calculation:

$$D_{Xi} = (X_{ih} - X_{il})^2/36.$$

It has become evident that the considered order of obtaining experimental data and appraising the main sensor's errors based on these data is a time-consuming process. Although this time can be even more significant if the sensors are produced by big consignments, the selective quality control widely used by plants manufacturing mechanical sensors (pressure gauges, weights, etc.) is unacceptable for evaluation of the different zirconia-based gas sensors. All manufactured sensors must be put through rigorous quality control tests. Descriptions of some of these tests are given in chapter 7. The optimal solution for this complex task can be the development of fully computerized automated test stations. Figure 6.9 illustrates one of the examples of such a test station, specifically designed in KASTEC for testing various gas sensors.

FIGURE 6.9 Example of computerized, automated test station for testing various zirconia-based gas sensors.

Consequently, the need for routine sensors testing has been widely supported by different industries consuming zirconia gas sensors and may be described as follows [13]:

- Making sure that manufactured gas sensors meet the requirements of the relevant national and international standards.
- Determining that zirconia gas sensors are reliable and perform satisfactorily in both normal and extreme conditions.
- If failures *do* occur: providing feedback to manufacturers indicating the nature of the failure, which is vital for R&D work.
- Consumer information: statistical analysis determines which type of zirconia gas sensor is the most reliable.
- Important information and feedback for risk engineering: statistical information aids to determine which type of zirconia gas sensors to install in different applications and why.

REFERENCES

1. BS EN 61207-1, *Expression of Performance of Gas Analyzers*, London, British Standards Institute, 1994, 34.
2. Kovarich, R.P. et al., Highly accurate measurement of oxygen using a paramagnetic gas sensor, *Meas. Sci. Tech.* **17** (2006) 1579–1585.
3. Rabonovich, S.G., *Measurement Errors and Uncertainties: Theory and Practice*, New York, Springer, 2000, 316.
4. Zhuiykov, S. et al., High-temperature NO_x sensors using zirconia solid electrolyte and zinc-family oxide sensing electrode, *Solid State Ionics* **152–153** (2002) 801–807.
5. Miura, N., Nakatou, N., and Zhuiykov, S., Impedancemetric gas sensor based on zirconia solid electrolyte and oxide sensing electrode for detecting total NO_x at high temperature, *Sens. Actuators B: Chem.* **93** (2003) 221–228.
6. Miura, N., Nakatou, N., and Zhuiykov, S., Development of NO_x sensing devices based on YSZ and oxide electrode aiming for monitoring car exhausts, *Ceramics Int.* **30** (2004) 1135–1139.
7. Badwal, S.P.S., Ciacchi, F.T., and Giampietro, K.M., Analysis of the conductivity of commercial easy sintering grade 3 mol % Y_2O_3–ZrO_2 materials, *Solid State Ionics* **176** (2005) 169–178.
8. Zhuiykov, S., ZrO_2-based partial gas pressures sensors in vacuum: Analysis main components of errors, *Vacuum Tech. & Coating* **2** (2006) 38–44.
9. Antropov, L.I., *The Theoretical Electrochemistry*, Moscow, Science Publishing, 1975, 540.
10. Zhuiykov, S., Electron model of solid oxygen-ionic electrolytes used in gas sensors, *Int. J. Applied Ceram. Tech.* **3** (2006) 401–411.
11. Rozanov, L.V., *Vacuum Equipment*, Leningrad, Chemistry, 1975, 336.
12. Tretjakov, U.D., *Chemistry of Non-stoichiometric Oxides*, Moscow, Chemistry Publishing, 1974, 364.
13. Zhuiykov, S., Fire detectors: Evaluation, statistical analysis and quality control by audit sample testing, *Fire & Safety* **2** (2005) 4–12.

7 Organization and Planning of Testing Zirconia Sensors

7.1 MAIN PRINCIPLES OF TESTING ZIRCONIA SENSORS

Figure 7.1 broadly represents the zirconia gas sensor development chart covering the range of activities from the scientific concept of measuring the selected gaseous component to the engineering developments required to deliver the devices, instruments, and control systems. Researchers working in the application environment (i.e., delivering scientific and engineered solutions) are most commonly exposed to the industrial sensor needs and interests. At the same time, the existence of integrated technology solutions facilitates the transfer of information to staff working on the development or improvement of appropriate sensing mechanisms and devices. In detail, the main stages of zirconia gas sensor development can be considered as follows.

7.1.1 SENSING MECHANISMS OF ZIRCONIA GAS SENSORS

These are fundamental mechanisms and combinations of interactions at the TPB that allow the detection and quantification of the selected gaseous component (i.e., gas concentration or partial gas pressure) with the required accuracy level. A sensing and interaction mechanism is also known as a *transduction mechanism* that converts

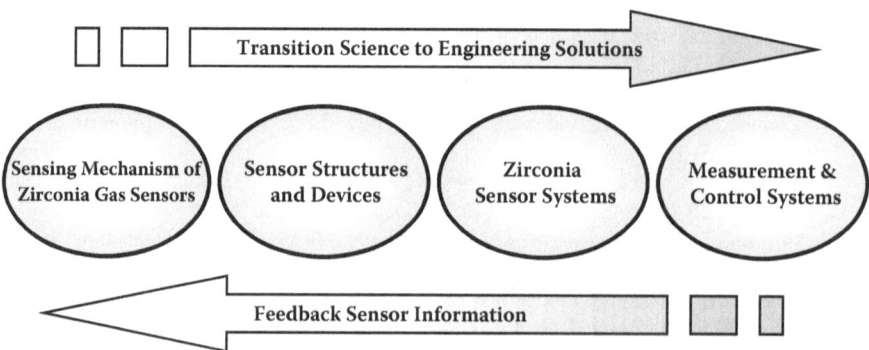

FIGURE 7.1 Zirconia gas sensor development chart.

the presence of target gas at certain temperatures to some form of usable output signal (*emf*, or current).

7.1.2 Sensor Structures and Devices

These zirconia-based devices can detect the presence and/or quantify the concentration of a family of chemicals or specific gases exposed to the sensors. The sensors will rely on one or more sensing mechanisms and produce a signal that indicates the chemical's presence and/or concentration.

7.1.3 Zirconia Sensor Systems

Typically, this is a zirconia-based gas sensor with associated components (electronics and user input-output devices) that can convert the sensor's output signal to usable information on a single target gaseous component in a measuring environment. These may be stand-alone instruments or a sensor module for incorporation in a more comprehensive system. Zirconia sensor systems usually include electronic devices/circuits plus software (embedded and discrete) that provide functions including signal processing, data acquisition, communications, device/system control, and so on.

7.1.4 Measurement and Control Systems

These are integrated systems that are developed to meet specific client application requirements and can include the following: single or multiple sensors, signal processing and data analysis functions, data monitoring and recording, and control functions. In research and development and in the application of technologies, statistical expertise is required to plan data collection activities and interpret the results of experiments. Uncertainties in the data must be quantified, constrained, or modeled to exercise proper control of a system or to ensure correct decisions and conclusions.

It is evident that conducting even the simplest tests of the zirconia gas sensors at high temperatures entails certain expenses, and the efficiency of experiments is determined as a ratio of the positive effect obtained during experiments to the test expenses. Therefore, the strategy of planning the sensors' tests is usually focused on the development of those algorithms that provide the maximum efficiency of the tests.

Structure of the test planning and execution of the zirconia gas sensors should cover the following main tasks:

- *Select the aim of the experiment.* The aim is essential, even if it looks very simple. It reflects all aspects of the planning. The aim of an experiment consists of the answer to the question of whether zirconia sensors can measure the selected gas concentration under certain environmental conditions (high temperatures, presence of humidity and other gases, etc.). The answer usually is yes or no. The test expenses in this case are insignificant. For example, the aim of the experiment can be the assessment of applicability of the zirconia sensor to the frequent changes in working conditions of the measuring environment. In this case, the final result will be assessed as a combination of all parameters involved in the experiment. There is no demand for optimal strategy of the test and the rational distribution and usage of the technical resources available.

- *Performance evaluation methodology.* Methods of performing tests, collection of data, and subsequent analysis vary widely with the intended application. However, it is important in all cases that these methods be carefully planned from the outset in terms of objectives, data format and data-processing methods, and scope of tests to ensure the workload is reasonable and the intended results are achieved. If the experiment is not well planned, then the number of tests can become excessive compared to labor resources, or alternatively the test data will become unmanageable or be in a format that makes posttest processing impossible without specialized software. The approach to executing tests and analyzing the results is dependent on the application and can be well bounded with careful planning.

- *Planning.* Planning experiments is one of the main stages of zirconia sensors' testing. There are many methods of planning experiments [1–5]. Their complexity usually depends on the diversity of the sensors' applications; and as far as zirconia sensors' testing is concerned, the most commonly used methods can be presented as follows:

    Dispersion analysis. The task is formulated like this: it is necessary to propose such a scheme of the sensor testing, which would allow factorizing the summary dispersion on the constituents. This method has been widely used in testing the different zirconia gas sensors.

    Factor experiment. The task is formulated like this: it is necessary to estimate linear effects and interaction effects at a large number of the independent variables. In this case, the variation of several factors (variables) takes place, which ultimately increases the efficiency of the experiment. The significant effects assessment is usually done afterwards by the dispersion analysis.

    Assessment and identification of mathematical model. One of the possible task formulations that comprises providing identification of the parameters of the developed mathematical model and verification of the adequacy of the model to the real work function of the sensor obtained during experiments. The number of factors, the volume of essential measurements at the different test conditions, as well as the extract volume should be determined during the planning of experiments.

- *Equipment selection.* During secondary equipment selection (measuring electronic devices, etc.), it is not necessary to obtain equipment with the maximum possible accuracy level because the trivial results could be obtained even on the unique apparatus. The accuracy of the electronic measuring devices should correspond to the required trustworthiness of the results. It should also be remembered that ultra-high accuracy is inseparably connected with extra expenses on equipment and on the time of test reiteration. The reasonable compromise should be found in each particular case depending on the sensor application.

• *Data processing*. At present, data processing always takes place during experiments. In order to obtain the estimated appraisal of experimental data, mathematical processing is used for further preparation of the test results. Mathematical processing allows obtaining the accurate results from an excessive quantity of inadequately precise data. If the volume of extracts is significant, then the experiment should be planned so that the dependent variables would be the input data of the processing algorithm. It should also be noted that in some cases, the estimated type of data processing, an adequate mathematical model, and embedded software correspondent to the developed model, can be the dominating prerequisite at the experiments planning stage.

• *Analysis*. During analysis of the test results, the decisions are made in accordance with the aims of the experiment. For example, the decisions can be a continuation or interruption of the experiments, changing the test direction or volume of data for experimental processing, and so on.

Generally speaking, all experiments can be divided into active and passive. An active experiment is usually carried out based on the test program approved beforehand, when the test officer sets the boundary and the number of the independent variables, determines the order of setting of these values and the number of measurements, and so on. The test officer, to some extent, totally controls the situation during an active experiment.

During a passive experiment, the test officer acts as an observer, who can only register the values of independent and dependent variables.

Based on the fact that the zirconia sensor tests are predominantly active experiments, let us consider some important aspects of planning such active experiments, which have practical significance for industry.

7.1.5 SELECTING THE NUMBER OF INDEPENDENT VARIABLES (FACTORS)

This task can be formalized only with analysis of the a priori information determined by the experience and intuition of the researcher. During study of combination, the interconnected complex processes occur at high temperatures on the TPB; even the simple list of all potential factors influencing the output sensor signal can be significant, as has been shown in Chapter 6. The planning and execution of an experiment with consideration of all these factors will lead to the substantial increase of time and cost of the experiment without any significant improvement of its efficiency. Consequently, the first criterion of the variables selection should be "commonsense" criterion, based on a priori analysis of the physical, chemical, and electrochemical processes on the TPB. In the next stage, the estimated appraisal of the factors' significance can be fulfilled by setting the series of one-factor experiments, where one factor varies in each individual test only and all others have the lowest values within the range of their changes. In the simplest case, the independent variables vary only on two levels, corresponding respectively to the maximum and minimum values of their changes. As a result of these experiments, the selective values of regression coefficients β_i by each ith factor are determined. The dispersion of the regression coefficients is decreasing proportionally to the number of tests in a one-factor experiment.

Having known dispersions of the independent variables σ_i^2, the contribution of each of them to the total resulting dispersion can be appraised as follows:

$$D_i = \beta_i^2 \sigma_i^2. \tag{7.1}$$

Then the test officer can make a decision about taking into account or rejecting the appropriate factor, after calculation of the values of the relative contribution of the ith variable into the total dispersion D:

$$\gamma_i = \left(D_i/D\right) = \left[\beta_i^2\sigma_i^2 \Big/ \sum_{i=1}^{i=n} \beta_i^2\sigma_i^2\right]. \tag{7.2}$$

The decision-making process described above is acceptable only to the linear tasks. However, considering zirconia gas sensor testing, where the volume of main testing is connected to appraisal of the sensor metrological characteristics, the quasi-linear tasks usually take place. Therefore, despite the restriction by linearity as well as a certain level of subjectivism, this way is relatively easy and clearly excludes variables, the insufficient value of which at comparison γ_i is unquestionable.

The distribution laws of values of the influential factors are accepted to be uniform at calculation of σ_i^2. The value of the total dispersion D is equal to 0.1433. The multicriteria active experiment has to be done for more accurate determination of valuable dominating factors. The results of this experiment are used in dispersion analysis. Then the residual dispersion, stipulated by the error of experiment, is calculated as follows:

$$D_{res} = D_0 - \sum_{i=1}^{n} D_i. \tag{7.3}$$

And, taking into account the ratio D_i/D_{res}, criterion of statistical significance of the Fisher dispersion can be applied for further calculation. In (7.3), D_0 is the total dispersion of the results relative to the total average value, D_1 is the dispersion of the average results value by all r levels of the first factor, and D_n is the dispersion of the average results value by all r levels of the nth factor. More detailed explanation of the dispersion analysis theory can be found in [6, 7].

Carefully calculated reduction for number of the significant factors leads to a less expensive and still efficient experiment, allowing in the receiving the same volume of useful data. At the same time, unfounded exclusion of the significant factors leads to the increase of dispersion of the results, engendered by the unconsidered factors and in some cases to wrong interpretation of data obtained during experiment.

7.1.6 Determination of Experimental Data Volume

It is essential to select a limiting number of independent variable levels at the planning stage of an active experiment. It is also required to select an end rational number of experimental measurements. If a small volume of experimental measurements is planned, then the aim of the test may not be achieved. In contrast, if the measuring data volume is too big, the cost of the experiment has risen and its duration and data processing are taking too much time.

The selection of experimental measurements begins with the determination of the boundary conditions of the independent variables. These values are often given by the sensors measuring conditions and are determined by the specific research tasks. The selection of the intermediate values of independent variables (or intervals between values) is determined by the character of experimental function, by the reproduction means of the selected function, and by the required accuracy of measurement.

Two points are sufficient in the case of the strictly linear experimental function. These points correspond to the boundary values of the independent variable. All other intermediate values (if necessary) can be calculated during further data processing.

However, if the sensor function is not strictly linear, especially on boundaries of the measuring range, and the test officer possesses good knowledge about its character, then the values of independent variables can be determined from the required accuracy of measurement.

Let us consider the example of the potentiometric zirconia oxygen sensor working at high temperatures. The measured sensor function represents a quasi-linear regression corresponding to the Nernst equation with a certain accuracy level, and a part of this function is shown in Figure 7.2. If the reproducibility error Δy is given, and the sought sensor function is restored by the experimental measurements of the output *emf* connected by the linear segments, then the intervals between the values of the experimental points are found as distances between abscesses of the intersection points of the experimental sensor function and the line segments. In this case, the maximum difference between them should be equal to the allowable measuring error (interpolation errors). Sometimes the intermediate values of the independent variable are selected in the way that the segments of the experimental functions ΔS are equal to each other, as shown in Figure 7.3. At such a selection, the maximum error can be on the segment with the biggest curvature — and, in the case of the same curvature, on the segment with the bigger value of the first derivative [8]. The length of the ΔS segment is determined from the maximum allowable measurement error.

It is necessary to take one more circumstance, connected to the relevant accuracy of measurement, into consideration at the selection of intervals between experimental measurements. More experimental points should be selected, that is, the intervals between points will be minimized on those parts of the experimental quasi-linear sensor function where the set relative accuracy of the independent variable is moderate.

In those cases where the errors of measurement lead to the uncertainty of the results, it is desirable to repeat measurements, that is, set the value of the independent variables several times, and to average the results obtained. The resulting dispersion changes in accordance with the well-known correlation $\sigma^2 = \sigma_i^2/n$, where σ_i^2 is the

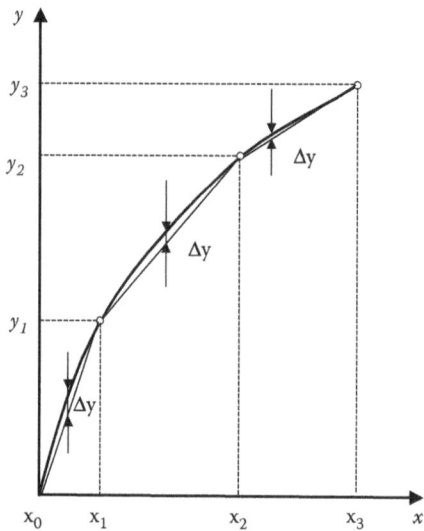

FIGURE 7.2 Quasi-linear regression of zirconia sensor corresponding to the Nernst equation.

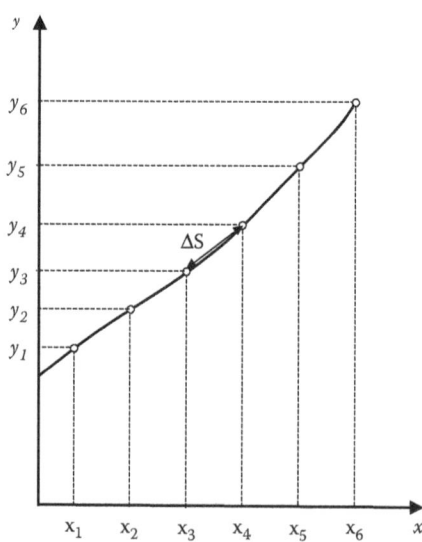

FIGURE 7.3 Segments selection from the equality condition for the lengths of experimental quasi-linear regression of the zirconia sensor.

dispersion of each individual ith result and σ^2 is the dispersion of the average from n measurement results. It is obvious that averaging the measurements makes the experiment more expensive and prolonged. Therefore, if the required and really achievable accuracies of measurement are known and can be appraised by σ and σ_i values, respectively, then $n = \sigma_i^2/\sigma^2$.

The reiteration of measurements can be connected with different drifts of the sensor characteristics caused by aging and the like. Further consideration of the situations given above can be found in the literature of the theory of experiment [9].

7.1.7 SEQUENCE OF EXPERIMENT

Sequence of experiment can be divided into two main types. In the first type, one of the boundary values of the independent variable is set, and then consecutive transfer from one experimental point to another takes place until the second boundary value will be achieved. A plan of such experiment is called a *consecutive plan*.

In the second type, the value of the independent variable alternates randomly. Such a plan of experiment is usually called a *randomized plan*.

The consecutive plan is widely used in many sensor experiments, especially in tests where the sequence of experiment itself is a peculiar parameter. Examples can be the sensor tests, where the sensor function is accompanied by hysteresis.

The randomized plan is also suited to many sensor experiments. Its main arguments are based on the fact that the external (controlling) conditions of experiment can vary in time and some undetected faults can influence the value of the independent measuring variable. The main concept of randomization is based on the fact that the systematic influential factors, which are hard to control with certain accuracy, should become the accidental factors for their statistical control. However, the randomization may be unnecessary in the complex experiments, when the establishment of the fixed experimental conditions requires substantial extra time and the accidental sequence of transfer from one test condition to others brings even more time expenditures.

7.1.8 DATA PROCESSING

Basically all methods of mathematical statistics can be used for data processing of the zirconia sensor tests. At data processing, the average value of some measurement is used for estimation of the real value of the variable. However, it is the most effective appraisal only for those distribution laws which are close to the normal one.

The following correlation is used for dispersion evaluation:

$$D = \frac{\sum_{i=1}^{n}(x_i - \bar{x})^2}{n-1}, \tag{7.4}$$

where n is the number of tests, x_i is the value of the measuring variable on the ith measurement, and $\bar{x}$ is the average value.

Grouping of experimental data at the histograms building can lead to the displacement of the calculating characteristics. The Sheppard corrections are used for the removal of this displacement [10].

Method of the least squares is used at processing of the functional experiments data. In the case of the one independent variable (one-factor experiment), the type of the regression should be appraised by the experimental measurements plotted within the y-x coordinates. The most frequently used dependences in practice are as follows:

$$y = kx^b; \quad y = c + kx^b; \quad y = k_1x^{b1} + k_1x^{b2} + \ldots + k, \tag{7.5}$$

where k, b, and c are coefficients, which can be positive and/or negative and can be more and/or less than unity. Quite often, the logarithmic scales are used in data processing:

$$\lg y = \lg k + b \lg x. \tag{7.6}$$

The regression coefficients k_i can be found from the distribution of the measured parameters. To find the regression coefficients, it is necessary to work out a task: select the real function that minimizes the total square deviations of the measured values from the values of the selected function. The task solution consists in the solution of the following system of equations determining the extremes of private derivatives of the total square deviations function:

$$\sum_{i=1}^{n} [y_i - \varphi(x_i, k, k_1, \ldots, k_{n-1})]^2 . \tag{7.7}$$

In general, this system can be expressed as follows:

$$\begin{cases} \sum_{i=1}^{n} [y_i - \varphi(x_i, k, k_{1,\ldots,}k_{r-1})](\partial\varphi / \partial k)_i = 0; \\ \\ \sum_{i=1}^{n} [y_i - \varphi(x_i, k, k_{1,\ldots,}k_{r-1})](\partial\varphi / \partial k_1)_i = 0; \\ \\ \ldots\ldots\ldots\ldots\ldots\ldots\ldots\ldots\ldots\ldots \\ \\ \sum_{i=1}^{n} [y_i - \varphi(x_i, k, k_{1,\ldots,}k_{r-1})](\partial\varphi / \partial k_{r-1})_i = 0, \end{cases} \tag{7.8}$$

where n is a number of measurements.

If the measurement number n is less than the number of the searching coefficients r, it is impossible to find a solution for (7.8); at $n = r$, the system (7.8) has only one solution. However, at $n > r$ there is a possibility to have several systems and consequently to get a row of the searching functions. In the last case, the obtained data should be averaged to increase the formalization accuracy of the experimental work function.

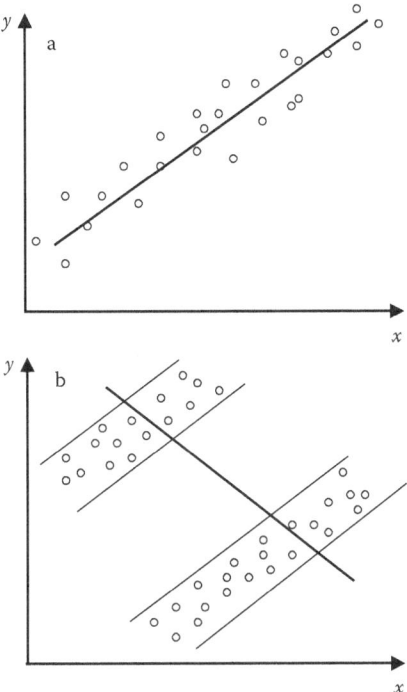

FIGURE 7.4 Examples of the measurement results: (*a*) even distribution; and (*b*) uneven distribution.

Using the method of the least squares does not always provide an ultimate positive outcome. In some cases, it can lead to the wrong conclusion and an incorrect reflection of the existing appropriateness. For example, it is possible at inhomogeneous distribution of the measurement results (Figure 7.4). The selected sensor function (Figure 7.4, *a*) is correctly reflecting the dependence between the input and output parameters at the even distribution of the measurement results. On the contrary, at uneven distribution, which is basically provided by two different independent dependences, the formal usage of the method can lead to the false results (Figure 7.4, *b*). Therefore, it is mandatory to analyze the obtained test results first in order to select a valid algorithm of data processing.

During processing of experimental data, the regression line should also be taken into account. Sometimes, the regression lines (lines 1 and 2 in Figure 7.5) do not correlate to each other owing to the difference in the correlation coefficient k. Let's assume that the equation is determined by correlation $y = k + k_1 x$ for the regression line 1 and $y = k' + k_2 x$ for the regression line 2, respectively. The ratio k_1/k_2 determines the value of the square of the correlation coefficient between y and x. If we consider that the real correlation corresponds to the central line of the dissipation field, as a dotted line in Figure 7.5, then the error caused by usage of only one of the regression lines can be calculated as follows:

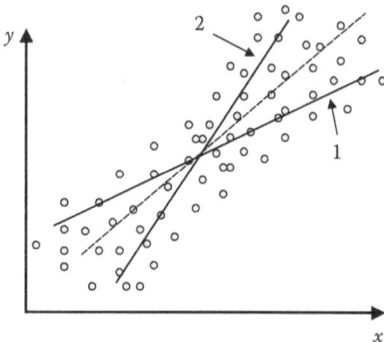

FIGURE 7.5 Graphical determination of the regression line.

$$\gamma = \frac{\Delta k}{k_{ave}} = \frac{k_2 - k_1}{2} \times \frac{2}{k_1 + k_2} = \frac{1 - r^2}{1 + r^2} . \tag{7.9}$$

If the error is more than 10%, then the correction should be implemented. For example, if $r = 0.7$, then $\gamma = 33\%$; coefficient k_1 should be increased by 33% and coefficient k_2 should be decreased by 33%, respectively.

The linear models use at data processing of the multifactor experiments

$$y = a_0 + a_1 x_1 + a_2 x_2 + \dots + a_n x_n, \tag{7.10}$$

or models of the following type:

$$y = a_1 x_1^{b1} x_2^{b2} \dots x_n^{bn}, \tag{7.11}$$

which can be brought to the linear models by finding the logarithm:

$$lg\ y = lg\ a + b_1\ lg\ x_1 + b_2\ lg\ x_2 + \dots + b_n\ lg\ x_n. \tag{7.12}$$

The appraisal of coefficients a_i and b_i can be done afterwards by the least squares method.

Coefficient of multiple correlations is used as a level of linear connection between the measuring variable and a composition of the independent variables [11]:

$$r = \sqrt{1 - \frac{\sigma_{res}^2}{\sigma_y^2}} , \tag{7.13}$$

where σ_y^2 is the dispersion of the measuring variable y, and σ_{res}^2 is the residual dispersion of deviation of the measured values y_i from the corresponded values, determined by the model.

The residual dispersion characterizes the errors of experiment provoked by the undetected factors, and the errors caused by the approximate character of the model. The relative value of these errors can be expressed by $\gamma = \sigma_{res}/\sigma_y$.

The value of γ depends on the number of considering factors. If the considering factors are sized by the significance (by the value of introducing effects), then at the small number of detected factors the value of γ will be significant because of the substantial influence of undetected factors. The residual dispersion and γ value are dropping at the increase of the detected factors number, then are rising again due to some k number of the detected factors. It can be explained by the fact that the regression coefficients, corresponding to the small influencing factors, determine with very little trustworthiness and, consequently, have big dispersion themselves. This situation can be used for selection of the dominating factors. For this purpose, all variables have to be centered and standardized beforehand:

$$t_{yi} = \left(y_i - \overline{y}\right)/\sigma_y; \quad t_{ji} = \left(x_i - \overline{x}_j\right)/\sigma_j \; ; \tag{7.14}$$

where t_{yi} is the ith count of the dependent variable; t_{ji} is the ith count of the jth independent variable; $\overline{x}$ and $\overline{y}$ are the average values of the independent and dependent variables, respectively; and σ_y and σ_j are the average square values of the dependent and independent variables, respectively.

Thus, the regression equation with β-coefficients should be compiled as follows:

$$t_y = \beta_1 t_1 + \beta_2 t_2 + \ldots + \beta_k t_k, \tag{7.15}$$

where $\beta_j = b_j \sigma_j / \sigma_y$ are the β-coefficients.

Sizing of the independent variables is proceeded by the value of β-coefficients. Then the error of γ can be appraised by consistently decreasing the number of β-coefficients. The total sum of dominating factors can be determined by the minimum of γ. It is obvious that the described procedure is very time-consuming. However, the calculations can be substantially simplified, if the independent variables vary only on two boundary levels and their standardization is done so that the boundary values would be equal to -1 and 1, respectively:

$$x_{j min} = \frac{x'_{j min}}{\overline{x} - \Delta x_j}; \quad x_{max} = \frac{x'_{j max}}{\overline{x} + \Delta x_j} \; ; \tag{7.16}$$

where $\overline{x} = {}^1/_2 (x_{j\,max} + x_{j\,min})$; $\Delta x_j = {}^1/_2 (x_{max} - x_{j\,min})$; and $x_{j\,max}$ and $x_{j\,min}$ are maximum and minimum boundary values of the jth factor.

If the plan of experiments is rich (number of measurement $n = w + 1$, where w is the number of factors) and symmetrical (sum of the values of each independent variable, including the repeatable values during experiment, is equal to zero), and the variables themselves are orthogonal (i.e., independent), then the regression coefficients can be calculated independently from each other by the following equation [11]: $b_j = \sum_{i=1}^{n} x_{ij} y_i / n$ with dispersion $\sigma^2(b_j) = \sigma^2 y/n$.

7.2 PLANNING OF EXPERIMENTS

The main purpose of the zirconia gas sensor is to transform information about the measuring gas concentration (partial pressure) at the presence of a significant number of the influencing factors. The main quality index of the sensor is the trustworthiness of transforming information, and the main criteria of the trustworthiness are the sensor's error and its reliability. Therefore, the testing which has been dedicated to the evaluation of the sensor's error and reliability is the biggest part of the experimental work during development of the sensor.

At the planning of appraisal of the sensor's error experiments, the test officer usually knows the list of the influencing factors and their boundary conditions. From the theory of experiment point of view, the optimal test setup would be the multifactor experiment at which both all influencing factors and measuring gas concentrations would be varied within all ranges with following appraisal of the dispersion of the output signal. However, it is impossible to implement the optimal test setup in practice.

The first restriction is an inability to use the randomization principle. It is explained by the fact that the majority of the zirconia gas sensors are based on using operational principles possessing hysteresis effects (hysteresis of the electrochemical reactions at the TPB, hysteresis of the setting-required temperature of the sensor, etc.).

The second and main restriction is the absence of the test equipment which would allow varying all influencing factors on the different levels. For example, for the zirconia sensors measuring O_2, C_xH_y, CO, CO_2, and NO_x concentrations in vehicle exhausts, it is impossible to set simultaneous deviation of the measuring temperature and vibration, which is usually present during vehicle acceleration on a country road.

The first restriction results in the consecutive plan of experiments, and the second one in resolution of the multifactor experiment on some sequence of single-factor and/or two-factor experiments.

The main groundwork of planning experiments for the sensor's error determination (metrological tests) is based on appraisal of the statistical model of error, which can be calculated on the basis of a mathematical model of the zirconia gas sensor with distributed parameters (considered in Chapter 2 for the NO_x sensor). In accordance with this model, the sensitivity of the zirconia sensor to the measuring gas concentration and to the influencing factors can be sequentially determined. It is assumed that all influencing factors are independent of each other and the contributions of each factor to the total dispersion are divided into two components: dependant and independent from the measuring gas concentration (multiplicative and additive) constituents. Owing to such division, sensitivities to the influencing factors are determined by the two-factor experiment with simultaneous variation of each factor and the measuring gas concentration.

Having done the determination of all sensitivities and appraised the dispersions of the influencing factors by one of the methods presented earlier in Chapter 6.4, the total dispersion of the measurement result has to be calculated.

Other types of sensor tests, such as checking the vacuum-tight joints of the sensor, a vibration test, and so on, have usually been planned as one-factor experiments.

The decisive importance at the planning of the zirconia gas sensor testing belongs to the sequence of the tests. In the first place, those verifications and tests should be done which do not affect the efficiency of the sensor. Examining the sensor's polarity, inputting impedance at various temperatures, checking the vacuum-tight joints, checking the initial and background levels of the output signals, and so on usually need to be done at this stage.

Then, the zirconia sensor should be tested at the normal working conditions: for example, examinations of its sensitivity at different temperatures, verification of the work function hysteresis, determination of the measuring concentrations range, and so on. After that, all other metrological characteristics of the sensor have to be appraised.

The most important and substantial, by the time and volume of work required, are those tests where the determination of main metrological characteristics, at which the evaluation of the influence of various external factors on the sensor's sensitivity is performed, should be done. A cross-sensitivity test to other gases, impact of the presence of humidity on the sensor output signal test at different temperatures, a vibration test, a thermoshock test, and other tests imitating the extreme working conditions of the sensor will follow afterwards.

7.2.1 DEVELOPMENT OF THE INDUSTRIAL PROTOTYPE OF THE SENSOR

Tests at this stage become more specific and involve the first level of sensor proving. This may include tests to verify elements of sensor design, which are difficult to do in the lab, for example methods of compensation for thermal effects (thermal gradient between SE and RE). It may also involve the tuning of detection algorithm performance, assessing practicality of packaging or development hardware, or methods.

7.2.2 PRODUCT VERIFICATION

At this stage is a more rigorous and long-term planned testing to verify the total sensor functionality, based on the product specification. Ideally this phase runs for *at least* a year on a final version of the sensor design to allow estimation of the main sensor's characteristics as well as its so-called long-term stability. It allows for final product adjustments to account for characteristics unforeseen at the development stage that may occur under unusual environmental conditions. Unfortunately, the long-term stability concept has been misunderstood in the majority of pure academic publications. As far as industry is concerned, the long-term stability starts only after one year of operation. This is much more of a warranty issue for the manufacturing companies rather than the reliability of the sensors. From my personal industrial experience, the well-designed and well-made zirconia-based gas sensors provide long-term stability from 3 to 10 years of operation [12, 13]. Thus, despite a significant number of published research works so far, there is a lack of reliable zirconia-based gas sensors available on the market at present.

7.2.3 TRAINING

Site-training applications may in themselves extend from initial research investigations to final product applications. For example, high-level technical personnel from the end user commonly may be asked, or may request, to assess the potential of research concepts to their applications, long before a product exists. The purpose is to help steer the future development beginning at the engineering level, in order to meet market expectations. At later development stages, customers find it advantageous to witness the product in field operation to verify it meets their anticipated needs. They may also want to perform some of their own tests using installed zirconia sensors on-site. Other manufacturers or users may find it useful to have their sensors evaluated against an application in order to assess suitability or establish necessary modifications required before wider use. Finally, the deployed zirconia-based sensors on-site provide training via a hands-on learning experience to users and installers prior to receiving hardware at their own facility.

7.3 RELIABILITY TESTING OF ZIRCONIA GAS SENSORS

The reliability testing of zirconia gas sensors is related to the most complex type of experiments owing to some very specific requirements to such tests. Special methods have been developed for their planning and appraisal in the mathematical statistics. Not only quality of appraisal but also the cost of testing to a considerable extent depend on the correct usage of these methods. Moreover, the reliability testing is the most important link in the chain of experiments providing the zirconia sensors reliability at their design and manufacturing stages.

The following consideration of planning of the zirconia gas sensors' reliability testing is related to reliability in general without its division into the metrological and mechanical reliabilities. Therefore, the destruction of any sensor's elements as well as the drop out of the allowable limits of any of the sensor's technical and metrological parameters can be considered a sensor failure.

At the planning of the reliability testing, the main attention should be focused on the practical recommendations based on the generalization of the established experience in experimental appraisal of the sensors' reliability.

The conception of reliability consists of such properties of the zirconia gas sensors as an ability to repair, faultlessness, preservation, and longevity. As usual, the reliability tests are carried out only with the aim to determine or to control the faultlessness of the sensor. The use of the faultlessness characteristic is very convenient for appraisal of the gas sensors' reliability because the zirconia-based sensors related to the unrepairable type of the factory-made goods and their reliability is determined by only one accidental value — time of the faultless work t.

If the distribution law and the density of distribution of this accidental value $f(t)$ are known, then it is possible to evaluate the probability of the faultless work P (which is used as the sensor reliability index) during any arbitrary time frame τ:

$$P_\tau = 1 - \int_0^\tau f(t)\partial t .$$ (7.17)

Consequently, the task of experimental appraisal of the zirconia gas sensors' reliability is based on determination of the distribution law of the faultless time or the parameters of distribution law, if its type is known.

Unfortunately, the distribution law of the faultless time for the zirconia-based gas sensors is unexplored. Assumptions that the distribution law is close to the exponential, normal or others, sometimes used during the planning of the tests are frequently ungrounded. Therefore, it is necessary to take into account that the distribution law of the accidental value $f(t)$ is unknown during the selection of the reliability appraisal method.

If various quality indexes such as, for example, the transference coefficient, mass, and so on are possible to determine accurately for each sensor in the batch, then the reliability index — probability of the faultless work within the set time frame — is impossible to determine accurately and individually for each sensor. It can be determined either for only the batch of sensors or for the total sum of sensors. Based on the fact that the number of sensors in the total sum is usually huge, the selection number of samples, called an *extract*, is evaluated. This testing is called *audit sample testing* [14, 15]. The reliability appraisal of sensors by the results of audit sample testing is usually spreading on the total sum of the sensors.

The obtained result of the reliability appraisal has an accidental value, and its trustworthiness is characterized by the principal probability γ or by risks of suppliers α and customers β because of the selective character of testing. It is obvious that the bigger the extract volume (i.e., the number of sensors to be tested), the better the trustworthiness of reliability appraisal of the total sum of sensors. However, the costs of sensor testing will also increase with expansion of the extract volume. Therefore, one of the main tasks at the planning of reliability experiments is the selection of a minimum number of audit samples and minimum duration of the tests at which the required reliability can be achieved and can be controlled with the requested accuracy.

The reliability tests are divided into the determining and control tests. The task of reliability determination, especially with such highly reliable products as sensors, is much more complicated than the task of control. This fact has been reflected at the planning of testing. Planning of the determining experiments possesses the reference character and is coming to the approximate determination of the extract volume and duration of the tests, proceeding from the expected reliability of the sensors and the set trustworthiness of appraisal. Duration of the determining tests can substantially exceed duration of the control tests of the same zirconia sensors.

During the control tests, the answer to the question of whether the testing of zirconia sensors corresponds to the required reliability level should be obtained. The simple answer should be yes or no. Furthermore, at the planning of the control tests, the extract volume and the duration of tests can be precisely determined, which is very suitable for production. Therefore, the control testing has been the most widely distributed at the appraisal of the sensors' reliability among manufacturers.

There are several methods of the sensors' reliability appraisal at the control testing: method of single extract, two-step method, method of the sequential analysis, and method of the truncated sequential analysis [16–18]. The most convenient for the sensors' reliability appraisal are the method of the truncated sequential analysis and the method of single extract. The method of single extract is simpler but leads to a bigger volume of testing compared with the method of the truncated sequential analysis [17].

The minimum number of periods of the sensors' work V_{min} can be determined by using main correlations of the sequential analysis [16]. This period is defined as a time of the faultless sensors' work corresponding to the required reliability level (the duration of each period should be equal to t_p):

$$V_{min} = \frac{ln\,\beta/(1-\alpha)}{ln(1-q_{01})/(1-q_0)} \, , \qquad (7.18)$$

where β is a customer's risk or the probability that the taken hypothesis $P \geq P_0$ is wrong and hypothesis $P \leq P_0$ is right, α is a risk of suppliers or the probability that the rejected hypothesis $P \geq P_0$ is right, $q_{01} = 1 - P_{01}$ is the maximum allowed value of the failure probability within time t_p, and $q_0 = 1 - P_0$ is the failure probability within time t_p.

The number of samples in the extract yields

$$n = V_{min}/m, \qquad (7.19)$$

where m is the number of the working periods with duration for each sample. The minimum number of failures within V_{min} periods, at which the sensors do not correspond to the required reliability level K, can be determined as follows:

$$K = \frac{ln\dfrac{1-\beta}{\alpha} - V_{min}\,ln\dfrac{1-q_{01}}{1-q_0}}{ln(q_{01}/q_0) - ln(1-q_{01})/(1-q_0)} \, . \qquad (7.20)$$

If at $V_{min} = nt_p$ no failures have been detected, then the sensors are considered to be corresponding to the required reliability level.

At the detected number of failures $K_d \geq K$, all sensors are considered noncorresponding to the required reliability level. If $0 < K_d < K$, then the testing continues until level V_d will be achieved:

$$V_d = \frac{ln\dfrac{\beta}{1-\alpha} - K_d[ln(q_{01}/q_0) - ln(1-q_{01})/(1-q_0)]}{ln(1-q_{01})/(1-q_0)} . \qquad (7.21)$$

Sensors are considered corresponding to the required reliability level if, during testing from V_{min} to V_d, no failures are detected. In such a situation, the risk of suppliers will be increased. Nevertheless, this method has often been used on practice.

Despite the fact that the mathematics of the statistical methods for the control experiments of reliability testing is well developed, the sensors designer usually has difficulties with considering specific test methods for the reliability testing applicable to various zirconia-based gas sensors.

In general, these difficulties are connected to the requirement of making the reliability testing conditions as close as possible to the industrial test conditions of the sensors. First of all, in this situation the natural changes of various parameters of the measuring environment in time are unknown; second, it is impossible to imitate the simultaneous impact of the complex of factors corresponding to the real measuring conditions on the sensor output. Furthermore, the specifics of common requirements to the reliability testing can also be a reason for some difficulties.

Let us consider the common requirements to the organization and routine of the zirconia gas sensors' reliability testing in details.

The zirconia gas sensors' reliability testing can be carried out either as an independent type of testing or simultaneously with other tests focused on the determination of the sensors' technical characteristics. Planning of experiments is getting more complex in the last case. However, the cost of the tests will be lowered.

The plan of experiments should be composed like this: the impact of various factors on the sensors' workability should be carried out before the reliability testing. Such tests are a mechanical strength test, an impact of transportation on sensor characteristics test, and so on. The test setup should be as close as possible to the real industrial test conditions of the zirconia sensors. Usually, the maximum values of the influencing factors as well as the time of the faultless sensor work have been indicated at the design stage of the sensor development. Therefore, it is very difficult to set the routine of laboratory sensor tests in conditions close enough to the industrial test conditions owing to the absence of comprehensive information as well as multiple-components sensor work stations imitating precisely the working conditions of sensors.

Consequently, owing to the difficulties in the planning of test routines, all tests are often divided into laboratory, standard, and on-location tests. Generalization of the statistical data of all tests should be based on the appropriate processing and will ultimately provide appraisal of the sensors' reliability. At present, the following practice usually takes place during organization of the reliability testing at laboratory conditions: the sum of the calculated minimum number of periods of the sensors' work V_{min} can be divided into at least three cycles, and each of them has the sensor's work resource at the influence of each of the most significant influential factors. The value of these factors should correspond to value set at the design stage of the sensor development. The time impact of each of the factors should be equal to the time at which the sensor shows no faults. The sequence of the influencing factors at testing, if it is not specified specifically to the developed sensor, should be as follows: mechanical charges, high temperature, high humidity, temperature cycling, low temperature, and presence of different gaseous components. Generally speaking, the set of such testing has a formal character. Careful consideration of the real test conditions and investigation of the zirconia sensor susceptibility to the various influencing factors are imperative to the optimal planning of the sensor tests. Furthermore, it is also necessary to develop the reliability testing plan even in the development stage of the sensor design considering the long-term stability test results of similar sensors because the real reliability of the sensor is unknown.

REFERENCES

1. Cox, D.R., *Planning of Experiments*, New York, Wiley, 1992, 320.
2. Artukhin, E.A., Optimum planning of experiments in the identification of heat-transfer processes, *J. Eng. Phys. & Thermophys.* **56** (1989) 256–260.
3. Jeffwu, C.F. and Hamada, M., *Experiments: Planning, Analysis, and Parameter Design Optimization*, New York, Wiley, 2000, 664.
4. Buzzi-Ferraris, G., Planning of experiments and kinetic analysis, *Catalysis Today* **52** (1999) 125–132.
5. Ruggoo, A. and Vandebroek, M., Model-sensitive sequential optimal designs, *Computational Stat. & Data Anal.* **51** (2006) 1089–1099.
6. Scheffé, H., *The Analysis of Variance*, New York Wiley, 1959, 278.
7. Peiponen, K.E., Vartiainen, E.M., and Asakura, M., *Dispersion, Complex Analysis and Optical Spectroscopy: Classical Theory*, New York, Springer, 1998, 130.
8. Peiponen, K.E., Vartiainen, E.M., and Asakura, M., Complex analysis in dispersion theory, *Optical Review* **4** (1997) 433–441.
9. Roll, W.G., *Theory and Experiment in Physical Research*, Ayer, Stratford, NH, 1986, 510.
10. Kendall, M. et al., *Kendall's Advanced Theory of Statistics: Volume 2A—Classical Interference and Linear Model*, London A Hodder Arnold, 6th ed., 1999, 912.
11. Miles, J. and Shevlin, M., *Applying Regression and Correlation*, Thousand Oaks, CA, Sage, 2001, 272.
12. Zhuiykov, S., Zirconia single crystal analyser for low-temperature measurements, *Proc. Contr. Quality* **11** (1998) 23–37.
13. Zhuiykov, S., "*In-situ*" diagnostics of solid electrolyte sensors measuring oxygen activity in melts by developed impedance method, *Meas. Science and Techn.* **17** (2006) 1570–1578.
14. Zhuiykov, S., Fire detectors: Evaluation, statistical analysis and quality control by audit sample testing, *Fire & Safety* **2** (2005) 4–12.
15. Zhuiykov, S. and Kats, E., Audit sample testing of fire detectors: Helping the community, *Fire Australia* **5** (2005) 14–18.
16. Karger, D.W. and Murdick, N.L., *Quality, Reliability and Process Improvement*, New York, Industrial Press, 1985, 407.
17. Tobias, P.A., *Applied Reliability*, Boston, CRC Press, 1995, 422.
18. Crowder, M., *Statistical Analysis of Reliability Data*, Boston, CRC Press, 1994, 264.

Index

Bold page numbers indicate material in tables or figures.

A

Absolute error
additive, 230
dispersion and, 245, 247
mathematical modeling of, 232
measuring parameter and, 230
multiplicative, 230
quality and, 227
reduction error and, 228
of sensitivity, 230, **231**
sensor function and, **231**, 245
uncertainty and, 230
of the zero, 230, **231**
Absolute temperature, 5, 98, 139, 237
Accidental errors, 232
Accuracy, factors affecting, 227
Activation energy
of adsorbed gas molecules, 52
of ionic conductivity, 5, 172
in NO sensors, 66
in O sensors, 52
transference, 5, 25
Active experiments, 256
Additive absolute error, 230
Additive sensitivity, 233
Adhesive joining, 197, 207
Adsorption
action radius of forces, 51
activation energy of, 52, 66
additives and, 149, 152, 221
catalytic activity and, 62
cross-sensitivity and, 152
current and, 30, 33
diffusion and, 31, 33, 45
dissociative, 44, 51, 64–65, 234
emf and, 30
in error analysis, 234, 236–237
gas concentration and, 33, 44, 51
gas pressure and, 237
in impedance-based sensors, 121
ionization and, 30

isotherms, 33, 51
in mixed-potential sensor models, 62
in NO sensors, 62, 64–66, 121
in O sensors, 33, 50–52, 54, 62
permeability and, 30
polarization and, 36
probability of, 52, 237
rate of, 78
resistance and, 121
surface topography and, 44, 51
temperature and, 51, 65, 152
in thin-film sensors, 64
time and, 52, 66
Aging
of alumina-doped YSZ, 137
charge carriers and, 12
cubic structure and, 12, 14
diffusion and, 12, 15
grain boundary and, 12
grain size and, 79
of hafnia-stablized YSZ, 14–15, **16**
ionic conductivity and, 11–15
oxygen partial pressure and, 15
phase boundaries and, 12–15, **13**
of platinum electrodes, 14, 98
of polycrystalline electrolytes, 12–15, 137
resistance and, 12, 14, **16**
of scandia-stablized zirconia, 14
temperature and, 11–15
vacancies and, 12
of YSZ, 12–15, **15**, **16**
of zirconium oxide, 137
Algorithm, modeling, 47–49, **48**, 82–84, **84**
Alumina
bismuth oxide and, 215
CTE of, 198
isolation layers of, 162
magnesium doping of, 145, 204–205, **206**
nanoparticles of, 216
nickel oxide and, 215
for plug-type sensors, 143, 198–201
zirconium oxide and, 137
Alumina-doped YSZ
aging of, 137
coefficient of thermal expansion for, 137, 138
conductivity of, 137, 155, 161

H